Video Engineering

Other McGraw-Hill Books of Interest

BARTLETT • *Cable Communications*
BARTLETT • *Cable Television Handbook*
BENNER • *Fibre Channel*
COOMBS • *Printed Circuits Handbook, 4/e*
FINK AND BEATY • *Standard Handbook for Electrical Engineers*
CHRISTIANSEN • *Electronic Engineers' Handbook, 4/e*
LUTHER AND INGLIS • *Video Engineering, 3/e*
JOHNSON • *Antenna Engineering Handbook, 3/e*
LAMPEN • *Wire, Cable and Fiber Optics for Audio/Video Engineers*
LENK • *Lenk's Audio Handbook*
LENK • *Lenk's Video Handbook, 2/e*
MEE AND DANIEL • *Magnetic Recording Technology, 2/e*
POHLMAN • *Principles of Digital Audio, 3/e*
ROBIN AND POULIN • *Digital Television Fundamentals*
SOLARI • *Digital Video and Audio Compression*
SYMES • *Video Compression*
TAYLOR • *DVD Demystified*
WHITAKER • *Standard Handbook of Video and Television Enginering, 3/e*
WHITAKER • *DTV, 2/e*
WILLIAMS AND TAYLOR • *Electronic Filter Design Handbook, 2/e*

To order or receive additional information on these or any other McGraw-Hill titles, in the United States please call 1-800-722-4726 In other countries, contact your local McGraw-Hill representative.

Video Engineering

Arch C. Luther
Andrew F. Inglis

Third Edition

McGraw-Hill
New York San Francisco Washington, D.C. Auckland Bogotá
Caracas Lisbon London Madrid Mexico City Milan
Montreal New Delhi San Juan Singapore
Sydney Tokyo Toronto

Library of Congress Catalog-in-Publication Data

Arch C. Luther
    Video engineering / Arch C. Luther, Andrew F. Inglis—3rd ed.
      p. cm
    Includes index.
    ISBN 0-07-135017-9 (hardcover)
    1. Television. 2. Digital video. I. Inglis, Andrew F. II. Title.
III. Series.
TK6630.I54 1999
621.388—dc20   99-10913
                                                                                                                    CIP

McGraw-Hill

A Division of The McGraw·Hill Companies

Copyright © 1999 by The McGraw-Hill Companies, Inc. All rights reserved. Printed in the United States of America. Except as permitted under the United States Copyright Act of 1976, no part of this publication may be reproduced or distributed in any form or by any means or stored in a data base or retrieval system, without the prior written permission of the publisher.

1 2 3 4 5 6 7 8 9 0 DOC/DOC 9 0 9 8 7 6 5 4 3 2 1 0 9

ISBN 0-07-135017-9

The sponsoring editor for this book was Stephen S. Chapman, the editing supervisor was Caroline R. Levine., and the production supervisor was Pamela A. Pelton.

Printed and bound by R. R. Donnelley and Sons Co.

 This book is printed on recycled, acid-free paper containing a minimum of 50% recycled, de-inked fiber.

Information contained in this work has been obtained by The McGraw-Hill Companies, Inc. ("McGraw-Hill") from sources believed to be reliable. However, neither McGraw-Hill nor its authors guarantee the accuracy or completeness of any information published herein and neither McGraw-Hill nor its authors shall be responsible for any errors, omissions, or damages arising out of use of this information. This work is published with the understanding that McGraw-Hill and its authors are supplying information but are not attempting to render engineering or other professional services. If such services are required, the assistance of an appropriate professional should be sought.

Contents

Preface — xiii

Chapter 1. Video System Fundamentals — 1

    1.1 Introduction — 1
    1.2 Video Systems — 2
    1.3 Scanning — 4
    1.4 Resolution — 6
    1.5 Sampling, Pixels, and Quantizing — 7
    1.6 Choosing Scanning Standards — 11
    1.7 Worldwide Video Scanning Standards — 14
    1.8 Television Signal Waveforms — 17
    1.9 Digital Video Signals — 25
    1.10 Analog-Digital and Digital-Analog Conversion — 25
    1.11 Summary Comparison of Analog and Digital Systems — 26
    1.12 The Audio Component — 26

Chapter 2. Color Video Fundamentals — 29

    2.1 Introduction — 29
    2.2 The Properties of Color—A Summary — 29
    2.3 Color Specifications — 31
    2.4 The Primary Colors — 32
    2.5 Dominant Wavelength and Saturation — 33
    2.6 Luminance and Color Difference Components — 34
    2.7 The CIE Chromaticity Coordinates — 34
    2.8 Camera Colorimetry — 38
    2.9 Display Device Primary Standards — 41
    2.10 Colorimetry of Computer Displays — 43

Chapter 3. Introduction to Digital Technology — 45

    3.1 Digital — 45
    3.2 Number Systems — 45
    3.3 Analog-to-Digital Conversion — 47
    3.4 Serial and Parallel Formats — 57
    3.5 Binary Arithmetic — 58
    3.6 Digital Signal Processing — 60
    3.7 Digital Video Transmission — 63
    3.8 Digital Error Protection — 69
    3.9 Digital-to-Analog Conversion — 73
    3.10 Video Data Compression — 74
    3.11 Data Compression Technologies — 75
    3.12 JPEG and MPEG Compression — 80

Chapter 4. Elements of Image Quality — 87

    4.1 Overview — 87
    4.2 Basic Image Quality Criteria — 89
    4.3 Image Defects — 90
    4.4 Characteristics of Human Vision — 92
    4.5 Image Definition — 94
    4.6 Aperture Response — 95
    4.7 Sampling and Aliasing — 99
    4.8 Limiting Resolution — 102
    4.9 NTSC Television System Aperture Response — 106
    4.10 HDTV System Aperture Response — 111
    4.11 Image Definition—NTSC Systems — 113
    4.12 Image Definition—HDTV Systems — 114
    4.13 Visual Perception of Broadcast and HDTV Images — 116
    4.14 Image Gray Scale — 118
    4.15 Signal-to-Noise Ratio — 123
    4.16 Image Defects (continued from Section 4.3) — 126
    4.17 Comparison of Film and Television Picture Quality — 130

Chapter 5. Audio Technology for Video — 137

    5.1 Significance of Audio to Video — 137
    5.2 Natural Sound — 137
    5.3 Audio Systems — 139
    5.4 Analog Audio Signals — 141
    5.5 Digital Audio Signals — 143
    5.6 Audio Signal Processing — 146
    5.7 Music — 147

Chapter 6. Analog Video Systems — 151

    6.1 Introduction — 151
    6.2 Standardization — 151
    6.3 Video Signal Formats for Color — 152
    6.4 Analog Color TV System Configurations — 154
    6.5 Composite Signal Components — 155
    6.6 Subcarriers — 156
    6.7 The NTSC System Subcarrier — 156
    6.8 The PAL System Subcarrier — 160
    6.9 Comparison of NTSC and PAL — 163
    6.10 Color Signal Distortions and Artifacts — 164
    6.11 Subcarrrier Crosstalk — 165
    6.12 Analog Component Formats — 167
    6.13 Multiplexed Analog Component Systems — 167

Chapter 7. Digital Video Systems—DTV — 171

    7.1 Introduction — 171
    7.2 DTV System Objectives — 173
    7.3 The ATSC DTV System — 174
    7.4 The Picture Formats Layer — 176
    7.5 The Compression Layer — 178
    7.6 The Transport Layer — 180
    7.7 The Transmission Layer — 183
    7.8 ATSC HDTV Audio — 189
    7.9 Digital Video Broadcasting Standards — 190

Chapter 8. Digital Video Systems—Computers — 195

    8.1 Introduction — 195
    8.2 The Personal Computer — 195
    8.3 Software — 201
    8.4 Computer Video Formats — 203
    8.5 Computer Video Standards — 209
    8.6 DVD on Computers — 210
    8.7 Digital HDTV on Computers — 210

Chapter 9. Video Cameras — 211

    9.1 Introduction — 211
    9.2 Photoconductive Storage Tubes — 216
    9.3 Charge-Coupled-Device (CCD) Sensors — 217
    9.4 CCD Imagers — 222

viii Contents

9.5 CCD Imager Performance	227
9.6 Camera Chain Packaging	234
9.7 Camera Optical Components	236
9.8 Camera Signal Processing Circuits	238
9.9 Digital Camera Electronics	242
9.10 Camera Operating Controls	245
9.11 Camera Specifications	246
9.12 Video Film Camera Systems	246
9.13 Digital Still-Image Cameras	249

Chapter 10. Professional Video Recorders — 253

10.1 Introduction	253
10.2 Principles of Magnetic Tape Recording	256
10.3 Video Head Design	257
10.4 The Helical-Scan Principle	259
10.5 Tape Transport Configurations	261
10.6 Magnetic Recording Tapes	263
10.7 Tape and Record/Playback Head Performance	264
10.8 Analog Recording	268
10.9 SMPTE Type C Recorders	271
10.10 ½-inch Cassette Recorders	279
10.11 Digital Recording	280
10.12 Digital Recording Standardization Policy	281
10.13 The D-1 Component Format	283
10.14 The D-2 Composite Format	283
10.15 The D-3 Composite and D-5 Component Formats	284
10.16 The D-6 HDTV Component Format	285
10.17 Betacam SX	285
10.18 DVCPro	288
10.19 Computer Hard Disk Video Recording	288
10.20 Optical Video Recording	291
10.21 Trends in Professional Video Recorders	293

Chapter 11. Home and Semiprofessional Video Recorders — 295

11.1 Introduction	295
11.2 Comparison with Professional Recorders	295
11.3 Recording Drum Configurations	299
11.4 Home VCR Record Formats	299
11.5 Automatic Scan Tracking	305
11.6 Record Information Density	305
11.7 Color-Under Signals	307
11.8 S-VHS and Hi-8	309

11.9 Digital Home Recorders	310
11.10 Camcorders	312
11.11 Semiprofessional and Professional Camcorders	315

Chapter 12. Video Postproduction Systems — 317

12.1 Introduction	317
12.2 Electronic Editing of Videotape	318
12.3 Tasks of Postproduction	323
12.4 Time Code	326
12.5 Digital Postproduction	327

Chapter 13. Television Receivers and Video Monitors — 333

13.1 Introduction	333
13.2 Receiver Configuration	333
13.3 Tuner Section	335
13.4 IF Amplifier	337
13.5 Detection	340
13.6 Sync Separation and Scanning Generation	340
13.7 Decoding Signal Formats to RGB	340
13.8 Color Display Devices	342
13.9 Special Receiver Features	345
13.10 Receiver Performance Criteria	347
13.11 Professional Picture Monitors	349

Chapter 14. Digital Video Display Systems — 351

14.1 Introduction	351
14.2 Computer Video Subsystem	351
14.3 Video Overlay	356
14.4 Digital TV Displays	358
14.5 Computer Display Monitors	358
14.6 Flat-Panel Displays	359
14.7 DTV Display Processing	365

Chapter 15. Interactive Video Systems — 369

15.1 Introduction	369
15.2 Ingredients of Interactivity	370
15.3 Communicating with the User	371
15.4 Video-Based User Interfaces	371
15.5 Computing Capability for Interactivity	374
15.6 Multimedia	376
15.7 Interactive Television	381

Chapter 16. Terrestrial Broadcasting Systems — 385

    16.1 A Brief History and the Future Outlook — 385
    16.2 Allocations and Assignments — 386
    16.3 The Standard Analog Broadcast Television Channel — 390
    16.4 The Visual Carrier — 391
    16.5 The Aural Carrier — 393
    16.6 Elements of a TV Transmitting System — 393
    16.7 FCC ERP/Antenna Height Limitations — 394
    16.8 Television Transmitters — 394
    16.9 Representative Transmitter Power Ratings — 399
    16.10 Transmitter Performance Standards — 399
    16.11 Transmission Lines — 402
    16.12 Waveguides — 406
    16.13 Properties of VHF and UHF Radiation — 407
    16.14 Antennas — 411
    16.15 TV Station Service Areas — 418
    16.16 Digital Broadcasting — 421
    16.17 DTV Transmission Standards — 423
    16.18 Display Formats for HDTV (DTV) Systems — 424
    16.19 DTV Channel Assignments — 425
    16.20 DTV Station Coverage — 426
    16.21 DTV Channel Utilization Alternatives — 427
    16.22 DTV Transmitters — 427

Chapter 17. CATV Systems — 429

    17.1 A Brief History of CATV — 429
    17.2 Cable System Elements — 429
    17.3 Cable Television Channels — 430
    17.4 Coaxial Cable — 432
    17.5 Line and Bridging Amplifiers and Equalizers — 434
    17.6 Trunk Circuits — 436
    17.7 Reverse Path Circuits — 437
    17.8 Video Performance Standards — 437
    17.9 Head Ends — 438
    17.10 Set-Top Converters — 439
    17.11 Pay-TV Systems — 441
    17.12 Wireless Cable — 442
    17.13 Digital Transmission and Fiber-Optic Cable — 445
    17.14 Future Cable Systems — 447

Chapter 18. Satellite Video Communications — 451

    18.1 Satellites and Television — 451

18.2 Satellite Communication Systems Elements	452
18.3 The Geosynchronous Orbit	453
18.4 Satellite Classifications	456
18.5 Satellite Construction	457
18.6 Earth Stations	461
18.7 Transmission Formats	466
18.8 FM Performance Characteristics	466
18.9 Digital DBS Services	472

Chapter 19. Fiber-Optic Transmission Systems — 479

19.1 Overview	479
19.2 Fiber-Optic Construction and Operation	480
19.3 Single and Multimode Fibers	481
19.4 Attenuation and Dispersion	482
19.5 Signal Formats and Modulation Modes	484
19.6 Path Performance, Digital Format	486
19.7 Transmitters for Fiber-Optic Systems	487
19.8 Receivers for Fiber-Optic Systems	489
19.9 Passive Optical Components	491

Chapter 20. Video on the Internet — 493

20.1 Introduction	493
20.2 Internet Architecture	493
20.3 Accessing the Internet	497
20.4 Connecting	498
20.5 The World Wide Web	500
20.6 Internet Video Architectures	503

Chapter 21. The Future of Video Systems — 505

21.1 Introduction	505
21.2 Technologies in Video Engineering	505
21.3 Video Systems	509
21.4 Communications	511
21.5 New Video Systems	512

Glossary — 519

Bibliography — 543

Index — 551

Preface

The advance of video technology makes necessary another edition of this book only *three* years after the previous one—wow! The digital revolution that was becoming visible in 1996 is now upon us and it is only a matter of time before all video and audio equipment will be digital. Along with digital technology come the Internet and the World Wide Web, which are rapidly developing another important channel for video and audio communication. In addition, notable advances have occurred in nearly every other aspect of video technology.

The spectacular progress of electronic technology has made electronics a participant in nearly all aspects of modern life. From television or personal computers to automobiles or kitchen appliances, electronics is everywhere. Nowhere are the advances more evident than in the electronic reproduction of pictures—video. That is the subject of this book and, in the three years since the Second Edition was written, so much has happened in video engineering that a Third Edition has become mandatory. This Third Edition, therefore, includes most of the subject matter of the Second Edition and is greatly expanded with descriptions of the technical developments of the past three years.

The same electronic progress applied to personal computers has made digital technology inexpensive and eminently useful to video systems. The result is higher performance, higher reliability, and lower cost. The newly developed digital HDTV system is an excellent example—high-resolution moving pictures with surround sound are transmitted in the same 6-MHz bandwidth that handles analog television and by using extra channels that could not previously be utilized for analog broadcasting because of interference. The result is an unprecedented merging of television and computer technologies.

The potential audience for this book has grown because analog video engineers need to learn about digital technology and digital engineers entering the video field need to learn about the analog principles that underly all video systems—even digital ones. And, of course, new video engineers need to learn about both. This book treats analog and digital video equally in the perspective required by future video engineers.

In-depth coverage of all the subjects embraced by video engineering would require many volumes of great detail, far beyond what is practical in one book. This book focuses on the fundamentals of video engineering in all aspects, and details are presented only as examples. Complete coverage of details is left to other writers. This concept is supported by end-of-chapter references and the Bibliography.

Another reason for adhering to fundamentals is that the details of a technology change much more rapidly than its fundamentals. This causes rapid obsolescence of books that go into great detail in a fast-moving field such as video engineering, which is unfair to readers unless they actually need the detail. It is also unfair to writers who will find that their books quickly go out of print!

In addition to a complete updating of the entire book to deal with the changes in technology, two new chapters have been added: Chapter 5 to more fully cover the audio field as it relates to video, and Chapter 20 to cover the Internet.

It is with great sadness that I must report the passing of my co-author, Andrew F. Inglis, in July 1998, before work began on this edition. I have known Andy ever since he came to RCA in 1953—he was a boss, a mentor, and a friend. I especially enjoyed working so closely with him on the Second Edition of this book, and experiencing his wonderful family. As I was working on this edition, Andy's words were constantly before me—he remains a co-author in absentia. Andy will be sorely missed by us all.

In addition to the authors, a project such as this book involves the contributions of many people, including those who helped with the First Edition, the Second Edition, and now the Third Edition. I especially want to acknowledge the following for their suggestions, comments, or reviews to the various editions: Stanley Basara, Sidney Bendell, Walter Braun, John Christopher, Peter Dare, Marvin Freeling, Luigi Gallo, Charles Ginsburg, Jukka Hamalainen, Norman Hurst, Anthony Lind, Robert Neuhauser, Koichi Sadashige, John Smiley, and Larry Thorpe.

Video engineering is a field that has an exciting future. This book will give its readers a full view of that potential.

Arch C. Luther

Chapter 1

Video System Fundamentals

1.1 Introduction

Everyone knows what video is—it is another word for television. For more than 40 years, television has been in our homes; it is an integral part of modern life. News, information, and entertainment are all received through "video."

However, an engineer would take a broader view of the definition of video—he or she would equate it with the use of a cathode-ray-tube display or other equivalent means of electronic display. Since this book is about video engineering, the engineer's viewpoint will be used:

Video is the technology for electronic capture, storage, transmission, and reproduction of images and motion pictures.

That definition embraces much more than television. In particular, it includes the display technology used in personal computers (PCs), which now are present in a large percentage of our offices, schools, and homes. These two spheres of usage of video are the hubs from which many other applications have grown.

For many years, the two spheres—television and computer—although they both used video, were artificially separated by the fact that TV was analog and computers were digital. That era is now past—analog and digital technologies now are used wherever they suit and are cost-effective. However, because of continued advances in capability and cost reduction of digital video technology, the trend is for digital techniques to replace many video functions that previously were analog. The signal path of television equipment for analog systems is rapidly becoming all digital. In fact, the new HDTV standards are all digital. The transition to digital television is accelerating. Because digital video systems will be discussed extensively throughout this book, the key features of digital video are discussed in the next section.

2 Chapter One

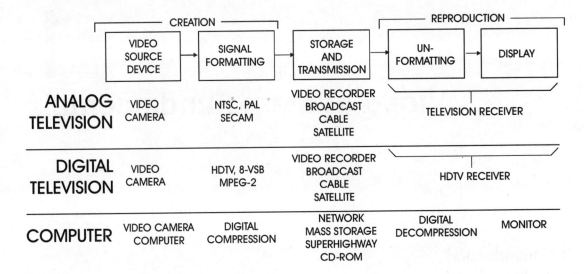

Figure 1.1 Elements of a video system.

1.2 Video Systems

Video, as shown by Figure 1.1, requires a *system* that includes the elements of creation, storage, transmission, and reproduction. The figure shows the most common names for each of the parts of a system for both television systems and computer systems. Each of these system parts will be covered in later chapters.

Creation is the process of generating an electrical video signal, either from a live scene using a live camera, from motion picture film using a telecine camera, or artificially by a computer. Video cameras are covered in detail in Chapter 9. In signal formatting, the video signal is processed to enhance it and to conform it to a signal standard for use later in the system.

1.2.1 Analog and Digital Systems

Existing television broadcasting systems are *analog*, meaning that signals are transmitted as a continuous-valued representation of the picture and sound. When existing standards were developed, this was the only technology available. Today, through the developments of the computer industry, *digital* video technology is available where the picture and sound representation is by a stream of values having only two levels—one and zero.

Digital signal representation has a number of important advantages.

- Signal performance is determined at the point of creation of the digital signal, usually by converting from an analog signal because most signal generation devices are analog. If desired, the rest of a digital system can be designed to be completely *transparent*, meaning that it provides a perfect reproduction of the

incoming signal. This eliminates the ever-present concern in analog systems of accumulating signal degradation from cascaded processing, transmission, or storage steps.
- Although digital circuits are conceptually more complex than analog circuits, advances in integrated circuit (IC) technology have made the hardware for digital processes smaller, more reliable, easier to design, and less expensive than equivalent analog hardware.
- Certain processes, such as signal delay and dynamic video special effects are easier to accomplish in the digital domain. This has caused digital products to be designed to implement these processes in otherwise analog video systems. The early development of digital video systems started this way.

Key digital video concepts are described in Chapter 3.

1.2.2 Video Signal Standards

Signal standards in a video system are extremely important because video signals pass through equipment of many different types and different manufacturers. Each equipment must know exactly the form of the signal it receives and must output a similarly precise signal. Otherwise, the many parts of a system could not work together. Signal formats are covered in Chapters 2 to 8.

Broadcast television systems have historically been standardized in terms of a signal radiated by a transmitter for reception and direct display. This has provided a solid basis for the design of receivers, but as the industry has grown and become more diverse, there is also a need for a signal-generation standard. This is called a production standard and it defines the video signal at its point of creation or assembly. Such a standard must be more demanding than a transmission standard because the signals will be stored, broadcasted, or presented in many formats, and the standard must allow for degradation that may occur in those processes.

In digital video systems, broadcast standardization is still in terms of a radiated signal, but receivers can have great flexibility in how they display the signal. This is because receivers can contain digital processing to convert a transmitted signal into a different format for display. As a result, digital transmission standards provide a multiplicity of signal formats.

1.2.3 Storage

Storage is another important part of a video system. The home video cassette recorder (VCR) is a typical video storage device for recording and replaying of video and audio. Video recorders of many types, both analog and digital, are covered in Chapters 10 and 11. Computers have their own forms of storage, including RAM memory and hard disks. These are covered in further detail in Chapters 3 and 8.

Because of the capability to store and replay video, one can assemble a video program from stored sequences taken at different times or different places. The result is stored as a complete program. This process is called *editing*. It allows program creation to be divided into two steps: the initial capture of video and audio, called *production*, and the editing and program assembly process, called *postproduction*. They are covered in Chapter 12.

1.2.4 Transmission

Transmission is the process of sending a video signal from one physical location to another. Usually it refers to transporting video in real time, although shipping tapes or disks is a valid alternative if the time of arrival of the video is not critical. The largest real-time video transmission system is the broadcast television system, although computer network systems are catching up. Broadcast transmission is covered in Chapters 16, 17, and 18. Computer transmission systems are covered in Chapter 3.

The reproduction part of a video system receives a signal and, by knowing its standard, unformats it to a form suitable for driving a display device such as a cathode-ray tube (CRT) or a solid-state display panel. The viewer sees the result. Displays are covered in Chapters 13 and 14.

1.2.5 Video System Types

Figure 1.1 is an abstraction of a video system that emphasizes the common elements in all systems. Most of the time in this book, real systems are discussed—not abstractions. Three major system types that appear throughout the book are:

Television (TV)—The broadcast systems used around the world, including NTSC (U.S. and Japan), PAL (most of Europe and China), and SECAM (France and Russia) color video systems (see Chapter 2). These systems are called *standard-definition TV* (SDTV).

High-definition TV—(HDTV) The emerging higher-performance video system included in the new digital TV (DTV) systems

Computers—The many forms of video used by personal computers

Electronic reproduction of video is not complete without sound. One has only to turn off the sound on a TV news program to appreciate how much information is carried by the sound channel. From an artistic point of view, the sound contributes as much as the picture to the overall psychological and emotional effect of a video presentation. Sound for video systems is described in Chapter 5.

Often the viewer is aware of only part of a video system, such as when viewing a television receiver, but the other parts of the system must exist (or have existed) for video to work.

1.3 Scanning

Natural scenes are three-dimensional (x, y, and time), but they can be converted to two dimensions using, for example, photography. A photograph presents millions of pieces of information at once, because all the points in the picture exist simultaneously. But an electrical signal has only one value at a time, so to represent a picture with a single electrical signal, one must use a technique that converts the picture to a one-dimensional signal vs. time. That process, known as *scanning*, is similar to the way one reads a page of English text. The reader starts at the upper-left corner of the page, reads across a line, moves down to the next line, and continues reading lines until the bottom of the page is reached.

Scanning is used by all video systems in the process of converting optical images into electrical signals. The invention of scanning was a major breakthrough and an indispensable first step in the development of video technology. Like many basic technical developments, it did not have a single inventor but appears to have occurred to a number of engineers simultaneously. Note that theoretically a picture could be reproduced electrically by creating a separate signal for each point (pixel) in the picture; however, the physical impracticability of having the many thousands of parallel channels that this approach requires precludes its use except inside of sensor devices such as CCDs (see Chapter 9.)

1.3.1 Scanning Principles

Scanning operation in a television camera is illustrated in Figure 1.2. An image of the scene is optically focused on the sensitive surface of the camera's imager, where scanning takes place. In the scanning, an electronic sensing spot moves across the image in a pattern called a *raster*. The sensing spot converts each image point that it sees into an electrical voltage. Thus, the voltage output from the imager varies as the spot passes over different brightness areas in the image. The actual mechanism for moving the spot is different between tube imagers and solid-state imagers, but the raster pattern and the electrical output are virtually the same. Both television and computer video systems use a rectangular raster pattern. Scanning begins at the upper-left of the image and the spot moves rapidly in a horizontal direction across the image to form a scanning line. It then snaps back (called horizontal *retrace*) to begin another line. However, the spot also has a slower vertical motion downward, so that—after retrace—the spot is just below the previous line, ready to scan a new area of the image.

When the vertical motion reaches the bottom of the image, both horizontal and vertical retrace actions occur, bringing the spot back to the upper-left starting position. A complete scan of the image area is called a frame.

Scanning is also used to reproduce the image in a display device. In an analog video system, the display scanning process must be exactly the same as at the camera except that the display's moving spot is a spot of light whose intensity is controlled by the electrical signal from the camera. In a digital video system, the display scanning is not necessarily the same as at the camera (see Section 14.4).

Frames must be repeated often enough so that the viewer's eye will not see the moving spot or the lines in a reproduced image. This is possible because of persistence of vision in the eye—a suitably fast frame rate will produce an apparently continuous image without any visibility of the scanning motion.

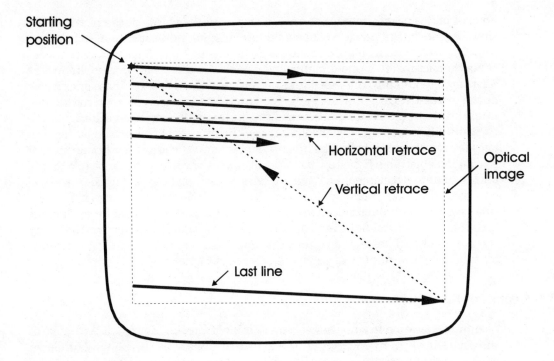

Figure 1.2 Scanning.

1.4 Resolution

The *resolution* of a video system describes the system's ability to reproduce fine detail. The amount of resolution required in a video system depends directly on how the output of the system will be viewed. It is not necessary (or desirable) to have more resolution than the viewer will be able to see. The principal parameter of the viewing situation is the *viewing ratio*, which is the ratio between the distance from the viewer to the display and the height of the display screen. For analog television in the home, optimal viewing ratios are from 4:1 to 7:1, whereas for computer screens, the ratio can be as low as 1:1. HDTV systems are designed for viewing ratios around 3:1.

In television, resolution is expressed as the maximum number of alternate black and white lines that can be distinguished in a dimension equal to the height of the picture (both black and white lines are counted). This number is expressed in *TV lines of resolution* (TVL), which can be measured in any direction on the image, although horizontal and vertical measurements are most common. The TVL measure of resolution is not often used with computers, but the way computer people express resolution is closely related.

When discussing resolution, there may be confusion between the image patterns used to measure resolution and the direction of resolution involved. For example, a pattern of vertical lines is used to measure horizontal resolution and a pattern of horizontal lines measures vertical resolution. To reduce the confusion, this book will call the image features that affect resolution in the horizontal direction *horizontal detail*. Similarly, image features that affect vertical resolution are called *vertical detail*.

The vertical resolution—the number of horizontal black and white lines in the image (vertical detail) that can be distinguished or resolved in the picture height—is a function of the number of scanning lines per frame (but it is not equal to the number of lines as explained in Section 7.8). The horizontal resolution—the number of vertical lines in the image (horizontal detail) that can be distinguished in a dimension equal to the picture height—is determined by the signal bandwidth in analog systems, but in a digital system it is a function of many factors including data rate, degree of data compression applied, and others.

Resolution and its significance in evaluating the performance of video systems is described more fully in Chapter 4.

1.5 Sampling, Pixels, and Quantizing

Although an image in nature is a continuous analog function in space and time, the process of scanning creates a noncontinuous output in the form of lines and frames. This is known as *sampling*, which is a fundamental technique used to some extent in all video systems. This section discusses sampling and the related subjects of pixels and quantizing.

1.5.1 Sampling

The purpose of sampling is to reduce the infinite content of an analog moving image to a manageable number of values that can be handled by a video system. For example, it can be shown that 30 frames per second is a sufficient rate for the display of image motion, so there is no need to send frames any faster. Similarly, the human eye's resolving power and the viewing conditions allow the number of scanning lines to be set to a finite value. The values in the analog image that exist in space between the lines and in time between the frames can safely be eliminated.

A second aspect of sampling makes possible a digital video system. Digital systems convey information as a series of numbers. By means of sampling in three dimensions (frames, lines, and horizontal), a series of discrete values is created to which a digital system can assign numbers. The process of assigning numbers is called *quantizing*, which is described below.

Although the parameters that are subject to sampling are not necessarily electrical (space and time), the analysis of sampling is often carried out by the use of electrical analogy. For example, frequency is the analog of TV line number. The powerful mathematical tools that have been developed for electric circuit analysis are then available. This analysis must be carried out in both the time and frequency domains, because both are needed for a full description of the sampling process.

1.5.1.1 Sampling analysis—time domain

The sampling process in the time domain is shown by Figure 1.3. In Figure 1.3(a), the sampling rate is more than twice the signal frequency and in Figure 1.3(b) it is less than twice the signal frequency. A sampling rate of twice the highest signal frequency component is known as the *Nyquist frequency*. Similarly, the highest signal frequency that can be sampled without distortion for a given sampling frequency is known as the *Nyquist limit*. Sampling at lower rates than twice the highest signal frequency is known as sub-Nyquist sampling. Sub-Nyquist sampling generates a new frequency component equal in frequency to the difference between the sampling and signal frequencies [see Figure 1.3(b)]. This effect is called *aliasing*.

For vertical detail, the spatial sampling rate is determined by the number of scanning lines in the raster and ideally this would equal twice the highest line number of the vertical detail components in the optical image presented to the scanning imager. There would then be no scanning line aliasing. In modern cameras, this is usually not the case and vertical aliasing may be visible in scenes that have a strong vertical detail pattern.

A solid-state imager is inherently sampled both vertically and horizontally because its sensitive surface is divided into an array of cells that are scanned in a raster pattern to obtain the electrical output. The numbers of cells in the horizontal and vertical directions are primary determinants of the imager's resolution performance. However, the output signal is still analog in the sense that the output from each cell can have any value at all, depending on the optical brightness that falls on the cell.

1.5.1.2 Sampling analysis—frequency domain

The pulses generated by sampling have characteristics similar to an RF carrier that is amplitude modulated by the video signal. The frequency spectrum of the pulses consists of a strong carrier component at the sampling frequency with sidebands that correspond to the frequency components of the video signal. (The sampling pulses themselves are rich in harmonics and the pattern of carrier and sidebands is repeated at harmonics of the sampling frequency.) The resulting spectrum for the baseband and sampled signals are shown in Figure 1.4. Figure 1.4(a) shows the spectrum when the sampling rate is more than twice the highest video frequency and thus meets the Nyquist criterion.

Figure 1.4(b) shows the effect of a sub-Nyquist sampling rate. In this case, there is an overlap between the frequency components of the video and the sidebands of the sampling signal. This overlap cannot be removed by subsequent signal processing and it is the source of aliasing. The various cases of aliasing are discussed in Section 4.7.

1.5.2 Pixels

In a digital system, a picture is made up of a series of digital values that represent individual points along the path of scanning the image. Thus, the image is divided

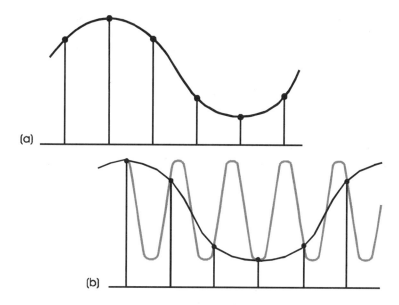

Figure 1.3 Sampling, viewed in the time domain: (a) sampling rate more than and (b) less than twice the signal frequency.

into discrete points, called *pixels*. If one examines a digital image closely, he or she will probably see little squares or rectangles of color—these are the pixels. In a well-designed system, the pixel pattern will be invisible at normal viewing ratios. The process of dividing an image into pixels is *horizontal sampling*, which means that a series of equally spaced readings of the image color have been taken during the scanning process. The output from sampling is a series of individual values, rather than a continuously varying signal.

The resolution of a digital video system is determined by its pixel counts, horizontal and vertical. This is given by two numbers in the form: (horizontal) × (vertical). A typical system (computer VGA) will have 640 × 480 pixels. The usual computer display shows all the pixels; none are lost to retrace intervals or to overscanning of the display.

1.5.2.1 Pixels in Analog Systems

It is convenient to use the concept of pixels even with analog video signals that have not been horizontally sampled. That is done by defining the width of a pixel to be one-half cycle of the highest video frequency in the system and the height of a pixel to equal the height of one scanning line. (Note that the scanning lines are a sampling process for vertical detail.) Thus, the number of pixels per line PPL is:

$$PPL = 2\frac{B_\mathrm{W}}{f_\mathrm{H}}(C_\mathrm{H}) \tag{1.1}$$

where B_W is the system bandwidth in hertz, f_H is the horizontal scanning frequency, and C_H is the fraction of the horizontal scanning interval devoted to

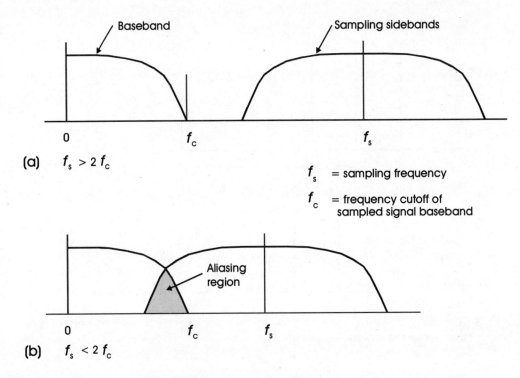

Figure 1.4 Sampling, viewed in the frequency domain.

signal transmission (active line factor).

For the NTSC television system with a bandwidth of 4.2 MHz, the pixels per line are:

$$2 \times \frac{4{,}200{,}000}{15{,}734} \times 0.84 = 448$$

The pixels per frame would then be:

$$448 \times 480 = 215{,}040$$

The number 480 is the number of scanning lines that are actually visible. The rest of the 525 total lines are blanked out for the vertical retrace. Pixels per frame is a measure of the maximum information content of a single frame for a system.

1.5.3 Quantizing

To create a fully digital signal, a second process, called *quantizing*, is necessary. In this process, each sampled value is forced to take a number corresponding to the closest position in a series of equally spaced values. This series is determined by the number of bits per pixel in the digital standard being used. For example, a system that uses 8 bits per pixel is capable of a maximum of 256 (2^8) values. For more discussion of quantization, see Section 3.3.3.

1.6 Choosing Scanning Standards

Scanning is one of the properties of a video system that must be standardized. The choice of the number of scanning lines requires a tradeoff among the conflicting requirements of image resolution, signal bandwidth or data rate, and image flicker. Before examining this tradeoff, each of the requirements and a few other contributing factors will be discussed.

1.6.1 Image Flicker and Frame Rate

In a video display, the action of scanning will become visible if the vertical-scan rate is too low. The common name for this effect is *flicker*. The visibility of flicker vs. vertical-scan rate is discussed in detail in Chapter 4, but it is sufficient to say here that it depends in complex ways on the characteristics of the human eye, the image brightness, and the viewing conditions. For television-style viewing, it has been determined that the vertical-scan rate ought to be 50 Hz or higher. For computer-style viewing, it may need to be even higher.

In the early days of television, it was important to consider the relationship between the vertical-scanning rate and the local power frequency because of serious problems in filtering all power-frequency interference out of a television receiver. As a result, early monochrome television systems synchronized vertical scanning to the power frequency—60 Hz in the United States, 50 Hz in Europe, etc. However, when color television was introduced, other considerations mandated that vertical scanning could not be locked exactly to the power line, but the frequency relationship was still kept close. These measures insured that power-line interference, if present, would cause a stationary or slowly moving pattern in the picture, thus minimizing the visibility of such interference.

The *frame rate* is the rate at which all the lines in the picture are reproduced. In a simple scanning system, the frame rate is the same as the vertical-scan rate. However, in television systems, a technique called *interlaced scanning* is used to make the frame rate lower than the vertical-scan rate.

1.6.2 Interlaced Scanning

As will be explained shortly, higher frame rates lead directly to higher bandwidth requirements. In television systems, interlaced scanning permits the frame repetition rate, and hence the bandwidth, to be reduced by one-half with very little increase in flicker. This is done by scanning only half of the total lines in each vertical scan and designing the system so that scan lines on alternate vertical scans fall between the lines of the previous vertical scan.

Thus, each frame takes two vertical scans (*fields*), with the even and odd lines scanned on alternate fields. The perception of large area flicker is based on the field rate, but the bandwidth requirements are based on the frame rate, which is one-half the field rate. The use of interlaced scanning in television to reduce flicker is analogous to the use of a shutter in a film projector that displays the film with two light flashes per frame.

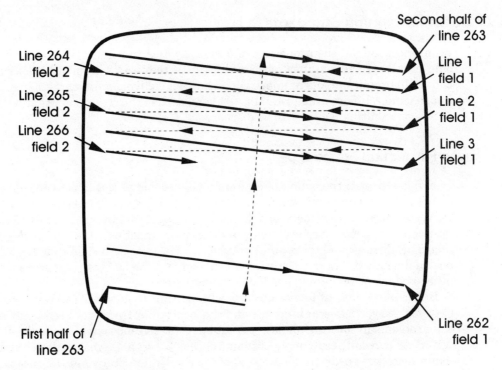

Figure 1.5 Interlaced scanning.

To generate interlaced scanning, each frame must have an odd number of scanning lines as shown in Figure 1.5. The vertical-scan frequency is chosen so that half of the scanning lines are contained in each field. This causes the first line of alternate fields to begin in the center of the picture, and the lines are interleaved between fields. In the system used in the United States there are 525 scanning lines in each frame and 262½ lines in each field.

In common with all techniques used in television to reduce the bandwidth requirements, there is a price for the benefits of interlaced scanning. The technique is not fully effective because of the tendency of the scanning lines to *pair* in the receiver, i.e., to fail to interlace. This is a receiver design problem. When pairing occurs, the effective number of scanning lines and the vertical resolution are reduced by one-half, thus eliminating the advantage of interlacing. Pairing can be eliminated by careful design of the display scanning circuits.

A more fundamental problem is small area flicker. Fine vertical detail is repeated at the slower frame rate, and a sharp horizontal edge that is reproduced only by an odd or even field will appear to flicker. Another problem with interlaced scanning is the greater visibility of motion effects because the number of scanning lines is halved for moving objects.

Small area flicker and motion problems are not particularly noticeable in most standard broadcast transmissions because the sharpness of the edges and the fineness of detail produced by the camera signal are not sufficient to generate visible spurious signals at normal TV viewing distances. They become a significant problem in HDTV systems in which the images are necessarily sharper and the detail finer because of smaller viewing ratios. As a result, the issue of interlacing is being reconsidered for HDTV (see Chapter 7).

1.6.3 Sequential or progressive scanning

The problems of interlaced scanning described in the preceding section can be greatly reduced or eliminated by the use of sequential or *progressive* scanning, where every line is scanned on every field. This is an important consideration to DTV because of the desire to reduce all kinds of picture impairments. The penalty, of course, is that twice the bandwidth is ordinarily required. However, the digital compression used in DTV will allow sequential scanning to be available as an option for broadcasting DTV. It may also be used in DTV production systems or other systems that employ wideband transmission modes such as fiber optics.

Sequential scanning is commonly used in computers and it can be used in television receivers that have a digital frame memory. The memory stores the broadcasted interlaced frames and outputs the lines in a progressive-scan format for display. DTV receivers will always have a frame memory and that allows the receiver display scanning to be decoupled from the scanning assumed in the transmitted signal. That means, for example, that a DTV receiver could employ sequential scanning for its display even though the incoming signal was interlaced.

1.6.4 Aspect ratio

Motion pictures have a wider screen than television or computers. This is determined by the system parameter called *aspect ratio*. For standard NTSC or PAL television and computers, the aspect ratio is 4:3 (1.33:1), which is the ratio of the width to the height of the picture. Wide-screen movies have a 16/9 (1.77:1) or higher ratio. At the time television standards were set, movies also were close to 4:3, but in the intervening years they have created new standards. Television developers are doing the same thing with DTV, where an option for 16:9 aspect ratio is provided.

Computer displays generally use 4:3 aspect ratios. This is not to be the same as television, but because it is convenient at the small viewing ratios involved to have a display width that can be viewed without much movement of the head. Compatibility with television is not important—computer CRTs are different than television CRTs because of the need for higher resolution and flatter screens called for by the close viewing conditions. Nevertheless, it is likely that in the future when wide-screen TV receivers are commonplace, that computer screens will also become wider.

1.6.5 Bandwidth and Data Rate

Bandwidth is a measure of an analog system's ability to transmit a range of frequencies. The scanning process generates both high frequencies and low frequencies, requiring a bandwith of at least 4.2 MHz (for NTSC television). Spectrum space for broadcast systems is limited, and it is desired to keep the bandwidth as low as possible. In a digital system, the equivalent parameter is data rate, which measures the speed at which the system can handle or transmit data.

1.6.6 The Tradeoff

As an example of the tradeoffs in choosing scanning standards for a video system, the following equation describes how the factors discussed above combine (for analog television) to establish a system's bandwidth requirement:

$$BW = \frac{1}{2} \frac{AR \; FR \; N_L \; R_H}{C_H} \qquad (1.2)$$

where BW is the system bandwidth in hertz, AR is the aspect ratio, FR is the frame rate in frames per second, N_L is the number of scanning lines per frame, R_H is the horizontal resolution desired in TVL, and C_H is the fraction of time occupied by active horizontal scanning (after subtracting the horizontal blanking period).

The factor ½ is the ratio of the number of cycles to the number of lines resolved. The aspect ratio in the equation is required because the horizontal resolution is measured in lines per picture height.

Choosing the optimum combination of BW, FR, N_L, and R_H was a major concern of the original TV system designers, since it involves a complex series of tradeoffs. From the above it can be seen that a wide bandwidth is required to resolve fine detail while maintaining a high enough picture repetition rate to avoid objectionable flicker. This explains the huge spectrum requirements of television systems.

1.7 Worldwide Video Scanning Standards

Scanning standards are in use for television systems and computers.

1.7.1 Television Scanning Standards

Television scanning standards adopted by various countries around the world are listed in Table 1.1. The diversity of standards is the result of chronological, political, and technical factors that affected the decision differently in different countries. In the current move toward HDTV standards, there is a concerted worldwide effort to minimize the number of different systems.

Some of these TV standards are more than 50 years old. Since TV receivers are purchased by consumers, who are generally nontechnical and have limited finan-

Table 1.1 Worldwide television scanning standards

System	Aspect Ratio	Inter-lace	Frames/ Second	Total/Active Lines	Lines/ Second	Bandwidth (MHz)
United States						
Mono	4:3	2:1	30	525/480	15,750	4.2
Color NTSC	4:3	2:1	29.97	525/480	15,734	4.2
Color HDTV	16:9	no	60	750/720	45,000	6.0[1]
Color HDTV	16:9	2:1	30	1125/1080	33,750	6.0[1]
United Kingdom						
Color PAL	4:3	2:1	25	625/580	15,625	5.5
Japan						
Color NTSC	4:3	2:1	29.97	525/480	15,734	4.2
France						
Color SECAM	4:3	2:1	25	625/580	15,625	6.0
Germany						
Color PAL	4:3	2:1	25	625/580	15,625	5.0
Russia						
Color SECAM	4:3	2:1	25	625/580	15,625	6.0
China						
Color PAL	4:3	2:1	25	625/580	15,625	6.0

1 Digital transmission. Scanning standards shown are typical—other variations are possible.

cial resources, it has been important to set a single standard within a country and hold it unchanged for many years. This allows the lowest-cost receivers to be built and the consumers' investment in them pays back over a long useful life. The situation is quite different in the computer market.

1.7.2 Computer Scanning Standards

Until very recently, personal computer (PC) users were required to have considerable technical knowledge about their system. Although this has surely limited the market for PCs, the value of the technology has been great enough that large numbers of users have made the investment in learning their systems. This has caused the standards situation to be very different from TV.

The personal computer is inherently a programmable device. With software, one can theoretically program a PC to do anything. Although the software might be able to do anything, there are limits on what the hardware can do. The thrust in hardware, then, has been to make it as powerful and flexible as possible—

Table 1.2 Computer scanning standards

Resolution Mode	Color Mode	Frames/s	Lines/Frame	Lines/s	Data Rate (MB/s)
640 × 480 VGA	4bpp	60	525	31,500	9.2
640 × 480 SVGA	8bpp	60	525	31,500	18.4
640 × 480 SVGA	RGB-16	60	525	31,500	36.8
640 × 480 SVGA	RGB-24	70	525	36,750	64.5
1024 × 768 SVGA	8bpp	70	800	56,000	55.0
1280 × 1024 SVGA	4bpp	70	1100	77,000	45.9

consistent, of course, with a selling price that users can afford. Over the years of PC development, hardware capabilities have grown at a prodigious rate. For example, the processing power of the microprocessor chip that is the heart of a PC has increased by a factor of 2 every two years! That kind of growth has made product life in the PC market much shorter than it is in television.

Turning to the effect this growth environment has had on video display technology, computer displays are generally designed to operate with a variety of scanning standards, covering a range of line and field frequencies as well as a range of color formats. Often a single PC will operate with different scanning at different times, simply by changing its software. The display hardware must be able to accommodate such versatility. Thus, there is much less need for a single country-wide scanning standard for computers.

However, computers need to share data with one another and the data requirements are different when scanning standards are changed. Although data can theoretically be converted between standards with software, the process is not always perfect and it uses computer resources that may slow down other operations. Therefore, the industry has made an effort to limit the number of different standards, but there is no reason to have only one. Table 1.2 lists some of the most popular scanning standards for IBM-compatible PCs.

The letters "VGA" in the table stand for *video graphics array*, which is a hardware video display standard originally developed by IBM and widely used. VGA defines many different resolution and color modes, although the one given in the table is the highest one. "SVGA" stands for *super VGA*, an enhanced version of VGA that supports higher-resolution and higher-color modes. SVGA has become the current standard because it supports video modes that do a better job of reproducing realistic images.

The color mode column in the table tells how many colors the format can reproduce at once. "4bpp" refers to 4 *bits per pixel* and can only display 16 (2^4) colors on the screen at once. This is extremely limiting for displaying realistic images and is the reason that plain VGA displays are phasing out. An 8bpp color mode displays up to 256 colors at once, which is still too small a number for real pictures. However, an ingenious technique called *dithering* allows remarkably good reproduction of many images using only 8bpp (see Section 7.4.2.2).

The other color modes (RGB-16 and RGB-24) support up to 65,536 or 16,777,216

Video System Fundamentals 17

colors and give excellent reproduction of realistic images. These modes are sometimes called 16bpp and 24bpp modes. Color modes are discussed more completely in Chapter 6.

Notice that there is a range of horizontal and vertical scanning frequencies in the table and that the numbers are all higher than television numbers. This is because all the formats listed use progressive (noninterlaced) scanning.

The data rates listed in the table represent the raw data that must be supplied to update a display at the frame rates indicated. These rates normally occur only within the video display hardware of the computer. Data rates handled by the microprocessor and the storage of the computer are many times lower because of a variety of strategies for data rate reduction, called *compression*. This is covered in Chapter 3.

1.8 Television Signal Waveforms

The units of most television systems communicate by means of a composite signal, which is a special format that contains all the information needed to convey a picture (brightness, color, synchronization) on a single cable. The composite signal is standardized by specifying its waveform, which is a view of how the signal varies with time. Figure 1.6 shows the waveform diagrams for the brightness or *luminance* part of the NTSC color television system used in the United States.

It is important to remember that these waveform standards were set 50 years ago, when all television technology was analog and the hardware capabilities were not what they are today. It is remarkable that any standard in electronic technology should last so long! The standards contain features that are specific to the problems of the era when they were developed; some of these are not as important today. However, the following discussion explains the standards in terms of the environment in which they were developed. Later chapters will give the modern views of some of these issues.

Views (a) and (b) of Figure 1.6 show the luminance waveform in the vicinity of the vertical retrace period of the scanning. The signal interval that accommodates vertical retrace is called the *vertical blanking interval* (VBI) and it takes about 7.5% of the frame time. The word "blanking" means that the video voltage during the VBI has a range of values that will always appear as black on a television receiver. This was necessary to insure that a video display will show no spurious information while vertical retrace is taking place. View (c) shows some details at the end of the VBI. View (d) shows details of the vertical synchronizing part of the VBI. Finally, view (e) shows details of the *horizontal blanking interval*, which is the time allowed for horizontal retrace of the scanning. Horizontal blanking takes 14 to 18% of a line time, so the total of blanking intervals is about 25% of a frame. That time is not available for picture transmission.

1.8.1 The dc component of a video signal

The drawings in Figure 1.6 show the video signal in a black-positive polarity, with peak white in the picture represented as 7.5% signal, blanking level at 75% signal,

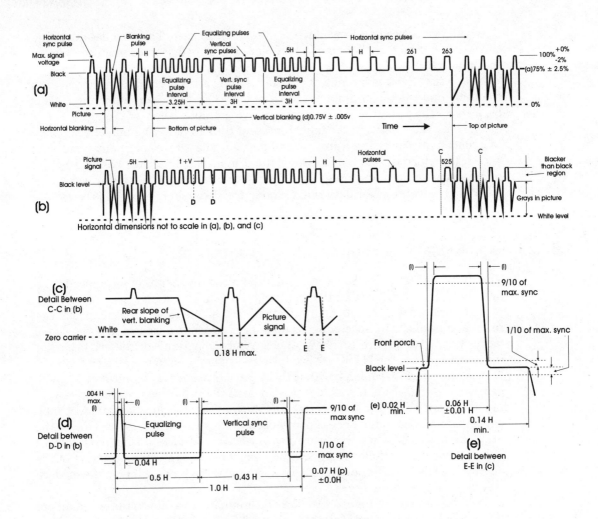

Figure 1.6 NTSC luminance video waveform drawings.

and synchronizing pulses at 100% signal. This is the format at the output of a television transmitter, where the synchronizing pulses produce the highest power output. Notice that the average voltage of the signal will vary with the average brightness of the picture-dark pictures will have a higher average voltage than bright pictures. This means that there is a direct-current (zero-frequency) component in a video signal.

It is not common practice to transmit the dc component throughout an analog video system. At most places in the system, video voltages are ac-coupled, which means that there is no absolute voltage consistently associated with any part of the signal. As shown in Figure 1.7, the absolute voltages of the signal vary with

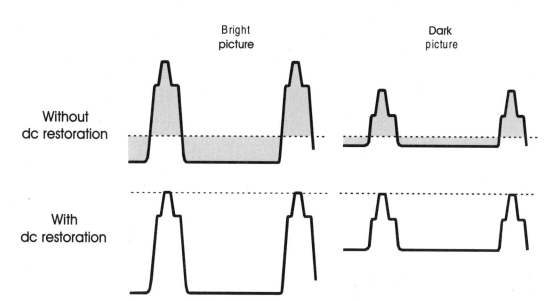

Figure 1.7 Dc restoration.

the average brightness of the picture. However, there are places where the dc component is required for correct operation. The TV transmitter is one place; other places are where any gray-scale operations take place and at the video display device. Fortunately, it is easy to restore the dc component wherever required by using a circuit that continually sets some level in the signal to a specific value. This is called *dc restoration* and it operates by setting all sync pulse tips to the same level as shown in Figure 1.7.

In a digital video system, the dc component is retained throughout the system because the digital data carries absolute values for the pixels. When a system involves conversion from analog to digital formats, dc restoration must be used at the conversion point to make sure the digital data does contain the dc component.

The video waveforms in Figures 1.6 and 1.7 are shown as black-positive, which applies to the output amplitude from a NTSC television transmitter. However, at most other places in the system, the signals are white-positive. This is simply a polarity inversion and has no other significance.

1.8.2 Synchronizing Signals

Television signals must carry the information necessary for the scanning of displays to be synchronized with the scanning that took place in the imager. This is done by adding *synchronizing pulses* (usually called *sync pulses*) to the horizontal and vertical blanking intervals. In order that the sync pulses do not interfere with the purpose of the blanking intervals, they go "blacker than black" so they represent the greatest signal excursion in the black direction. This also allows them to be easily separated from the rest of the video by simple clipping (passing of signals above a particular signal level). Proper clipping of the sync, called *sync separation*, requires dc restoration at the clipper.

Because the video content of the signal is constantly changing and could interfere with synchronization if it got into the sync circuits of a display or a receiver, the NTSC standard signal contains several features that facilitate clean sync separation. Providing for clipping of sync is one of them.

Another such feature is the *front porch* on the horizontal blanking interval. In Figure 1.6(e), the front porch is shown as a short inflection between the start of horizontal blanking and the beginning of the horizontal sync pulse. This makes sure that variation of picture content at the right-hand edge of the picture cannot smear over into the edge of the sync pulse and disturb the timing of the synchronization. In a signal channel that has poor high-frequency response, video signals can become slightly smeared to the right. The front porch provides a small barrier region to prevent that affecting horizontal synchronization.

There is a related problem for vertical synchronization except that it is caused by interlacing rather than video content. This gives rise to the features of the NTSC signal called *vertical serrations* and *equalizing pulses*, shown in Figure 1.6(a) and (b). In principle, vertical synchronization could be accomplished by providing a sync pulse and a front porch in the vertical interval, just like the horizontal interval. The vertical sync pulse could be wide enough that it could be separated from the horizontal pulses after clipping by simply narrowing the bandwidth (called *integration*) of the sync enough that the relatively narrow horizontal pulses are suppressed relative to the wider vertical pulse. Another clipping operation then separates the vertical pulse by itself. This is actually what is done in a TV receiver.

However, the wide vertical pulse would represent an interruption of horizontal synchronization. That is taken care of by making narrow slits (serrations) in the vertical pulse at horizontal sync times during the vertical pulse. The black-going edge of the serrations can be horizontal sync. At other times in the vertical interval, horizontal sync pulses could be present as normal. This way, horizontal sync continues through the vertical interval.

But, as shown by Figure 1.8, the serrations and the horizontal sync pulses cause the vertical sync to look slightly different on the odd and even fields of the interlaced scanning. That seems like a small effect, but it is great enough to cause almost 100% pairing of the interlaced lines. The solution to this problem is to make the serrations at twice-horizontal frequency and also to add twice-horizontal pulses (equalizing pulses) for a short period before and after the vertical sync. The result is that vertical sync separation by integration is greatly enhanced.

1.8.3 Spectral Content of Video

A monochrome video signal (or the luminance component of a color signal) covers the system bandwidth from dc to the highest frequency determined by the horizontal detail of the picture. However, the frequency spectrum of video is not continuous—it contains a fine-grained structure determined by the scanning frequencies.

Video System Fundamentals 21

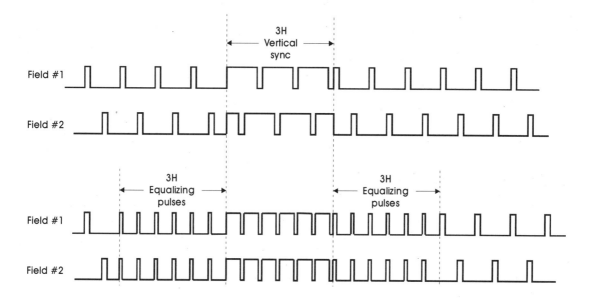

Figure 1.8 Correction of vertical sync errors with serrations and equalizing pulses.

This fine-grained spectral frequency structure, shown in Figure 1.9, results from the fact that video contains three periodic components, one repeating at the line rate, one at the field rate, and one at the frame rate—nominally at 15,750, 60 and 30 Hz for monochrome television. Each component can be expressed by a Fourier series, a sequence of sinusoidal terms with frequencies that are multiples (harmonics) of the repetition rate and having amplitudes determined by the waveform of the component. The summation of the amplitudes of the terms for the three periodic components over the entire video spectrum is the frequency content of the signal. The 30 and 60 and even the 15,750 Hz harmonic components are too closely spaced to be distinguished on a frequency scale of 4.2 MHz, and they are not disclosed on a conventional amplitude-frequency graph.

The nature of these harmonic components can be visualized by considering some special cases. (For the purpose of this analysis, the vertical blanking and synchronizing pulses are ignored.)

The simplest cases are all-white pictures, or pictures consisting of vertical bars. Each scanning line is identical, and the frequency components are harmonics of the line frequency with amplitudes as determined by a Fourier analysis of the waveform, including the horizontal blanking and sync pulses. The number of harmonics is large, usually extending to the limit of the video bandpass. In a 4.2-MHz channel it would be:

$$\frac{4.2 \times 10^6}{15,750} = 266$$

Another example is a picture in which the top half is white and the bottom half black. It generates two signal components that are sequential in time and periodic

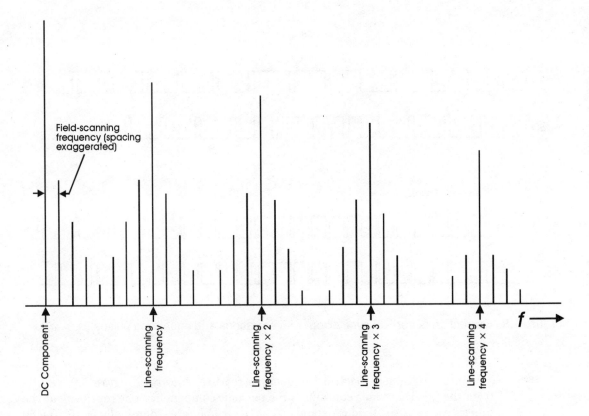

Figure 1.9 Fine-grained frequency spectrum of a video signal (field-scanning frequency components are shown with greater spacing for clarity).

at the line rate, one for the white area of the picture and one for the black. The spectrum of this signal will consist of a series of frequency components harmonically related to the line frequency and with their amplitudes determined by a Fourier analysis of the white and black waveform. In addition, however, the transitions from the black to white signals will generate another series of frequency components that are harmonically related to the field frequency, the rate at which the black-to-white transitions are repeated.

The relationship between the line and field frequency components is analogous to that between the carrier and modulating frequencies in an amplitude-modulated signal. The field frequency components cluster around the line frequency harmonics like sidebands around a carrier as shown in Figure 1.9 [1, 2]. The field frequency components convey the information necessary to reproduce line-to-line changes in the brightness of the image, either in time as the result of motion in the scene or in space as the result of picture content. Their number and amplitudes depend on the rate of change of brightness in the vertical direction.

In a stationary scene with typical brightness variations, most of the vertical components are contained in a band ±2 kHz on either side of each line rate harmonic for a total bandwidth of 4 kHz. This is less than half the 15,750-kHz spacing between harmonics. If there are moving objects or unusually rapid vertical brightness variations in the image, the clusters broaden, but seldom so much as to occupy more than one-half the spacing between harmonics. An exception can occur when the vertical variations in brightness are extremely rapid. Under this condition, the clusters can broaden so far that the field-rate components surrounding adjacent line-rate harmonics overlap. This results in aliasing. In general, however, the standard television signal is wasteful in its use of spectrum space as well as power, and half of the spectrum space is normally unused.

In summary, the frequency spectrum of the luminance component of a color signal or of a monochrome signal consists of a series of harmonics of the line frequency, each surrounded by a cluster of frequency components separated by the field frequency. The amplitude of the line-frequency components is determined by the horizontal variations in brightness while the amplitude of the field frequency components is determined by the vertical brightness variations.

1.8.3.1 Utilization of gaps in the video signal spectrum

The presence of gaps in the spectral content of luminance signals provides an important opportunity to transmit additional information without requiring more scarce bandwidth. These gaps have been utilized for the following purposes:

1. Compatible color systems. As will be described in Chapter 2, the frequency components of the color subcarrier and its sidebands, which also appear in clusters, are interleaved with the luminance frequency components. This minimizes the cross-talk between the luminance and chrominance signal components and makes it possible for the luminance channel of the receiver to utilize the full 4.2-MHz bandwidth of the video channel, even though it includes the color subcarrier. This is best accomplished by the use of a *comb filter* (see Section 6.11.3) in the luminance channel that removes the chrominance channel components without normally affecting the luminance.

2. Signal-to-noise improvement. A comb filter can also be used in the luminance channel of the camera to remove noise from the spectral regions between the line-scanning frequency clusters. Nearly half the noise can be removed by this technique without affecting the signal, leading to an improvement of up to 3 dB in the high-frequency signal-to-noise ratio.

1.8.4 Color Video Waveforms

The video waveforms shown in Figures 1.6 and 1.8 are for monochrome TV. Color TV requires the transmission of three such waveforms and that is exactly what is done in many computer systems that use the RGB color system. However, the use of separate signals for the colors is impractical for broadcasting, so they are

24 Chapter One

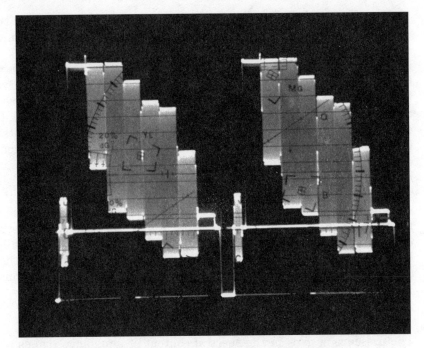

Figure 1.10 Waveform display of an NTSC color-bar video signal.

combined by an encoding process into a *composite* color signal.

NTSC or PAL composite color signals are designed to be viewable on monochrome displays—that is called being compatible. Compatibility is achieved by providing a luminance signal component that is exactly the same as a monochrome TV signal. Color information is added by encoding it on a high-frequency subcarrier that is superimposed on the luminance signal. Because the subcarrier frequency is high, it is not very visible on monochrome displays. There is more to it than that, which is explained in Chapter 5.

The basic sync and blanking waveforms are the same for color, but tolerances must be tightened and a color burst added for color subcarrier synchronization.

1.8.5 Viewing Video Waveforms

In the operation and maintenance of a television system, it is necessary to be able to view the video waveforms at different points. An instrument for that purpose is the waveform monitor (WFM), which is a special-purpose oscilloscope designed for viewing television waveforms. Figure 1.10 is a photograph of a WFM screen that shows the NTSC color signal for a color-bar pattern. The display is synchronized with the horizontal scanning and shows two lines. Vertical synchronization is also available in waveform monitors so that all the details shown by Figure 1.4 can be observed and measured.

1.9 Digital Video Signals

The concept of signal waveform that was just covered does not apply to digital video signals. Digital video being transmitted in real time is referred to as a *bit stream*, which is a continuously flowing sequence of digital bits. Observing it with an oscilloscope or waveform monitor, one sees a varying high-frequency signal, but the structure of an analog video signal is not visible. In order to see the structure of the digital video, its format must be known. Knowing the format, a computer or a dedicated processor can decode the bit stream and produce a form of video signal. The digital video format, then, is its signal standard.

Digital bit streams use the concept of a *header* to describe structure. A header is a block of data of a known format that contains information about the parameters and structure of the data that follows. It performs the functions of the sync pulses in analog video, but it actually does much more. There must always be a header at the start of a bit stream to identify it, its type, the details of the data's structure, and where the data begins. For example, a video data stream header may contain the name of the video, its resolution (pixel counts), its bits per pixel, details of any compression technique used, and other things. The header information tells the receiving system how to begin decoding the data.

There may be additional headers embedded within a bit stream. For example, a video stream will have embedded headers for every frame. Thus, the characteristics of the video stream can change on a frame-by-frame basis. A header will also contain a data count so that the system can tell when it gets to the next header. Since a digital system can be as precise as needed, it can count exact numbers of bits from header to header, even if there are millions of them.

The header concept supports one of the most powerful features of a digital system—it can be programmable to as much depth as needed. In standardizing for digital video, one must standardize on the content and layout of the headers, but that content can provide for nearly infinite variability of the data itself. For example, if the horizontal and vertical pixel counts are contained in the header, then a bit stream can be built for any resolution at all. The decoding system's software reads the header pixel counts and sets itself up to decode at the resolution required for the bit stream. Another bit stream can contain totally different numbers and the system will still work.

Carefully designed headers can provide for future options that didn't even exist when the header structure was initially set up. Therefore, a digital video system can improve in performance over time, simply by adding new software for new features. The new software is enabled when a bit stream calls for it. There is no need to expect that a detailed (and limiting) standard has to last for 50 years! Digital data formats are covered in detail in Chapters 3, 6, and 7.

1.10 Analog-Digital and Digital-Analog Conversion

Modern video systems can be all-analog, partly digital, or almost-all digital. The "almost-all" is because camera imagers and display devices are still analog. Even on a computer, "real" pictures start from an imager whose output is analog, and they are viewed on a monitor that uses an analog CRT. Except for an all-analog system, every video system includes conversion to digital and back to analog. These processes are called *analog-to-digital conversion* (ADC) and *digital-to-analog conversion* (DAC). Today, each one is done by a single integrated circuit.

The basic ingredients of ADC—sampling and quantizing—have already been explained. However, there are many considerations of sampling rates, filtering, digital formats, etc., that affect the performance of the conversion. These will be discussed in Chapter 3.

1.11 Summary Comparison of Analog and Digital Systems

There are many reasons for the trend to digital techniques in the design of video systems, even when the end result will be analog, such as for present-day TV broadcasting or cable systems. These will be covered in detail in Chapter 3, but here is a brief summary.

1. Digital systems can have perfect error correction so that transmission distortion, noise, or other degradation does not affect the information being transmitted. As a result, distortion does not accumulate as a system becomes larger.

2. Digital video is easily processed for delay, timing correction, signal enhancement, special effects, and storage with little, if any, quality losses.

3. Digital video can be compressed to remove inherent redundancy of information that results from the scanning process. Using these techniques, it can be transmitted with lower bandwidth requirements than analog video of equivalent subjective quality.

4. Although digital processes are often structurally more complex than their analog equivalents, they can be encapsulated into integrated circuit devices, resulting in lower-cost, higher-performance, and easier-to-build hardware.

5. The use of software-controlled (programmable) digital systems offers fantastic flexibility in signal processing. Furthermore, processing algorithms (techniques) can be changed or updated without the need to modify any hardware.

However, every comparison has its other side. As has already been explained, everything can't always be done digitally (at least not yet), so there are still analog imagers and display devices. The ADC process has limitations that result in the digital signal not being an exact reproduction of the analog input. Although the digital circuits themselves may well deliver perfect reproduction, there will be an accumulating analog distortion every time the signal goes through the ADC and DAC processes. Therefore, one must minimize the number of such conversions that cascade in a system.

1.12 The Audio Component

In the days of silent motion pictures, live music was played along with the showing of a movie so that the audience's sense of hearing was supported. Truly "si-

Table 1.3 Digital audio formats—PCs

Mode	bpp	Stereo	Bandwidth (kHz)	Sampling (kHz)	Data Rate (kB/second)
PCM speech	8	no	5	11.02	11.0
PCM music	16	no	10	22.05	44.1
CD-quality	16	yes	18	44.1	176.4
ADPCM speech	4	no	5	11.02	5.6
ADPCM music	4	yes	10	22.05	44.1

lent" movies were pretty unsatisfying because an important element of realism was lacking. The same can be said of video—sound is an essential ingredient of an effective and engaging presentation. All video systems must have sound (audio).

In analog television broadcasting, audio is transmitted by a separate transmitter operating at a fixed frequency offset from the video transmitter. In computers, audio is a data stream that can be handled separately or it can be interleaved with the video data. In DTV, audio is interleaved with the video. In all cases, the audio part of the system involves the same steps that were shown for video in Figure 1.1—creation, storage and transmission, and reproduction. Much of what has been said about video in this chapter also applies to audio, except that the numbers are grossly different. That makes audio a very different problem from video.

1.12.1 Audio Bandwidth

Video bandwidths are in the megahertz range, but audio bandwidths range only up to 20 kHz. However, the requirements for amplitude linearity and noise performance are much more demanding in an audio system than a video system. These requirements combine to make audio system design nearly as much of a challenge as video system design. In addition, audio poses another synchronization requirement—the audio must stay in sync with the video. Since the most demanding audio sync situation is when a person is talking on camera, audio-video synchronization is often referred to as lip sync.

Unlike video, audio cannot stand repetition of information (frames) or interruptions. Any such effects, however small, are sure to be heard.

1.12.2 Digital Audio

The use of digital audio in the mass market is more advanced than the use of digital video. This is primarily because of the audio Compact Disc (CD) that has taken over the prerecorded audio field from records and tapes. The audio CD provides extremely high performance in a conservatively designed system.

Digital audio also is now widely available in PCs. There, it has been handled in much the same way as digital video, using programmable formats, data compression, and the other digital features discussed in Section 1.7. The typical PC audio format uses less data than video, so it places a lower demand on the capabilities of the PC and it is therefore more often used. Table 1.3 lists some typical PC audio formats.

Notice that there is a wide range of data rates from minimal speech performance up to full CD-quality digital audio. The acronym "PCM" stands for *pulse code modulation*, which is what results from simple sampling and quantization. "ADPCM" stands for *adaptive differential PCM*, which is a compression technique. ADPCM can provide up to 4:1 compression, which is much less than can be achieved with video compression techniques. This is because audio has far less structure and redundancy than video. The programmable nature of PCs allows any of these audio formats to be used wherever they best suit the application. The computer will play audio at whatever quality level represented in the data it receives.

In DTV systems with audio data compression, a single audio channel may use only a small percentage of the channel data rate. For example, the ATSC DTV standard (see Section 7.3) supports up to 5-channel audio using only about 2% of the total data rate.

1.13 Summary

This chapter has introduced some of the fundamental concepts underlying video systems, whether analog or digital. Basic system architecture was covered, and the principles of scanning standards selection, sampling, signal formats, synchronization, and data conversion have been discussed. The rest of this book covers all of these subjects in greater detail.

1.14 References

1. P. Mertz and F. Gray, "A Theory of Scanning and Its Relation to the Characteristics of the Transmitted Signal in Telephotography and Television," *Bell System Technical Journal*, Vol. 13, July 1934.
2. P. Mertz, "Television–The Scanning Process," *Proc. IRE*, Vol. 29, October 1941.

Chapter 2
Color Video Fundamentals

2.1 Introduction

This chapter is devoted to colorimetry[1] and the properties of color that are the basis of color video technology.

2.2 The Properties of Color—A Summary

The term "color" as used in color video has a dual meaning—it is both a physical property of visible light and the perception of this property by human vision. Color as a physical property can be defined and measured in objective and precise terms. Its perception is subjective, it varies with individuals, and it depends on the surrounding environment so that it must be specified less precisely on the basis of an average or "standard" observer under "standard" conditions. A major task of video engineers is to establish and define measurable criteria for color images that coincide as closely as possible with their perception.

Each picture element in a color image has three basic properties. In objective terms they are luminance, hue, and saturation. The corresponding perceptual terms are brightness, color, and purity.

The term *luminance* has approximately the same meaning in color video as in monochrome (see Sections 1.8 and 4.4). As with monochrome, a color video system must have the capability of transmitting the luminance of each picture element, but in addition it must transmit and reproduce its hue and color purity or saturation.

---

1. Colorimetry originally meant the measurement of color with a colorimeter, but it now has the broader meaning of the science of color.

Table 2.1 Hue vs. wavelength

Hue	Approximate wavelength (nm)
Violet	400
Blue	450
Cyan	490
Green	520
Yellow	575
Orange	590
Red	640

2.2.1 Hue

The most prominent characteristic of color picture elements is their *hue*, the sensation created by visible light that is commonly described as its color (red, yellow, orange, etc.). The term "hue" is often used synonymously with "color."

Visible light is electromagnetic radiation having a spectrum of wavelengths[2] extending from approximately 400 to 780 nm (1 nanometer = 10^{-9} m)[3], and its hue is determined by its wavelength. The band centers of the wavelengths that produce the sensation of common hues are presented in Table 2.1.

This table is based on monochromatic light, i.e., light of a single wavelength. With the important exceptions of pure spectral colors resulting from the excitation of atoms as in a sodium vapor light or a laser, the colors that appear in nature are polychromatic, i.e., they are mixtures of many wavelengths. Each has a dominant wavelength that establishes the visual perception of its hue, but it may contain radiation components with wavelengths that extend over the entire visible spectrum. White or gray light results when the radiation at all wavelengths is present in approximately equal amounts.

Radiation with wavelengths just beyond the limits of the visible spectrum also has important applications in specialized video systems. Wavelengths in the range from approximately 780 to 25,000 nm are known as *near infrared*, while in the range of 100 to 400 nm they are known as *ultraviolet*.

2.2.2 Saturation

The appearance of any color can be duplicated by a mixture of white or gray light and a pure spectral color, the dominant wavelength, in the proper proportions

---

2. Wavelength rather than frequency is customarily used to specify position on the visible and neighboring regions of the spectrum. Wavelength and frequency are related by the equation: [wavelength (m)] [frequency (Hz)] = 3×10^9 m/s = velocity of light.

3. Scientists sometimes use the Angstrom unit (10^{-10} m) rather than the nanometer for specifying wavelength.

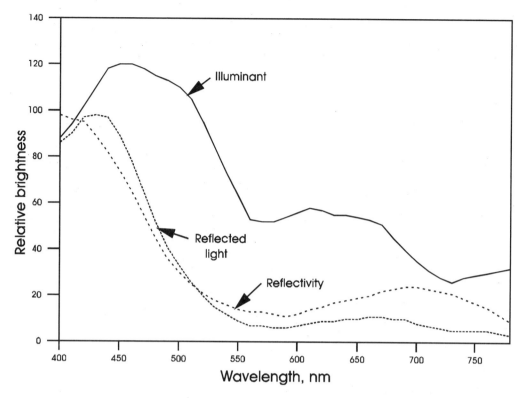

Figure 2.1 Spectral distribution of reflected light.

(see Section 2.5). The ratio of the magnitude of the energy in the spectral component to the total energy of the light defines its *saturation* or "purity." A pure spectral color has a saturation of 100 percent, while the saturation of white or gray light is zero.

2.3 Color Specifications

The three basic physical properties of color radiation—luminance, hue, and saturation—can be completely specified by its spectral distribution, a plot of its radiant energy vs. wavelength. Figure 2.1 shows the spectral distributions of incident light on a reflecting surface, the reflectivity of the surface, and the reflected light, which is the product of the first two.

A graph of amplitude vs. wavelength, however, is not always the most useful format for specifying color—numerical criteria are often more convenient—and the graph does not describe the relationship between the physical properties of color and their visual perception.

Scientists have studied the nature of this relationship for the past 400 years (Sir Isaac Newton carried out basic research on this subject in the early 17th century), and numerical criteria have been developed for specifying color that are stated in technical terms but are based on the subjective perceptions of thousands of viewers. Four of these criteria are commonly used in color television:

1. Primary color mix
2. Dominant wavelength and saturation
3. Luminance and color differences
4. Coordinates on the CIE Chromaticity Diagram

All of these criteria directly or indirectly specify three independent variables. This is consistent with the *trichromatic theory* of color, which states that the sensation of color results from stimulation of three sets of *cones* in the retina, each with a different spectral sensitivity. The perceived hue and saturation of a color is determined by the ratios of the amplitudes of the responses of the sets of cones to the color stimulation. Thus two colors will appear to have the same hue and saturation if they stimulate the same response from the retinal cones, even if the distribution of their spectral energies is different. (Their appearance is also affected by differences in the viewing environment, a fact well known to painters.)

2.4 The Primary Colors

A consequence of the trichromatic theory of color is that the hue and saturation of most colors can be duplicated by combining three *primary colors* in the proper ratio. This theory was one of the results of Sir Isaac Newton's work. It has since been developed further to meet the needs of the film and video industries.

There is wide latitude in the permissible spectral content of primary colors. The most important requirements are that they be independent, i.e., no one of the three can result from a combination of the other two, and that they be chosen so that the widest possible range of hues and saturations can be produced by suitable combinations of the three. Within these broad criteria, considerable variety is possible in their spectral characteristics.

Video colorimetry is basically different from that used in photography and painting or in the perception of color in real life. All of the latter are *subtractive* systems in which the picture or scene is illuminated by an external source of light such as sunlight, which includes components of many hues. The hue of the image is produced by the subtraction of color components by absorption, reflection, or transmission. Color television is an *additive* system since it produces hues by adding the primary color components. As a result, two types of primaries, subtractive and additive, have been defined (Table 2.2)

2.4.1 Subtractive primaries

The subtractive primaries are magenta (a blueish-red), yellow, and cyan (a greenish-blue). Each absorbs one of the additive primaries and reflects or transmits the other two. Since magenta has a reddish cast and cyan is bluish, it is popularly (and erroneously) stated that the primary colors are red, yellow, and blue.

Subtractive primary colors are used in the form of inks for color printing. Be-

Table 2.2 Primary colors

Subtractive Primaries	Reflects or Transmits	Absorbs
Magenta	Red and blue	Green
Yellow	Red and green	Blue
Cyan	Blue and green	Red
Additive Primaries	**Mixtures**	**Produce**
Red	Blue plus red	Magenta
Green	Red plus green	Yellow
Blue	Blue plus green	Cyan

cause practical ink pigments do not produce perfect primary colors, laying down all three primaries does not create a good black. This problem is solved in color printing by using four inks: cyan, magenta, yellow, and black. This combination is referred to as *CMYK* color (the K is for black).

2.4.2 Additive primaries

Additive primaries can be derived from the subtractive, for example, red can be produced by mixing magenta and yellow, the former absorbing green and the latter blue from the incident white illuminant, leaving red.

The luminance of the additive primary colors, R (red), G (green), and B (blue), and the amplitude of their electrical analogues, E_R, E_G, and E_B, are the basic color television parameters, and all other color specifications are derived from them. After gamma correction (see Section 4.14), it is customary to designate the amplitude of the electrical signals as E'_R, E'_G, and E'_B. The outputs of the camera and the inputs to the kinescope are the electrical analogues of the three primary colors in the scene.

Additive primaries were seldom used until the advent of color television because that is virtually the only imaging system that creates pictures by the addition of color components. Other hues, including the subtractive primaries, can be produced by combining the additive primaries. It is not intuitively obvious that yellow light is produced by mixing red and green, but it can be confirmed by superimposing red and green spotlights on a screen.

2.5 Dominant Wavelength and Saturation

It can be shown, both by theory and experiment, that the luminance, hue, and saturation of every color in nature can be duplicated by a mixture of white or gray light and monochromatic light in the proper proportions, no matter how complex its spectral distribution. The *wavelength* of the monochromatic component of this mixture that establishes its perceived hue is known as the *dominant wavelength*.

The electrical analogues of the dominant wavelength and saturation are the basis for the NTSC, PAL, and SECAM systems that are used for the transmission of color by all present-day broadcast systems.

2.6 Luminance and Color Difference Components

Colors can also be specified in terms of their *luminance* (Y) and *color difference* ($R-Y$ and $B-Y$) components where R and B are the amplitudes of the red and blue primaries. The amplitudes of the corresponding electrical analogues are E_Y, E_{R-Y}, and E_{B-Y} or, after gamma correction, E'_Y, E'_{R-Y}, and E'_{B-Y}. The luminance component, E_Y, is given by the equation:

$$E_Y = 0.299\, E_R + 0.587\, E_G + 0.114\, E_B \tag{2.1}$$

The coefficients of the primary components are based on the sensitivity of the human eye to each of the primary colors separately. The magnitude of E_Y is then proportional to the perceived *brightness* of the scene. The electrical analogues are the basis for *component* recording and transmission, a format that is widely used for video recording and other applications with high demands for quality (see Section 6.3.2).

2.7 The CIE Chromaticity Coordinates

The CIE (*Commission Internationale de l'Eclairage*) chromaticity coordinates are based on a system for describing colors in terms of the *color-matching functions*, x', y', and z', shown in Figure 2.2. These functions were developed on the basis of hundreds of observations, and are used to calculate the *tristimulus values*, X, Y, and Z:

$$X = \int_{380}^{780} L(\lambda) x'(\lambda)\, d\lambda \tag{2.2}$$

$$Y = \int_{380}^{780} L(\lambda) y'(\lambda)\, d\lambda \tag{2.3}$$

$$Z = \int_{380}^{780} L(\lambda) z'(\lambda)\, d\lambda \tag{2.4}$$

where $L(\lambda)$ is the spectral energy density, and x', y', and z' are the color-matching functions. The properties of the color matching functions are such that colors with the same tristimulus values will appear the same, even though their spectral distributions are different.

The *chromaticity coordinates*, x and y, are calculated from the tristimulus values:

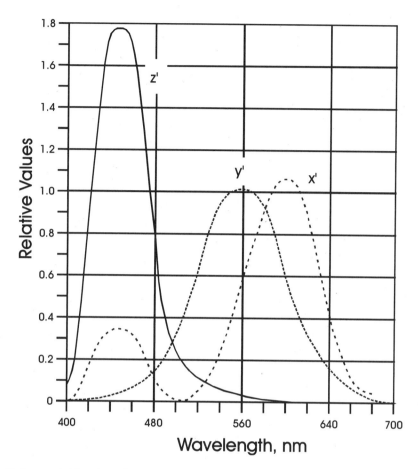

Figure 2.2 Color-matching functions.

$$X = \frac{X}{X+Y+Z} \tag{2.5}$$

$$Y = \frac{Y}{X+Y+Z} \tag{2.6}$$

(Since $x + y + z = 1$ by definition, it is unnecessary to calculate z.)

The concept of chromaticity coordinates was a major milestone in the development of imaging system technology. It was developed during the late 1920s and early 1930s, and the coordinates of the colors in the visible spectrum were first calculated and published in 1931. It is an invaluable tool for the analysis and design of color television products and systems. It is also a link between the objective physical nature of color and its subjective perception.

The plot of the chromaticity coordinates of visible light is known as the *CIE Chromaticity Diagram*, Figure 2.3. The color-matching functions and the

36 Chapter Two

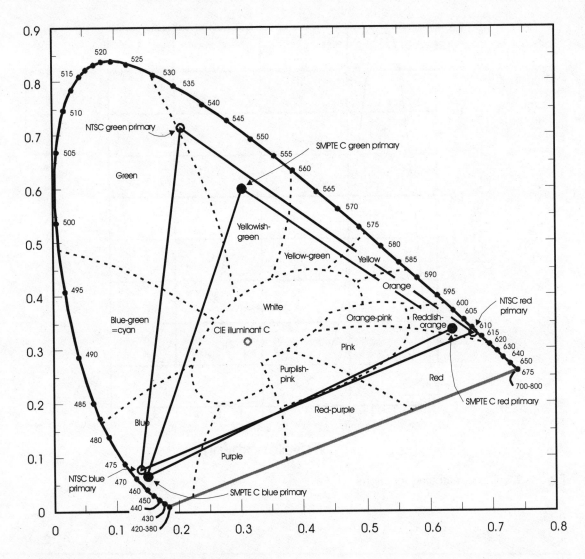

Figure 2.3 The CIE Chromaticity Diagram.

Chromaticity Diagram are known collectively as the *CIE 1931 Standard Observer*. A supplement to the 1931 paper, the *CIE 1964 Supplementary Standard Observer*, added data for viewing angles greater than 2°.

The following features of the Chromaticity Diagram should be especially noted:

1. The coordinates of spectral (monochromatic) colors are located on the horseshoe-shaped curve around the periphery of the diagram. Their wavelengths are indicated on the curve.

2. The coordinates of white light are located in the center of the diagram. The area on the graph corresponding to "white" light is rather large because there

Table 2.3 Chromaticity coordinates of standard illuminants

Designation	Source	x	y
Illuminant A	Tungsten at 2856K	0.4476	0.4074
Illuminant C	Daylight	0.3101	0.3516
Illuminant D_{65}	Daylight (revised)	0.3127	0.3290

is an almost infinite variety of white hues. The three sets of coordinates for white light that have been established by the CIE are tabulated in Table 2.3. The spectral power distributions of these *illuminants* are shown in Figure 2.4.

Table 2.3 includes a reference to *color temperature*, a fifth criterion for specifying the color (see Section 2.3) of illuminants. The use of color temperature as a specification results from the fact that the spectral distribution of radiant energy from an incandescent body (known as a black body or Planckian radiator) is determined solely by its temperature, usually expressed in Kelvin (K). As the temperature of a body is raised, its color changes from red to white, a fact reflected in the common terms "red hot" and "white hot."

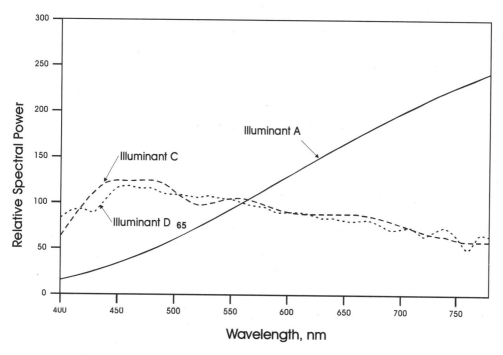

Figure 2.4 Spectral distribution of standard illuminants.

The spectral distribution from other illuminant sources, e.g., fluorescent lights, is different from incandescent sources and the concept of *correlated* color temperature was introduced. This is the temperature of a Planckian radiator whose perceived color most nearly matches that of the source being defined. The use of color temperature as a specification has usually been limited to illuminants and has become less common as nonincandescent sources have assumed greater importance.

3. All visible colors are located within the area bounded by the spectral curve and the line (the alychne) joining its red and blue ends.
4. Every color whose coordinates lie on a line joining a point at the white center of the diagram to the outer spectral curve have the same hue. The dominant wavelength for the color is determined by the point at which the line intersects the spectral curve. Its saturation is indicated on its location along this line, ranging from zero at the center to 100 percent at the spectral curve.
5. The coordinates of a primary set are located at the corners of a triangle. All colors that can be reproduced by a given set of primaries have coordinates within the triangle. Two primary sets for color display devices are shown in Figure 2.3; they are:
   - The NTSC/FCC standard primaries established in 1953.
   - The SMPTE C standard primaries established in 1982.

   These primary sets are described in Section 2.9.
6. The specification of color primaries is one of the most important applications of chromaticity coordinates (see Sections 2.8 and 2.9 below).

2.8 Camera Colorimetry

The camera generates three signals that are the electrical analogues of the red, green, and blue visual components of the scene (see Chapter 9). The camera's colorimetry is specified by graphs of spectral response vs. wavelength for each color channel. These curves are called *camera taking curves* or, simply, *camera primaries*.

2.8.1 Camera primaries

These are determined initially by filters in an optical color splitter, but the curves can also be modified electrically in a camera matrix circuit by a process known as *matrixing* (see Section 9.7). With matrixing it is possible to achieve a spectral response that would be impossible with optical filters alone.

The amplitude of each color signal is determined by the product of the spectral content of the scene and the colorimetry of the channel integrated over the entire spectrum. For example, the amplitude of the red signal, E_R, is given by Eq. (2.7):

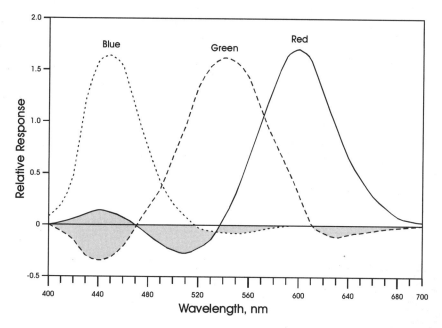

Figure 2.5 NTSC/FCC standard camera primaries.

$$E_R = R_C \int_{380}^{780} L(\lambda)R(\lambda)d\lambda \qquad (2.7)$$

where E_R = output voltage of the red channel, R_C = constant relating electrical output of the channel to luminance, λ = wavelength, $L(\lambda)$ = luminance of picture content, and $R(\lambda)$ = spectral characteristic of red camera primary.

The camera primaries must be compatible with the primaries of the display device, and this has led to the need for the establishment of display-device primary standards to provide the basis for the design of camera colorimetry.

The spectral content of the scene illuminant also affects the choice of primaries. If a scene is illuminated by incandescent light and the camera primaries are adjusted for daylight such as illuminant C, the reproduced image will appear reddish. Compensation for this effect is achieved by readjusting the camera matrix circuits, and a switch can be provided for these circuits with two positions, one for illuminant C and one for incandescent light.

The camera primaries recommended by the NTSC and adopted by the FCC in 1953 were based on compatibility with the NTSC recommendations for display device primaries and illuminant C. Their spectral responses are shown in Figure 2.5.

These primaries contain negative lobes that cannot be produced optically. This problem has two solutions. One is to ignore the negative lobes and the sidelobe in

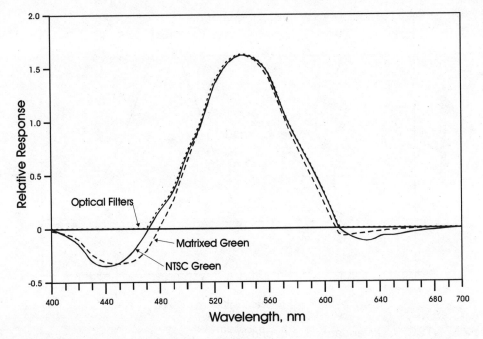

Figure 2.6 Green primaries—effect of matrixing.

the red primary, as shown in the cross-hatched area in Figure 2.5. This is a compromise that produces satisfactory results but with some distortion. The other is to employ matrix circuits to create the negative lobes electrically (Figure 2.6).

Another departure from the NTSC/FCC standards is required by industry practice, described in Section 2.9, of employing receiver primaries that deviate considerably from the original FCC specifications in order to provide greater screen brightness. This requires a corresponding modification in the camera primaries for the best results.

2.8.2 Matrix circuits

Electric matrix circuits convert an array of coefficients of independent variables into another array through a mathematical transform. They can be used to alter the spectral response of camera primaries, either to provide negative lobes that cannot be produced optically or to compensate for changes in the scene illuminant.

The operation of these circuits is illustrated by the example of the NTSC green primary shown in Figure 2.6. Optical filters cannot produce the negative values, but they can be produced electrically by a matrixing operation:

$$E_{GM} = E_G - 0.2\,E_B - 0.05\,E_R \tag{2.8}$$

Figure 2.6 shows that the matrixed signal gives a close approximation to the NTSC primary response.

Table 2.4 Chromaticity coordinates of different display primary sets

	NTSC/FCC		SMPTE C		240M		274M			
							Specified		Interim	
	x	y	x	y	x	y	x	y	x	y
R	0.67	0.33	0.635	0.340	0.630	0.340	0.67	0.33	0.640	0.330
G	0.21	0.71	0.305	0.595	0.310	0.595	0.21	0.71	0.300	0.600
B	0.14	0.08	0.155	0.070	0.155	0.070	0.16	0.07	0.155	0.060
Color Gamut	Wide		Narrow		Narrow		Wide		Narrow	

This is a simple example involving only the addition and subtraction of linear functions. More complex circuits make use of nonlinear elements so that the matrix coefficients are dependent on the signal level. With these systems, the operator can "paint" the picture in accordance with esthetic judgments (this is tricky and requires a great deal of skill). Also, matrixing makes possible the precise matching of camera color characteristics, which is required when several cameras are to be used together on the same scene.

2.9 Display Device Primary Standards

The colorimetry of receiver display devices is determined by the characteristics of the phosphors used in the picture tube and is specified by their CIE coordinates.

Since the number of receivers far exceeds the number of cameras, receiver display device performance and cost are the most important factors in the selection of primary coordinates. It becomes the responsibility of manufacturers of cameras to design their colorimetry to be compatible with receiver display devices. Because of their impact on system performance and cost, the selection of display primaries has been the subject of extensive research. Standards have been established and modified from time to time in response to changes in system objectives and component capability (particularly picture tube phosphors).

The coordinates of five important display device primary sets, existing and proposed, are tabulated in Table 2.4.

2.9.1 NTSC/FCC primaries

The coordinates of display device primaries for television broadcasting, as proposed by the NTSC and adopted by the FCC in 1953 as part of its specifications for color broadcast standards, were plotted in Figure 2.3.

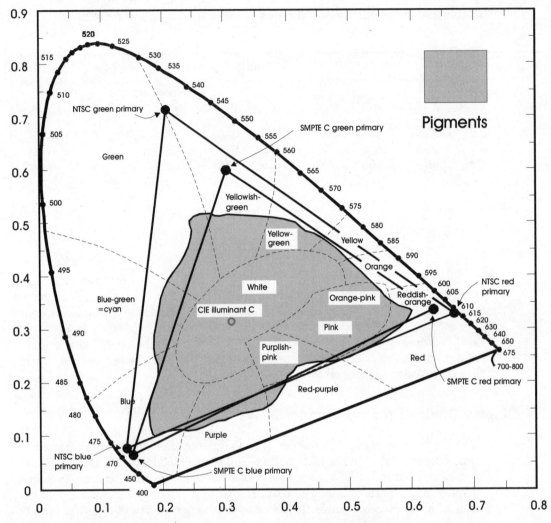

Figure 2.7 Color gamuts of dyes and paints. From K. Blair Benson (ed.), *Television Engineering Handbook*, McGraw-Hill, New York, 1986.

2.9.2 SMPTE Standard C primaries

With growing experience in receiver manufacture in the years after 1953, it was found that greater screen brightness could be obtained with primaries that differed from the original NTSC recommendations. Customers preferred brighter pictures, and many receivers were manufactured with revised primaries, even at the expense of a small loss in color fidelity. In recognition of this reality, the *Society of Motion Picture and Television Engineers* (SMPTE) issued a revised primary color standard for studio monitors in 1982 that was known as *SMPTE Standard C*. (Receiver manufacturers were not required to adopt this standard, but it was hoped that the use of a common standard for monitors would aid in standardizing the colorimetry of broadcast signals.)

The coordinates for Standard C are plotted in Figures 2.3 and 2.7. The plots show that the added brightness is obtained at the expense of a more limited gamut of green, yellow, and red hues but with a slightly greater gamut of blues. The compromise is not serious, however, as illustrated in Figure 2.7, which compares the gamut of spectral responses of typical paints and dyes with Standard C. The color gamut of the television system (the area within the triangle) compares favorably with the color gamut of paints.[4]

2.9.3 SMPTE 240 (Appendix) standard primaries

The color primaries included in the SMPTE HDTV standard 240M essentially duplicate the SMPTE C standard with the restricted color gamut. It was hoped, however, that it would not be necessary to limit the gamut indefinitely because of the current limitations of kinescope phosphors. The 240M standard, therefore, includes an Appendix that suggest a future improvement of the color gamut by means of signal processing. The coordinates of 240M and 274M (Appendix) are tabulated in Table 2.4.

2.9.4 SMPTE 274M standard primaries

The SMPTE Standard 274M is specified for use with the ATSC HDTV system (see Chapter 7), which is supposed to provide optimum performance without significant compromises. With the passage of time, however, it did not appear realistic to assume that HDTV phosphors could be developed that would combine high brightness with a wide color gamut, i.e., that would meet the wider gamut requirement [1]. Accordingly, SMPTE 274M specifies wide gamut primaries, but it allows for interim primaries that are virtually identical to the SMPTE C and 240M standards.

2.10 Colorimetry of Computer Displays

The considerations of computer display colorimetry are much the same as for television except that color information in a computer may need to be output by means other than the CRT display. The most important other use of color is in color printing, a subject that is beyond the scope of this book.

Computer motion video usually originates from a standard TV camera and therefore it has standard TV colorimetry. However, still images may also origi-

---

4 It should be noted, however, that the *Pointer colors*, a list of real surface colors developed by M. R. Pointer in 1980, includes a larger gamut of 576 colors. The standard C primaries will represent less than half of these. (See Section 2.8.3 for the efforts of industry to specify a standard for a wider color gamut.)

nate from image scanners or they may be computer generated—these sources are not necessarily tied to TV colorimetry.

A computer is capable of built-in color correction—through software it can do color balancing, matrixing, color gamma control, or other kinds of color processing. That means it can theoretically correct for different display primaries or, for example, make the display show images as they will appear when printed on paper with a specified set of CMYK inks. However, color correction of digital images requires a lot of CPU activity, so it is applicable only to still images. The extra processing would unacceptably slow down motion video playback.

2.11 Reference

1. Grand Alliance Expert Group on Scanning Formats/Compression, *Report on Colorimetry*, July 1995.

Chapter 3

Introduction to Digital Technology

3.1 Digital

Digital means "comprising digits or numbers." *Digital video*, then, is video represented by a series of numbers. This chapter explains how images that contain smoothly varying brightness, hue, and saturation values can be conveyed by a sequence of numbers and why it is so valuable to do that.

3.2 Number Systems

To most people, a number is one or more of the arabic numerals: 0, 1, 2, 3, 4, 5, 6, 7, 8, or 9. Each numeral in this system is a *digit* that can have one of ten possible values. No more than ten values are represented by a single digit. It is a *base-10* number system. Of course, more than ten values are obtained by combining multiple digits, such as in the number "25." The "2" represents a value of 20 (2 multiplied by the base 10), since it is in the second position to the left. Similarly, in the number "325," the "3" represents 300—the 3 is multiplied by the base squared. By combining enough digits, where each one to the left is multiplied by a successively higher power of the base, any integral value can be represented.

There is nothing magic about the base-10 for a number system. Any number can be used as the base—some common bases and their names are:

    base-2   binary
    base-3   ternary
    base-8   octal
    base-10 decimal
    base-16 hexadecimal

An analog system allows an infinite number of values to be possible and valid; it can be thought of as a number system of base infinity. In such a system, one cannot distinguish between valid values and errors because all values are valid. In a system of lower base, however, circuit values that are not exactly equal to an integral digit value are clearly in error and are prohibited. In electrical transmission of numerical values, the process of *quantization* is used to force all output to be exactly integral values.

3.2.1 Binary number systems

Quantization effectively eliminates transmission errors that are less than the reciprocal of the base. That means the lower the base value, the more robust is the transmission capability of the system. So a base-2 (*binary*) system is the most tolerant of transmission errors. However, as the base is lowered, it takes more digits to reproduce useful numbers and this makes any processing or transmission more complex.

In the early days of video and computing, different choices were made regarding the tradeoff between robustness and complexity based on differing needs. In video, the inherent redundancy in the scanning process and the characteristics of human vision provided considerable tolerance for errors, so the simplicity of an analog system was an easy choice. However, in computers, the requirement for complete elimination of errors at almost any cost shifted the tradeoff to a binary number system even though it was much more complex.

Today, with solid-state integrated circuits (ICs), the complexity of binary circuits has moved inside IC chips and most system designers now think it is easier to build binary hardware than analog hardware. The complexity issues of binary systems have been taken over by IC designers so that system designers now work at a much higher level and don't have to worry about the details of fundamental binary processes. In video systems, the choice between binary or analog is now just a matter of whether there are ICs available to perform a process at appropriate speeds for the video performance required. Of course, this is oversimplifying things, but the conclusion that nearly all video systems are moving to binary technology is still correct.

The developmental energies of the computer and the IC industries have primarily been directed to binary number systems and little hardware is available for anything other than base-2 systems. For this reason, the word *digital* has become synonymous with *binary*. It is used that way throughout this book.

3.2.2 Binary numbers

A binary digit, called a *bit*, can have only two values: 0 or 1. It takes many more than two levels to make a useful digital video system, so multiple bits are required to specify each pixel value. A typical value is 8 bits per pixel—that gives 256 levels for a pixel value. Computer people call a group of 8 bits a *byte*. Binary numbers are represented as a series of digits (0 or 1). For example, the decimal number 5 would be 0000 0101 in binary. Further explanation of this is shown in Figure 3.1.

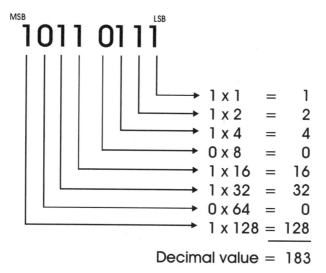

Figure 3.1 Binary numbers.

Working directly with binary numbers quickly becomes tedious, so there are several techniques to simplify the process. One that is widely used is *hexadecimal* (*hex*) notation that represents each 4-bit group as a single hexadecimal digit. Hex digits have 16 possible values, which is accomplished by using the numerals 0–9 and continuing with the letters A–F for values above 9. This is shown in Table 3.1. Because hex numbers that contain no digit values above 9 can become confused with decimal numbers, it is common to put the suffix "H" or the prefix "0x" on hex values. Some examples:

Decimal	Binary	Hex	Hex to Decimal
0	0000 0000	00H	(0×16+0)
5	0000 0101	05H	(0×16+5)
31	0001 1111	1FH	(1×16+15)
255	1111 1111	FFH	(15×16+15)
2550	0000 1001 1111 0110	09F6H	(0×4096+9×256+15×16+6)

Notice that the binary values above are shown in groups of 4 bits. This helps in the process of conversion between binary and hex.

A digital audio system usually requires more than 8 bits per sample—16 bits per sample (2 bytes) are often used. This gives 65,536 levels (2^{16}). Some audio systems use 20 or 24 bits/sample.

3.3 Analog-to-Digital Conversion

Conversion from analog to digital (ADC) may be required at several points in a system, depending on how much of the system is digital and what parts they are. An all-digital system performs ADC at the imager. (Imagers today all deliver

Table 3.1 Decimal, binary, and hexadecimal values

Decimal	Binary	Hexadecimal
0	0000	0
1	0001	1
2	0010	2
3	0011	3
4	0100	4
5	0101	5
6	0110	6
7	0111	7
8	1000	8
9	1001	9
10	1010	A
11	1011	B
12	1100	C
13	1101	D
14	1110	E
15	1111	F

analog outputs. The development of a truly digital imager could entirely eliminate the ADC process, and everything beyond that would be digital. Another system might use digital technology only for recording. In that case, ADC (and DAC—see Section 3.9) would be inside the recorder(s). Of course, if video is being generated from scratch by a computer, there is no ADC.

Section 1.5 explained how a video signal from an imager is sampled by the scanning process. However, the resulting video signal is still a series of analog values. To convert completely to a digital data stream, the analog values must be sampled and quantized. That subject was introduced in Chapter 1. But even that is not the whole story because there are many ways to translate sampled and quantized values into groups of bits—a process called *encoding*. This section expands on sampling, quantizing, and encoding as parts of the process of ADC or *digitization*.

Figure 3.2 shows functional diagrams of ADCs for composite and component analog video signals. The first element (optional) is an anti-aliasing filter. This is a low-pass filter with a cut-off below the Nyquist limit (see Section 1.5.1.1), i.e., below one-half the sampling rate. This eliminates the signal frequency components that could cause aliasing. It is desirable to have the sampling rate high enough so that there is a reasonable separation between the nominal cut-off frequency of the filter and the Nyquist frequency. That makes it possible to use a filter with a gradual cut-off without the excessive phase shift of sharp cut-off filters. In a 4:2:2 component format (see Section 3.3.1), the cut-off of the filter in the color difference channels is approximately one-half of that in the luminance channel in accordance with their reduced bandwidth. The anti-aliasing function in the ADC is followed by the sampling, quantizing, processing, and encoding functions.

Introduction to Digital Technology 49

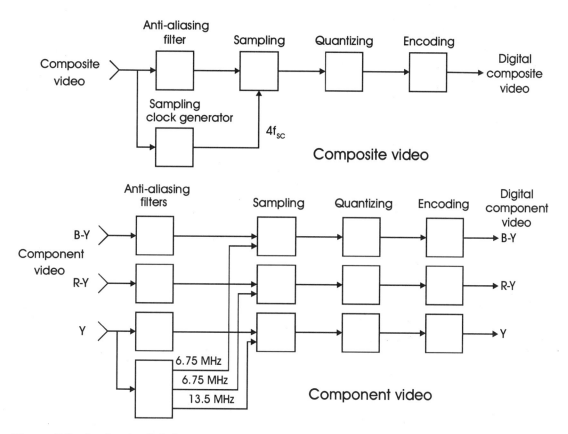

Figure 3.2 Analog-to-digital converters.

Detail specifications for the ADC process depend on where it is occurring in a system. That will determine what the analog input format is and what digital output format will be needed. The resulting digital data stream must conform to a format standard so that other systems will know what to look for in the data. It may be a proprietary format that applies to a specific situation, or it may be a public standard that is recognized worldwide. This is one of the major concerns of national and international standardizing bodies, including the SMPTE, IEEE, EBU (*European Broadcasting Union*), and ITU (*International Telecommunications Union*).

3.3.1 Horizontal sampling rate

The horizontal sampling rate or its reciprocal, the sampling interval, is a fundamental specification. It should meet a number of conditions:

1. It should meet or exceed the Nyquist limit to avoid degrading the horizontal resolution and to minimize aliasing (see Section 1.5). This requires it to be at least twice the upper limit of the analog signal frequency components or a rate

Table 3.2 SDTV bit rates and bandwidths

System	Bit rate, Mb/s	Bandwidth, MHz*
NTSC ($4f_{SC}$, 8 bps)	114.5	58
PAL ($4f_{SC}$, 8 bps)	141.9	72
Component**		
Luminance	108	54
Color-difference	54	27

* Based on one-half the bit rate, which is an approximation (see text)
** According to ITU-R Rec. 601-5, 4:2:2 format

of 8.5 MHz (2 × 4.25 MHz) for NTSC systems and 10.0 MHz for PAL systems. These are minimum rates, and it is standard practice to use rates that are somewhat higher than the minimum to permit the use of an anti-aliasing filter with a gradual cut-off.

2. It should be an integral multiple of the line rate. With this relationship, the sampling points on adjacent scanning lines are directly above each other (see Figure 3.3).
3. For digitizing composite signals that contain a color subcarrier, the sampling rate should be an integral multiple of the subcarrier frequency. This is not an absolute requirement, but signal processing and decoding are greatly simplified when the sampling rate meets this criterion.
4. For component signals, it should be the same frequency for 525-line 30-frame systems and 625-line 25-frame systems.

In response to these requirements, standards for sampling rates exist as shown in Table 3.2. (The bandwidth figures in the table assume simple binary transmission. With more sophisticated modulation, bandwidths can be less—see Section 3.7.2.)

3.3.1.1 Component sampling

Component sampling rates for SDTV are internationally standardized by the ITU in ITU-R Rec. BT.601-5. The sampling rate of 13.5 MHz was chosen for component signals because it is an integral multiple of both the NTSC and PAL line rates. It is not, however, an integral multiple of either the NTSC or PAL subcarrier frequency, but no subcarrier frequency is required anyway.

3.3.1.2 Composite sampling

For composite signals, sampling rates of $3f_{SC}$ or $4f_{SC}$ can be used. $3f_{SC}$ does not exceed the Nyquist frequency by a comfortable margin, and therefore a sampling frequency of $4f_{SC}$ is the most commonly used and is an SMPTE standard 244M.

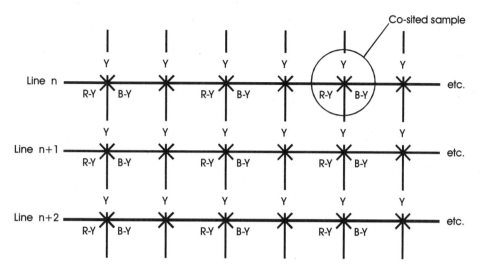

Figure 3.3 Location of sampling points, component signals.

3.3.1.3 Computer display sampling

In the computer industry, the choice of sampling rate depends not only on the input signal format as described in the list above, but also on the desired output display format. If a composite signal is being sampled, there is a need to decode the composite format because computers generally use a component format to describe the color value of each pixel. There are many variations of computer formats as explained in Chapter 8.

Sampling rate considerations for HDTV are discussed in Chapter 7.

3.3.2 Location of sampling points

The locations of the sampling points on individual lines for the ITU-R Rec. BT.601-5 component format are shown in Figure 3.3. The luminance and two color difference sampling pulses are synchronized so that the color difference points are *co-sited* with alternate luminance points.

In the formats for composite signals in Table 3.3, the sampling points have a specified relationship with the phase of the burst. For both PAL and NTSC at $4f_{SC}$, there are four sampling points for each cycle of the burst. For PAL, the sampling points are at the 0°, 90°, 180°, and 270° points of the burst waveform. For NTSC, the first sample is located at 57°, the I axis (see Section 3.14.1). The others are located at 147° (the Q axis), 237°, and 327° (Figure 3.4). This is known as *I-Q axis sampling*.

The sampling rate and phase must be synchronized with the line and subcarrier frequencies and phases in order to maintain these precise locations. For composite signals the synchronizing element is the subcarrier burst. The system element that generates the sampling frequency is called the sampling clock generator.

Table 3.3 Sampling frequencies

Composite standards	NTSC		PAL
	$3f_{SC}$	$4f_{SC}$	$4f_{SC}$
Bandwidth, MHz	4.2	4.2	5.5
Subcarrier (f_{SC}), MHz	3.58	3.58	4.43
Sampling frequency, MHz	10.74	14.318	17.72
Samples per total line	682	910	1134
Samples per active line	576	768	939
Bit rate, Mb/s (8 bps)	85.9	114.5	141.8
Component standards	525 lines, 59.94 fields ITU-R BT.601-5 4:2:2 SMPTE 125M		625 lines, 50 fields ITU-R BT.601-5 4:2:2
Luminance channel			
Bandwidth, MHz	5.5		5.5
Sampling frequency, MHz	13.5		13.5
Samples per total line	858		864
Samples per active line	720		720
Bit rate, Mb/s	108.0		108.0
Color-difference channels			
Bandwidth, MHz	2.2		2.2
Sampling frequency, MHz	6.75		6.75
Samples per total line	429		432
Samples per active line	355		358
Bit rate, Mb/s (8 bps)	54.0		54.0

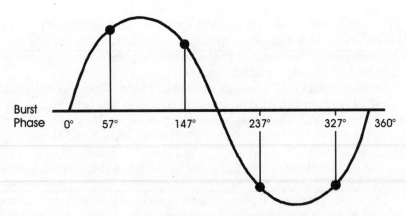

Figure 3.4 Location of sampling points, NTSC composite signals, $4f_{SC}$.

Introduction to Digital Technology 53

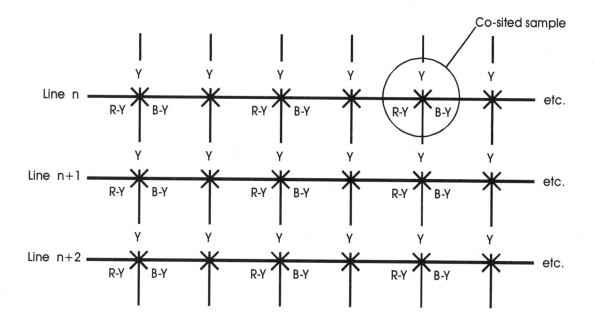

Figure 3.5 Quantizing.

3.3.3 Quantizing

The next step in the ADC process is to *quantize* the pulse stream resulting from the sampling process.

3.3.3.1 Quantized signal levels

Quantizing is depicted in Figure 3.5. The signal amplitude range is divided into discrete intervals, and a fixed or quantum level is established for each interval. This level is assigned to samples having an analog value that falls within the interval. The difference between the quantum level and the analog signal at the sampling point is *quantizing error*.

Most video quantization systems are based on 8-bit words with 2^8 or 256 discrete values. (2^{10} or 1024 levels are sometimes used for pre-gamma signals in cameras.) Not all of the levels may be used for the video range because of the need to use some of the values for data or control signals.

Figure 3.6 shows typical utilization of the 256 8-bit quantized levels for composite and component signals. The entire composite signal, including sync pulses, is transmitted within the quantized range of levels 4 to 200.

In component systems, sync pulses fall outside the quantized range and are not transmitted. They are unnecessary because line and field synchronization is

54 Chapter Three

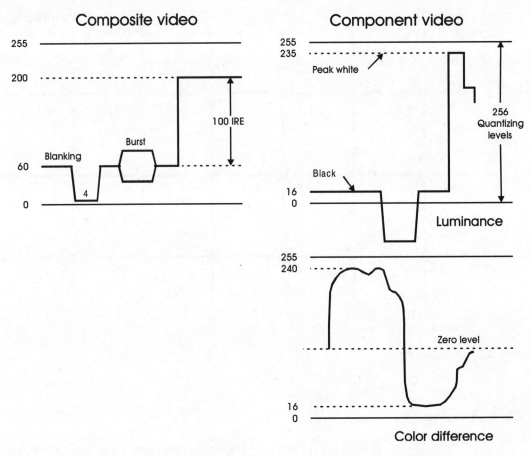

Figure 3.6 Assignment of quantizing levels for composite and component video.

provided by digital codes. The luminance signal occupies levels 16 to 235. The color-difference signals can be either positive or negative, and the zero black level is placed at the center of the quantized range because there are no negative numbers in the binary number system used for encoding quantized signals. Note that neither range includes the binary words 0000 0000 (0) or 1111 1111 (255), and they are available for synchronizing.

3.3.3.2 Quantizing errors

If the sampling rate is harmonically related to a signal frequency component, the result of quantizing errors is distortion of the waveform (see below). In the more unlikely event that the relationship is random, the result of quantizing errors looks like noise. The peak-to-peak signal-to-rms quantizing error ratio in dB is given by Eq. (3.1):

$$\frac{S}{Q_e} = 10.8 + 6.02n + 10 \log\left(\frac{F_S}{F_V}\right) \tag{3.1}$$

where: S is the peak-to-peak signal, Q_e is the rms quantizing error, n is the number of quantum levels as a power of 2, F_S is the sampling rate, and F_V is the video bandwidth. For $n = 8$, $F_S = 14.4$ MHz, and $F_V = 4.2$ MHz, $S/Q_e = 64.3$ dB.

The visible nature of the noise depends on the magnitude of n. If n is about 6 or greater, the noise will be random as from thermal effects. If n is less than about 6, it will produce a contouring effect in large areas of color.

Equation (3.1) assumes that the original analog signal is noise-free. With a noise-free input signal, the sampled and quantized waveform will be distorted when reconverted to an analog signal. For example, a smoothly rising signal amplitude will be reproduced as a stairstep with discontinuities equal to the quantizing intervals. If there is noise in the analog signal, however, that will partially or totally obscure the discontinuities. If the rms value of the noise exceeds one-third the quantizing interval, the discontinuities will disappear in the noise. The ratio, $S\backslash Q_e$ (in dB), under these conditions is given by Eq. (3.2):

$$\frac{S}{Q_e} = 6.02n \tag{3.2}$$

When $n = 8$, $S/Q_e = 48.2$ dB.

In some cases it may be desirable to trade off signal distortion for noise by deliberately introducing noise. This is known as *dithering*, an intentional rapid and random variation of some parameter of a signal that eliminates the visibility of quantizing discontinuities.

3.3.4 Encoding

The final step in ADC is to *encode* the quantized levels of the signal samples. The encoding and processing of digital signals has been the subject of intensive mathematical analysis. Hundreds of papers have been published on the subject as well as a number of excellent textbooks. A complete description of this subject is beyond the scope of this chapter. Rather, the purpose here is to describe the basics and define the principal terms of digital signal encoding as a foundation for a more extensive study of the subject.

3.3.4.1 Pure binary encoding

When the binary digits directly represent values multiplied by powers of 2, the encoding is called *pure binary*. The range of 8-bit pure binary encoding goes from 0 to 255 (256 levels). This is satisfactory to represent composite video waveforms, but it is not suitable to represent zero-center (bipolar) waveforms such as audio or color-difference signals. In that case, analog zero has to be offset to the center of the quantizing range (128). This is known as *offset binary* encoding, but it is not satisfactory because the offset values tend to interfere when the signal is processed.

56 Chapter Three

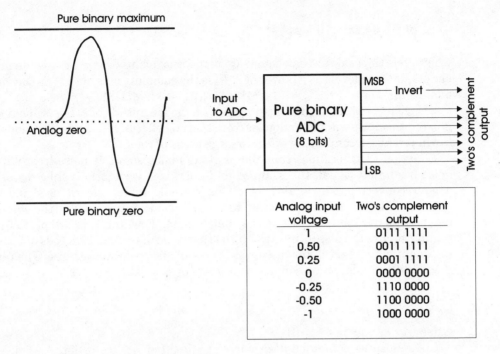

Figure 3.7 Two's complement encoding.

3.3.4.2 Two's complement encoding

A better strategy for bipolar waveforms is what is known as *two's complement* binary encoding, illustrated in Figure 3.7. In two's complement, the lower half of the pure binary range (when MSB = 0) is defined as negative numbers. That is accomplished by offsetting the analog signal so that it fits into the normal range of binary values (0–255 for 8 bits) with zero at the center, digitizing, and then inverting the MSB of the resulting digital values. Since zero remains at digital zero, this system of encoding does not interfere with processes such as addition (mixing) that depend on that.

The largest positive decimal number that can be represented in the two's complement code is $2^{(n-1)}-1$, where n is the number of bits. The absolute value of the largest negative number that can be represented is $2^{(n-1)}$. For 8-bit values, the range is from –128 to +127. The two's complement code for a positive number equal to or less than $2^{(n-1)}$ is the same as its pure binary code. The two's complement code for a smaller negative number, –X, is equal to the pure binary code for the positive number, $2^n-|X|$. The first digit in this code for positive numbers is always 0 while for negative numbers it is 1.

More sophisticated coding is often employed as a part of an error management system. Some examples of that are discussed in Section 3.8.

3.4 Serial and Parallel Formats

The bits in a digital code can be transmitted or recorded either serially over a single path, or in parallel with a separate path for each numbered bit in the code. Thus, the transmission (or recording) of an 8-bit byte by parallel transmission requires eight separate communication paths. The use of parallel transmission simplifies the terminal equipment, but the necessity for a multiconductor transmission path makes it impractical except for short distances such as the interconnections within a single equipment or within a studio. Serial transmission, therefore, is more commonly used for medium- and long-distance transmission.

In parallel transmission, a separate circuit is provided for each of 8 (or 10) bits. In serial transmission, the 8 (or 10) bits in each code, which are produced simultaneously, are stored in a shift register and read out sequentially to a single circuit.

The minimum bandwidth requirement of a single digital channel with simple encoding is nominally one-half the bit rate, corresponding to transmission of one bit on each half-cycle of the channel's highest frequency. (Advanced encoding and modulation can achieve greater than one bit per half-cycle.) An 8-bit parallel system would require eight separate channels, each with one-eighth the bandwidth of an equivalent serial channel. Each mode has advantages and disadvantages.

Video bit streams are usually created in a parallel format, and the cost and complexity of serialization equipment are avoided with parallel transmission. Thus, parallel systems are cheaper and simpler for local signal distribution. On the other hand, multiconductor cable is more expensive and bulky, and parallel transmission is more costly for long-distance transmission. Matching the delay times of parallel paths may be difficult in parallel systems, and there is the possibility of crosstalk between paths. Matrix switching between multiple inputs and outputs is difficult. Signal processing equipment is more costly and complex for the parallel mode. As a result, serial transmission is widely used where distances are great or for complex signal processing. This has led to the development of inexpensive serialization ICs and the cost of serialization today is minimal.

In the absence of data compression, the bandwidth requirements of either serial or parallel transmission, 54 MHz or higher for 8-bit SDTV composite or luminance signals, are a serious handicap to the use of digital long-distance transmission. The bit rates for 8-bit SDTV signals do not fall neatly into one of the steps in the standard 1.54 Mb/s T-1 communications hierarchy, which higher has levels at 44.7 and 274.2 Mb/s.

On the other hand, video compression technology is rapidly advancing and bandwidth requirements can be reduced for digital video transmission at all levels. For example, the compression in ATSC HDTV delivers high-resolution video at 19 Mb/s or less. By the use of advanced modulation methods (see Section 3.7.2) this data rate can be transmitted in a bandwidth of 6 MHz or less. In fact, many systems are being designed today for multichannel serial transmission of compressed digital video.

Bandwidth usually receives the major emphasis in digital transmission system design; this is because digital signals are considerably more tolerant to noise and amplitude linearity than analog signals. Digital interfaces are described further in Section 3.7

3.5 Binary Arithmetic

A branch of mathematics known as Boolean algebra (named for the English mathematician George Boole) deals with the relationships between and manipulation of binary numbers. It includes laws on deductive logic that are used for the analysis of complex relationships between many interacting components, for example, in the design of machine control systems.

3.5.1 Binary addition

In principle, the addition of pure binary numbers is similar to that of decimal numbers. Starting with the LSB column, the binary digits in each column are added. If the sum of the digits, n, in a column is 0 or 1, that number is entered as the total for that column. If n exceeds 1 and is an even number, 0 is entered as the total and $n/2$ is carried into the next column to the left. If n is an odd number, 1 is entered in the total column and $(n-1)/2$ is carried into the next column to the left. Two examples are:

	Binary number	Decimal equivalent		Binary number	Decimal equivalent
Carryover	111			1	
Binary A	1101	13		101	5
Binary B	1111	15		111	7
			Total	1100	12
Binary C	0011	3			
Total	11111	31			

Offset binary numbers can be added in the same fashion, but it must be remembered that the sum will include the sum of the offsets.

Addition of two's complement binary numbers is often necessary. Addition of two's complement numbers can be carried out with the same rules as with pure binary numbers.

	Binary number	Decimal equivalent
Carryover	101	
Binary A	0111	7
Binary B	0100	4
Binary C	1011	−5
Total	0110	6

When adding positive numbers, whether using a pure binary or a two's complement code, care must be taken that the sum does not *overflow* (exceed) the range available for the bit number of the system, i.e., 2^n for pure binary code and 2^{n-1} for two's complement. If the overload results from negative numbers in the two's complement format, some bits may actually be deleted. Some systems have overflow control circuits to avoid this effect.

3.5.2 Binary multiplication

Binary multiplication is not as simple as addition, but it is essential to many signal processing functions such as fading and mixing. Multiplication can be accomplished by repeatedly adding the bits in each multiplicand column in a shift register in accordance with the value in the multiplier. The sum for each column is weighted in accordance with its significance, ranging from the LSB to the MSB.

Binary notation	Decimal conversion
[101]	[4+0+1]=5
×[111]	[4+2+1]=7
=[100011]	5×7=[32+3+1]=35

This procedure requires a tremendous number of operations, but it is accomplished quickly and accurately with integrated circuits.

3.5.3 Logical operations

An important feature of computers is the ability to perform logical operations on or between binary values. A simple application of logic is found in the situation where a group of bits—say a byte—is used to represent eight individual yes/no *bit-flags* instead of a numerical value. This is often done in file header structures to define attributes of the file's data. Using the logic operations, the bits can be individually set, reset, or tested.

There are four basic logic operations: NOT, OR, AND, and exclusive-OR. Their behavior can be described by means of a *truth table*, which shows all the possible outcomes of the operation. Figure 3.8 shows truth tables for each of the four basic logic operations. The truth tables show the outcome of the operation as a matrix (shaded area) based on the possible values (0 and 1) of the 1-bit operands A and B. (The NOT operation has only one operand, A.)

3.5.3.1 NOT operation

This operation simply reverses the value of a bit, regardless of what it was.

3.5.3.2 OR operation

This function outputs a 1 when either or both of the input values is 1. It outputs 0 when both inputs are 0.

3.5.3.3 AND operation

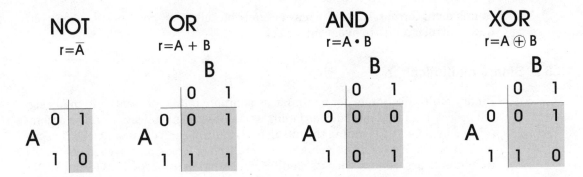

Figure 3.8 Truth tables for the basic logic operators: NOT, OR, AND, and XOR.

This function outputs a 1 when both inputs are 1. If either input is 0, the output is 0.

3.5.3.4 Exclusive-OR

This operation, often called XOR, is the same as OR except that it outputs 0 when both inputs are 1.

3.5.4 Bit-flag example

Returning to the example of a byte that contains 8 bit flags, the logic operators are used as follows to control the flags:

> The OR operator is used to set a bit flag. In logical operations, the flag bits are identified by their decimal values as indicated in Figure 3.1. Thus, to set the second flag from the right (0000 0010—whose value is 2), the flags byte is ORed with a byte whose value is 2.
>
> The NOT and AND operators are used to clear a flag. In this case, a byte must be created that has all bits 1 except the one to be cleared. This is done by using the NOT operator on the flag value and ANDing the result with the flags byte.
>
> The XOR operator has the property of reversing the value of a bit. Thus, if the flags byte is XORed with 2, the 2 bit will be flipped.

Various combinations of these operations can be used to modify several flags at once.

3.6 Digital Signal Processing

Digital video systems have the capability to perform signal conversion and processing functions that would be difficult or impossible with analog signals. After ADC, the signal consists simply of a series of numbers that can be manipulated with little or no loss of picture quality. This capability is enhanced by the ease with which the bit signals can be stored in memories for rapid retrieval as needed.

Digital processing can be done in special-purpose hardware that performs video processes on a signal in real time. This is the form of most digital devices intended to be embedded in video systems. However, it is also possible to do many processes in general-purpose programmable hardware (a computer) under control of software. This has exciting prospects because of the extreme flexibility that is available, but many video processing functions are still too complex to be handled by software in real time. That is changing every day as the power of computing hardware continues to advance, but most digital video systems today use a combination of hardware and software processing. This subject is discussed further in Chapter 8.

Four important signal processing functions that can be performed with digital signals are time-base error correction (Section 3.6.1), standards conversion, e.g., PAL to NTSC (Section 3.6.2), postproduction editing (Chapter 12), and data compression (Section 3.11).

3.6.1 Time-base error correction

Maintaining a stable time base for video signals and accurately preserving their original time relationships is a problem with videotape recorders because of timing instabilities[1] introduced by the mechanical systems involved. The time-base errors at the output of uncorrected helical scan recorders with their long recording paths on an elastic tape medium can be particularly severe.

An adequate degree of stability can be achieved for normal television viewing without time-base correction with consumer-type recorders that use the "color under" analog technique (see Chapter 11). However, more precise correction is required for professional recorders because of the necessity of performing editing and other postproduction functions on their outputs (see Chapter 12). These tasks require fully stabilized signals.

A number of methods have been developed for providing precise *time-base correction* (TBC) with the signal in analog form, using a variable electrical delay line as the basic element. Timing stability better than ± 2 nanoseconds (ns) has been achieved; that is approximately ±2.5° at the NTSC color subcarrier frequency. This is a satisfactory tolerance, but the range of correction required for helical-scan recorders is so great that long and expensive delay lines, covering several microseconds or more are required.

Time-base correction can be performed more cheaply and easily with digital techniques. The availability of low-cost, wide-range correctors has made helical-scan recorders standard in professional recording (see Chapter 10). The improvement in stability is quite impressive—timing errors of 10 μs or greater can be reduced to a few nanoseconds.

---

1. Time-base errors are frequency modulation of the scanning and subcarrier frequencies of the video.

The principle of operation of digital time-base correctors is straightforward. The signal from the replay process is converted to a digital bit stream, which is read into sequential locations of a memory in real time. The stored data are then read out at a rate that is precisely synchronized with the receiving system's clock. In effect, the memory acts as a buffer that absorbs the timing errors of the input signal. Precise control of the frequency relationship between input and output signals must be maintained so the memory buffer never overflows or underflows (runs out of data). This may require occasional dropping or repeating of a frame or line.

3.6.2 Standards conversion

Prior to digital television, *standards conversion* was accomplished with analog signals by the brute force methods of focusing a camera operating at the new standards at an image on a kinescope operating at the old. The picture quality of analog conversion systems was marginal at best. Standards-converted images suffered not only from the degradations that are inherent in the conversion of line and frame rates, but much worse was the cumulative degradation of the SNR, gray scale, and definition that are unavoidable in cascaded analog systems.

The use of digital processing does not necessarily solve the inherent problems of rate conversions, but it eliminates or greatly reduces the cumulative picture quality degradation of analog systems.

Digital standards converters use the same principle as time-base correctors—storage at the old standards and read-out at the new. The conversion process is considerably more complex, however, because the number of lines, fields, frames, and sampling rates must be changed without destroying time relationships of the events in the image. Also, it is not practical to convert the burst and subcarrier from one standard to another, for example, from NTSC to PAL. The first step in the conversion process must be to decode the composite signal to its luminance and color-difference components, thus eliminating the incoming subcarrier frequency. The line, field, and frame rates are then converted to the new standards, and then the signal components at the new standards are re-encoded to the new composite format.

The difference between the line period for 25-Hz, 625-line systems and 30-Hz, 525-line systems (64.0 vs. 63.55 μs) is so small that it is usually ignored. The change in the number of lines per frame from 525 to 625 and the number of frames per second, however, must be taken into account in the conversion process. Steady repeating or deleting scanning of lines or frames is not a satisfactory solution. Repeating or deleting lines will cause discontinuities in the image and repeating or deleting fields or frames will cause an appearance of jerky motion. The solution is to generate new signals by interpolation between lines and frames that approximate what the signals would have been had they been originally produced at the new standards. Digital circuits readily handle the arithmetic required for interpolation.

3.6.3 Postproduction editing

The art of postproduction film *editing*—the production of an edited master from a combination of short film or tape segments and special effects—was developed over a period of decades by the motion picture industry, and its practitioners are highly skilled artisans. The television industry tried for years to duplicate the complex postproduction functions performed routinely in film studios, but these efforts were only modestly successful with analog systems.

Digital formats can produce unlimited multiple generations of recorded material without significant loss of quality, incorporate precise computer control for switching and effects, and insert computer-generated program material. This has made digital video an extraordinarily powerful production and postproduction medium. With the improvements that have occurred in the performance of video cameras, video production techniques are now practical for the creation of high-quality programs.

The digitized analog signals used for editing can either be composite such as NTSC or PAL, or in component form, usually R, G, and B or Y, R–Y, and B–Y. Composite digital signals have the same basic problems of luminance-chrominance crosstalk as analog, and it is desirable, therefore, to perform production and complex postproduction editing functions with component signals. If necessary, signals are converted to composite formats only after completion of editing. Video editing systems are described in Chapter 12.

3.7 Digital Video Transmission

Transmission is the sending of video data from one point to another (point-to-point) or from one point to many others (broadcasting). Digital point-to-point transmission is widely used for connecting signal sources to signal-assembly centers via land lines, microwave links, and satellite links. Digital broadcasting of video is emerging, with the first example being the multichannel satellite services. However, with the digital standardization of HDTV, digital broadcasting is the wave of the future. This section discusses some of the special considerations of digital transmission of video, including packetization, modulation, error control, and synchronization.

3.7.1 Packetization

A digital bit stream must have a structure to facilitate synchronization and identification of the data. This is commonly done by dividing the data up into blocks called *packets*. Each packet begins with a header that identifies the packet and contains a unique pattern for synchronization (see Section 3.7.4), followed by a specified amount of signal data. Packets may be of any size, either fixed or variable (variable-size packets must have a size parameter in their headers), and there are arguments pro and con for each type.

Packetization allows interleaving of different types of data in a bit stream, such as video, audio, auxiliary data, or even multiple programs. The bit stream capacity (data rate) can be flexibly allocated to the different data types by varying the number and/or size of the packets for each data type.

Error handling also requires blocking of the data and it is convenient to choose a packet architecture that will support the error-handling methods used in the system. Error considerations usually lead to rather small packet sizes, such as a few hundred bytes.

A detailed example of fixed-length packetization for a digital video system is the Transport layer of the ATSC DTV system (see Section 7.6).

3.7.2 Digital modulation modes

Most transmission media are not inherently digital, but have many of the properties of analog systems. To efficiently transmit digital data, it is necessary to convert the data to a format that is compatible with the characteristics of the medium. This process is called *modulation*. An important objective of the choice of modulation method is to achieve an appropriate match to the bandwidth and noise performance of the transmission channel.

These two parameters can actually be traded for each other—for example, if better noise performance is available, the bandwidth for a given data rate can be reduced. There are two measures for this tradeoff:

1. The bandwidth requirement is expressed in terms of the bit rate that can be transmitted in a bandwidth of 1 Hz. This factor is sometimes called the *transmission speed*.
2. The noise performance of the system is expressed as the SNR required for a specified *bit error rate* (BER). A BER of 10^{-4} (1 error in 10,000 bits) is often used. Most systems require BER performance better than this; they achieve it by using *error protection* techniques. Error protection involves adding additional bits to the data (called overhead bits) and therefore the net data rate of the system becomes lower than the data rate of the raw channel (see Section 3.8).

Based on its bandwidth, a transmission channel is capable of transmitting a certain number of independent events per second. These events are called *symbols* and, depending on the modulation method, each symbol may represent one or more digital bits. When there is 1 bit transmitted by each symbol, the symbol has only two valid values—just like the bit it represents. A symbol that represents 2 bits has four valid values (2^2), and so on. Because multiple bits per symbol expand the number of valid values of a symbol, such modulation methods require higher channel SNR, but at the same time they are transmitting more data per second. This is how the tradeoff between bandwidth and SNR works—bandwidth can be reduced, but the necessary SNR increases.

All digital modulation modes establish conditions of amplitude, frequency, or phase (or a combination of these) to represent each of the values of a symbol.

Table 3.4 Performance of digital signal transmission modes

	OOK	QPSK	FSK	QAM
Constellation Diagram	(2 points on horizontal axis)	(4 points, one per quadrant)	(2 points on horizontal axis)	(16 points, 4×4 grid)
Bits/symbol	1	2	1	4
Transmission speed, b/Hz	0.8	1.9	0.8	3.1
Required E_b/E_a for BER=10^{-4}	12.5	9.9	11.8	13.4

Key:
 E_b = energy per bit
 E_a = white noise spectral density
 OOK = on-off keying
 QPSK = quaternary phase-shift keying
 FSK = frequency-shift keying
 QAM = quadrature amplitude modulation
SOURCES: J. D. Oetting, "A Comparison of Modulation Techniques for Digital Radio," *IEEE Transactions on Communications*, 23(12), 1979; Alladin Javed, "Signal Transmission Modes," in A. F. Inglis (ed.), *Electronic Communications Handbook*, McGraw-Hill, New York, 1988, Chap. 11.

There is a wide variety of possibilities, and a summary comparison of the speed vs. bandwidth relationship and the power requirements of four modes that span the range of possibilities is shown in Table 3.4. They are:

1. *On-off keying* (OOK) has 1 bit per symbol, represented by the two conditions of carrier and no-carrier.

2. *Quaternary phase-shift keying* (QPSK) transmits 2 bits per symbol using phase modulation of a constant-amplitude carrier with four possible phases spaced by 90°.

3. *Frequency-shift keying* (FSK) transmits 1 bit per symbol using frequency shift between two frequencies.

4. *Quadrature amplitude modulation* (QAM) is a combination of amplitude and phase modulation of the carrier that transmits 4 bits per symbol (16 values of the symbol).

The table also shows a *constellation diagram* for each of the modes; this diagram is a polar diagram showing the amplitude and phase of each of the valid carrier states. Among the four modes in the table, QAM can transmit the highest bit rate in a fixed bandwidth, but the power level must be high to achieve a satisfactory BER. Accordingly, it is often used in microwave systems where adequate power is available. QPSK presents an attractive compromise between power requirements and transmission speed, and it is frequently used in satellite system. The references cited in Table 3.4 describe other digital transmission modes. Modulation modes for magnetic recording are discussed in Section 10.12.3.

3.7.2.1 HDTV modulation modes

The ATSC DTV transmission system described in Chapter 7 is an example of a transmission system capable of digital HDTV. It uses multilevel *vestigial sideband* (VSB) amplitude modulation, referred to as *8-VSB* or *16-VSB* (the numbers give the number of amplitude levels involved). For example, the 8-VSB system transmits 3 bits per symbol using its 8 levels and the 16-VSB system sends 4 bits per symbol.

Vestigial sideband means that one of the amplitude-modulation sidebands is deliberately suppressed at the transmitter to save bandwidth. It can be restored before detection at the receiver by suitable processing. This technique is also used in analog TV broadcasting.

3.7.2.2 Trellis coding

The 8-VSB system actually sends only 2 bits of data per symbol; the data is expanded to 3 bits per symbol for transmission by means of a *trellis-coding* algorithm, an error protection strategy that can improve channel performance without increasing bandwidth. (The name comes from a diagram of this process that looks like a garden trellis.) In this approach, there are only four possible data states but eight possible symbol states, so not all combinations in the symbol are valid. The effect is to increase the differences between adjacent symbol states, which makes it easier to detect valid states.

At the receiver, detection is facilitated by the use of a *soft-decision decoder* (sometimes called a *Viterbi decoder* [1]), which operates on the channel signal before quantization to decide what a symbol value must be based on the context of adjacent symbols. Since the trellis coding has invalidated some symbol states or sequences, the decoder can correct many single-symbol errors.

With 8-VSB trellis-coded modulation, the DTV system achieves a data rate of 19.29 Mb/s in a 6-MHz channel, which is a transmission speed of 3.2 bits/Hz. It has a noise threshold of 14.9 dB. These performance numbers are not directly comparable to those in Table 3.4 because they describe the complete system including the error protection, whereas Table 3.4 shows raw modulated-channel performance at 10^{-4} BER.

The 16-VSB DTV system does not use trellis coding, so it achieves twice the data rate of the 8-VSB system. This is intended primarily for cable service, where the necessary higher SNR can be obtained. The noise threshold for 16-VSB is 28.3 dB. DTV modulation is described further in Section 6.6.3.

3.7.3 Component and composite signals

As described in Section 3.2, either component or composite analog color signals can be converted to a digital format. Of course, digitizing a composite signal cannot remove signal impairments that already exist caused by the composite format. The limitations of composite color signals are especially serious for postproduction editing that requires extensive use of multigeneration recording.

Introduction to Digital Technology 67

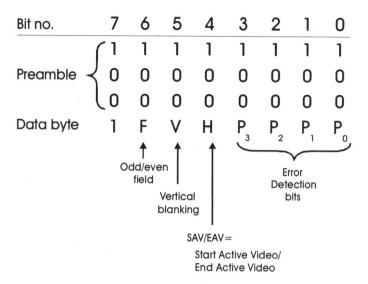

Figure 3.9 Serial synchronization pattern TRS-ID.

Component signals are more supportive of the advantages of a digital format—lower noise, lower distortion, greater flexibility for editing, and elimination of artifacts from luminance-subcarrier crosstalk. Accordingly, digital component formats are coming more used, particularly in recording (see Chapter 10).

In many system applications, however, it is necessary or desirable to transmit or record the signal in a composite form such as NTSC or PAL. As a result, industry standards are established for the frequency and location of sampling points for both composite and component signals (see Section 3.3 and Table 3.2).

3.7.4 Synchronization

The video bit stream from an ADC must not only convey the picture information but also needs digital words to synchronize it with the receiving device at the output of the system or with other processes occurring within the system. This is accomplished by the use of digital code words, which make it possible for the receiver to identify positively the following:

In serial systems, the beginning of each word

In component systems, the location of the beginnings of lines, the active regions of lines, fields, and frames.

The code words must have values that the decoder will not confuse with pixel values. Words 0000 0000 and 1111 1111, which are outside the quantizing range for pixels (see Section 3.3.3), meet this requirement.

Figure 3.9 shows a four-word synchronization pattern, known as *TRS-ID* (timing reference and identification signal), that is used for serial transmission. It is repeated for each line at the end of the horizontal sync pulse and synchronizes word beginnings, lines, fields, and frames. The first three words, the TRS, define the LSBs and MSBs in the bit stream. The fourth word, the ID, contains bit flags that identify odd and even fields, the beginning and end of the active line time, and the location of vertical blanking. Since each bit in the ID has only two states, the identification is limited to two choices such as the beginning and end of the active line time. In some cases, the ID has bits that are used for parity checking and error correction.

3.7.5 DTV transmission standards

Standards have been developed for various DTV requirements, including digitized composite and component signals, SDTV and HDTV, and serial and parallel transmission. Three examples age given in this section.

3.7.5.1 SMPTE 259M serial digital interface

The SMPTE *serial digital interface* for point-to-point transmission of 525- or 625-line Rec. BT.601-5 4:2:2 component or $4f_{SC}$ composite SDTV is also known as *SDI*. It uses coaxial cable of the same types that are already is use for analog video signals, which is a boon in converting existing facilities from analog to digital. SDI is based on 10 bpp video sampling and supports data rates up to 360 Mb/s.

3.7.5.2 SMPTE 125M component parallel interface

This standard provides a 10-bit parallel interface for point-to-point transmission of Rec. BT.601-5 4:2:2 component SDTV digital video. Twelve twisted-pair circuits in the cable include the ten bit channels and a clock channel. The twelfth pair is used simply for additional grounding. Y, C_R, C_B signals are multiplexed on each bit channel in the sequence $YC_R[Y]C_BY...$, where the three C_BYC_R values are the cosited samples and the [Y] value is the inbetween Y sample that has no cosited C_B and C_R samples. This means that the data rate on the cable is twice the 13.5 MHz sampling rate, or 27 MHz. Cable lengths up to 50 m are possible without equalization; longer lengths may be achieved with equalization.

3.7.5.3 IEEE 1394 serial interface

The two SMPTE standards just described are for one-direction point-to-point connections within a video facility. IEEE 1394 is different—it is a bidirectional *bus* standard that allows multiple units to be connected in daisy-chain fashion. The circuit operates as a computer network with messages being transmitted in packets at data rates up to 200 Mb/s or higher. 1394 cables have two twisted pairs for signal connections (one in each direction) and two other (optional) power wires. It is being used for connecting digital cameras to VCRs or computers in the home and field, and it should have further applications as the hardware cost falls with manufacturing volume.

3.8 Digital Error Protection

A major advantage of digital systems is their ability, within limits, to recognize and correct bit errors that are caused by noise or other interference in the system. If the number of errors exceeds the limits, an additional technique, *error concealment*, can be applied that reduces the visible effect of the errors. This technique is particularly important in video recorders and transmission systems.

3.8.1 Visual effect of bit errors

The visual effect of a single-bit error depends greatly on the significance of the bit. In an 8-bit PCM system, an error in the LSB will be 1 part in 256, which would have a barely noticeable effect. An error in the MSB, on the other hand, will cause a major disturbance to the signal. This range in sensitivities is considered in some error-protection systems, but it is not much of an issue any more because modern error-protection systems correct almost all single-bit errors anyway. Furthermore, systems that involve data compression may make all bits equally significant and a single-bit error can be highly visible. This means that error correction is more important in compressed-data systems.

Typical transmission systems can have *burst errors*, where bit errors occur in a group of adjacent bits in the bit stream. These are more difficult to correct directly and most systems use a strategy such as *interleaving* (see Section 3.8.3.3) to spread burst errors out into multiple error-protection blocks, where they will be seen and corrected as single-bit errors.

3.8.2 Error-protection system functions

Figure 3.10 is a functional diagram showing the essential elements of error-protection systems. The redundancy code is a bit or word added to the digital bit stream to make it possible to detect bit errors with a high degree of probability.

The retransmission request function is applicable only to systems that have two-way transmission and that can tolerate random delays in the data transmis-

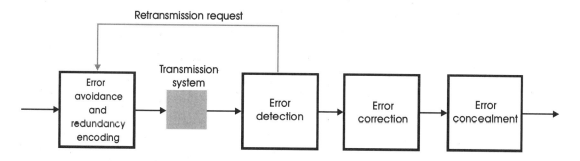

Figure 3.10 Error-protection system.

sion. The latter requirement usually precludes the use of retransmission for audio or video systems because it would interrupt the smooth reproduction of the audio or video.

Since bits have only two states, the correction of a bit error is straightforward once it has been detected—the bit value is simply reversed.

Error concealment, applicable primarily to recorders, is a last-resort process. It is used when error correction is ineffective, e.g., when a large number of bits are in error. Its principle is to reconstruct the missing or erroneous bits by interpolation of spatially and temporally adjacent bits. However, proper application of error concealment requires that the existence of an error be known, so a means for error detection is still required.

3.8.3 Encoding and error detection

A variety of redundancy codes have been developed to make it possible to identify bit errors with a high degree of certainty and probability. The simplest code is a single 1 or 0 bit, known as a *parity* bit, added to the end of each word as required to make the sum of the bits an even number (even parity) or an odd number (odd parity). Even parity code is illustrated for 2-bit words:

Signal words	Parity bit	Sum
0 0	0	0
0 1	1	2
1 0	1	2
1 1	0	2

If the sum of the bits in a word plus its parity bit after transmission or recording is an even number (0 is considered to be an even number), it is probable that none of the bits is in error, although there is a possibility that more than one are incorrect.

This simple parity check identifies single errors, but it does not point to the offending bit so that it can be corrected. It is particularly useful in situations where errors are most likely to occur one at a time, as in an electronic memory. It is not effective by itself for burst errors.

Error correction can be accomplished in parity systems by considering blocks of data as a two-dimensional matrix. For example, an 8-byte data block can be considered to be an 8×8 matrix of 64 bits. By assigning a parity bit to each row and column of this matrix—resulting in 2 bytes of parity overhead—single-bit errors can be located by looking at the intersection of the row and column that show errors. The offending bit can then be reversed to correct the error. This use of a two-dimensional error-protection block is known as a *crossword* or *product* code.

The simple row and column parity scheme described above works only for one error per block. However, more sophisticated codes have been developed that can operate with higher error rates. Two of the most common ones used in digital video are the *Hamming code* [2] and the *Reed-Solomon code* [3].

3.8.3.1 Hamming error-protection code

To generalize on the product code example above, a certain number of check bits are created from the data so that the pattern of failed check bits can be used to identify the data bits in error. The data bits of a block are transmitted unchanged and they are followed by the specified number of parity bits. This is known as a systematic code. In a Hamming code, the matrix for the generation of the check bits is designed so that the pattern of failed and unfailed check bits is an address to the failed data bits. In practice, Hamming codes are easy to implement in hardware and they can readily detect and correct single-bit errors per block. Additional errors can sometimes be detected, but they cannot be corrected.

A special form of the Hamming code is the *cyclic redundancy check* code (CRC), which is often used in data transmission. However, for digital video and audio, Hamming codes have been superceded by the much more powerful Reed-Solomon codes.

3.8.3.2 Reed-Solomon error-protection codes

One of the most widely used codes in digital audio and video is Reed-Solomon. It is used in CD audio, CD-ROM, digital video recorders, and DTV. It is a block coding scheme that can correct bursts of errors up to a limit governed by the amount of error-correction overhead that is chosen at design time. Although the processing is quite complex, it has been implemented in integrated circuits and is now easily included in system hardware.

3.8.3.3 Reed-Solomon as a product code

By defining a macro-block of data that contains a number of R-S blocks within it, a two-dimensional form of error correction is achieved that has much more power than a single R-S implementation. This is often used with the data from an entire TV frame or with the data from a tape track segment in a magnetic recorder. R-S coding is applied to the incoming data, after which the entire macro-block of data is stored in a memory arranged as a 2D matrix. The data are then read out from the memory by reading the memory matrix perpendicular to the way it was read in and a second R-S process is applied. For example, the data could be read into rows of the memory matrix and then it would be read out by columns. The first R-S process is called *outer* error coding and the second process is *inner* coding. At the receiving end, inner decoding is done first, the memory matrix transformation is done, and then the outer decoding is done to restore the original data. This scheme allows correction of large burst errors such as from tape dropouts.

In R-S coding, a data block is grouped into symbols, which (for example) might be bytes; a certain number of error-protection symbols are then calculated from the data. These are transmitted along with the data for use by the receiving R-S processor. The more error-protection symbols relative to the data block size, the more powerful the correction. Blocks are specified with two numbers, for example,

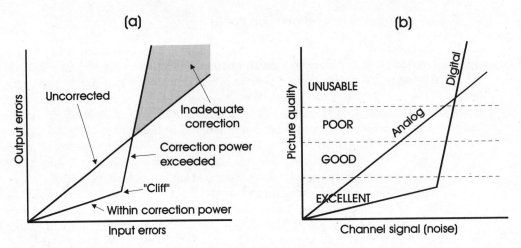

Figure 3.11 Error protection with high error rates.

a 208, 192 system has 192 symbols of data per block with 16 error-protection symbols.

3.8.3.4 Data interleaving

Although Reed-Solomon coding can handle multiple errors in one data block, transmission channels can sometimes have larger burst errors that would overload correction in a single block. This problem can be overcome after R-S encoding by *interleaving* the data between blocks so that the data for each block is not transmitted in normal sequence. Instead, it gets spread out over the whole interleave pattern. At the receiving end, the interleaving is removed before R-S decoding.

For example, if eight blocks were interleaved, then one-eighth of the data would be sent from each block in sequence, followed by the next eighth of each block, and so on. The result is that a single burst error in the channel cannot cause more than one-eighth of a block to be lost unless the channel burst gets longer than a whole data block. This process can be extended to deal with almost any size of channel interruption, but of course, the amount of processing and memory required increases with larger interleave patterns. In the inner-outer type of R-S coding, interleaving is done between the outer and inner coding.

3.8.4 Effectiveness of error protection

Figure 3.11(a) illustrates the performance of an error-protection system as the bit error rate is increased. In a system with no error protection, the output bit error rate increases with the input error rate (assuming that no additional errors are introduced by the system). Error correction is effective for input error rates up to a critical value. Above this level, sometimes called the "cliff," the error correction

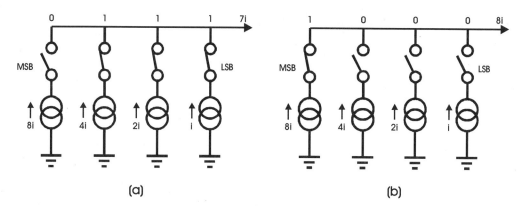

Figure 3.12 Digital-to-analog converter.

is not only ineffective but the output BER may actually be worse than without correction.

In comparing digital and analog systems, the cliff effect is sometimes presented as a disadvantage to digital systems. However, in a properly designed digital system, the cliff would not be reached until the transmission characteristics are so bad that analog transmission in that channel would also be unsatisfactory. This is illustrated in Figure 3.11(b).

3.9 Digital-to-Analog Conversion

Digital video signals must ultimately be reconverted to an analog format to be displayed. Figure 3.12 illustrates the principle of one type of *digital-to-analog converter* (DAC). A circuit is clocked so that each bit in a word, from the LSB to the MSB, controls a current flow that is proportional to its position in the word. The summation of these currents is proportional to the quantized level in the analog signal. The analog signal is then recovered by passing the output through a low-pass filter that removes the sampling frequency components.

An obvious problem with this system is the necessity of controlling the current in the MSB path with a high degree of accuracy. With an 8-bit code, the error in the MSB path must be substantially less than 1 part in 256 or it will overwhelm the effect of the LSB.

An alternative conversion system causes a constant current to flow into an integrator for a time interval that is determined by the position of the bit in the word.

The output of these conversion systems is a reconstruction of the pulse train that was created by the sampling process in the ADC. It is necessary to pass it through a low-pass filter having a cutoff at one-half the pulse rate in order to recover the original analog signal.

3.10 Video Data Compression

When digital video transmission or storage is viewed in terms of raw bit rates and analog bandwidth requirements, the numbers are so large that many tasks appear to be impractical. However, this view overlooks the potential of digital video and audio data to be *compressed*—that is, to be reduced. Because of the large amount of redundancy that exists in scanned video data, compression factors up to 100:1 or more are possible in some cases. This is the reason that all new video transmission systems will soon be digital.

It is an apparent paradox that digital video formats, which start out with more bandwidth than analog formats, can actually be used to reduce the bandwidth required for video signals. The technologies for reducing the data rate of video signals by data compression are now highly developed. It has been estimated that more than 40 companies worldwide have been engaged in this endeavor. Digital video compression can accommodate the growing requirements of computers, satellites, cable systems, and HDTV.

For the past several years, the video technical community has been engaged in two major development programs—HDTV and video compression—both directed toward more efficient use of spectrum space but with different objectives. The major purpose of HDTV development is to improve television picture quality with little increase in bandwidth requirements. The major purpose of video compression development is to reduce data requirements with a minimal loss of picture quality.

These developments have now come together in the ATSC DTV system that has been adopted in the United States, and the DVB project in Europe. Through MPEG video compression (see Section 3.12.2) and a creative digital transmission system (see Chapter 7), HDTV video and multichannel high-fidelity sound are transmitted over the same 6-MHz channels that now handle NTSC analog video.

Communications theory [4, 5] shows that under the conditions encountered in practical video systems, the rate at which information can be transmitted in a communications channel is proportional to its bandwidth. The name of the game in video data compression thus is to reduce the amount of information needed to send video pictures.

If video signals were random current or voltage variations, data reduction would not be possible without a major loss of picture quality. Fortunately this is not the case. Video signals are highly structured, and they are repetitive in time and correlated in space. Further, the rate of information transfer varies widely from frame to frame and in different areas of the frame. Areas of an image with rapid motion and a large amount of detail require more data than nonmoving areas with little detail. The role of data compression is to make use of these properties to reduce this rate with a minimum loss of picture quality. Some compression could be achieved with no loss in picture quality—this is called *lossless* compression and is used widely in the computer industry on all types of data. However, the degree of data reduction by lossless methods on video is usually 2:1 or less.

To achieve significant video compression, so-called *lossy* compression methods must be used. It should be emphasized, however, that any lossy data compression system will result in some reduction of picture quality—for example, loss of resolution at the edges of moving objects—and an objective is to make the defects least visible.

3.11 Data Compression Technologies

High-quality video compression requires extremely complex processing and effective techniques and hardware continue to be improved. A compression method is called an *algorithm*, which is a description of a mathematical process that performs the compression or the decompression required to restore the compressed signal to its original form. Many algorithms already exist for different purposes and different quality levels. Because of the tremendous diversity of existing algorithms and the vast potential that still exists for new development, standard-setting has been difficult. Good standards are in place, however, and they have left the door open for more improvements in the future.

3.11.1 Algorithms

As with other complex digital processing functions, the algorithm is the basic tool for specifying the mathematical operations that are to be performed for compression. Figure 3.13 is a diagram of some of the different techniques of compression that may be combined into an algorithm. Each of these techniques is potentially applicable to digital video at the pixel or line (spatial), or frame (temporal) levels.

3.11.2 Fixed and adaptive techniques

Figure 3.13 illustrates five common video compression techniques:
- Basic
- Interpolative
- Predictive
- Transform
- Statistical

This is by no means a complete list of techniques, but it is illustrative of the range of possibilities.

The techniques have two modes, "fixed" and "adaptive." In the fixed mode, their application is independent of image content. In the adaptive mode, their application depends on the content—for example, the system may operate differently for images with a great deal of detail or motion. The resulting algorithm processes the incoming digital signals continuously, and the output is a compressed bit stream.

3.11.3 Truncation

Truncation is a brute force technique that reduces the bit rate by dropping one or more of the LSBs. In the fixed mode, truncation has the obvious problem that it

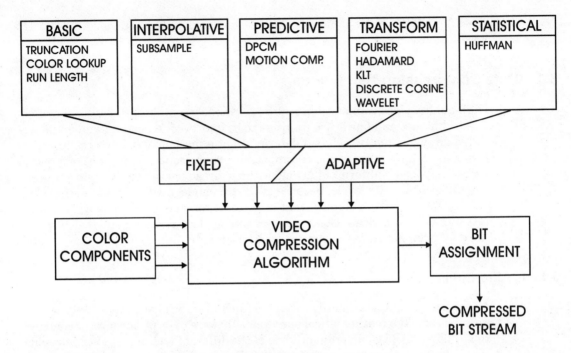

Figure 3.13 Video compression techniques. From [6].

lacks the ability to distinguish between the brightness levels represented by the LSBs, and the variations in image brightness will be discontinuous. This can be avoided by an adaptive transform that drops the LSBs only to the extent that the amount of picture detail permits.

The bit stream with variable word length is fed into a buffer, after which the missing LSBs are restored. A potential problem with this system is that the buffer can become overloaded if the picture contains an unusual amount of motion or detail.

3.11.4 Subsampling and interpolation

Subsampling and interpolation are widely used compression techniques, and they can also be used with samples that have not even been digitized. Figure 3.14 illustrates one example of the technique [7]. The sampling frequency is reduced by one-half (subsampling) and the samples on adjacent lines are offset by one-half the sampling interval. The values of the samples that are skipped as the result of subsampling are interpolated from the values of adjacent sampled elements, both horizontal and vertical. Two possible interpolation equations for the example in Figure 3.14 are:

$$i = \frac{b + c + d + e}{4} \qquad (3.3)$$

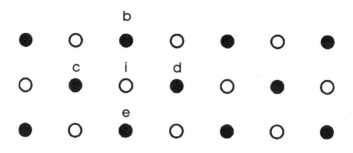

Figure 3.14 Spatial subsampling.

or

$$i = \frac{c+d}{2} + \frac{b-e}{2} \tag{3.4}$$

Equation (3.3) is the arithmetic average of all the adjacent sample values. This is sometimes called "A" interpolation. Equation (3.4) is the average value of the adjacent points on the same line plus the change in value of points on an adjacent line—a function of the high-frequency components in the signal. The choice between these formulas is largely subjective.

3.11.5 Prediction

Spatial *prediction* techniques are based on similarities between an object and its predecessor. Using the predecessor to predict the new object and then coding only their difference often results is less data needing to be transmitted.

One example of prediction is based on the fact that the difference in signal levels between adjacent pixels is usually small. If this is the case, the difference can be represented by a smaller number of bits per sample, say 4 instead of 8. This technique is called *differential pulse code modulation* (DPCM).

A DPCM system is overloaded when there is an occasional large difference between adjacent pixels as at a very sharp edge. This is called *slope overload* and results in a smeared edge.

DPCM also has the problem that the amplitude of each pixel signal is the sum of the amplitude of the previous pixel signal and the difference. If an error creeps into one of the pixels, it will be repeated until the level is reset. Because of these problems, DPCM is not widely used by itself, but it is a useful technique when combined with others.

Prediction can also be done temporally by examining adjacent frames and finding out what parts of a frame already exist in the preceding frame. These parts of the new frame do not have to be transmitted again. The technique is called *motion compensation* (see Section 3.12.2).

3.11.6 Transformation

Another possibility for video compression is to transform the values of a group of pixels into a different set, which may be transmitted with less data. After transmission, an inverse transform is performed to recover the original values.

Transforms have been the subject of extended investigation, their objective being to be effective and to operate speedily in real time. As a simple example, assume the values $A, B, C,$ and D are transformed to the values W, X, Y, Z by the equations:

$$W = A$$
$$X = B - A$$
$$Y = C - A$$
$$Z = D - A$$

$W, X, Y,$ and Z are transmitted. Since $X, Y,$ and Z are value differences, they are usually smaller than the untransformed originals, $B, C,$ and D and do not require as many bits per word.

After transmission, $A, B, C,$ and D are recovered by a reverse transform:

$$A = W$$
$$B = W + X$$
$$C = W + Y$$
$$D = W + Z$$

Adaptive transformation is another technique. The amount of information that must be transmitted for a particular portion of the image is proportional to the fineness of detail in that portion. A portion of the picture with little detail can be transmitted with very few bits, and this provides extra time for transmitting portions with high detail. A buffer is used to restore the original spatiotemporal relationship. If the entire picture has high detail, the buffer may become overloaded and the rate of information transfer reduced by increasing the quantum intervals. This, of course, reduces the image quality by introducing artifacts.

Transforms have been the subject of extended mathematical analysis, and many are in use. Some of the better known ones are the *Hadamard* transform, the *Fourier* transform, and the *discrete cosine* transform (DCT).

3.11.6.1 The discrete cosine transform

The DCT is a version of the Fourier transform that uses the fact that the input is a limited set of samples taken from a continuous waveform (thus, the word "discrete"). A Fourier transform produces both sine and cosine terms in its output in order to specify the phase of the frequency components. However, by performing the discrete transform on a block of samples and its mirror image, the sine terms can be made to cancel, leaving only the cosine terms. This is actually a simplification and it explains why the word "cosine" is in the name of the transform.

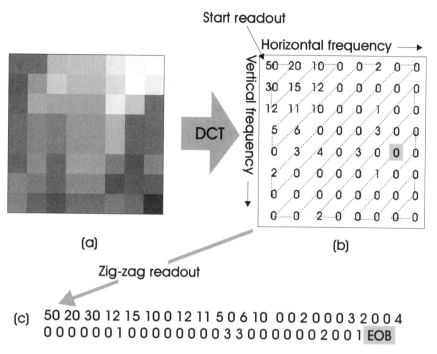

Figure 3.15 The discrete cosine transform.

The DCT is typically done on 8 × 8 blocks of pixels, so there are 64 total samples to be processed, as shown in Figure 3.15(a). The output of the transform is 64 new values representing coefficients for the spatial frequency components in the block, as shown in Figure 3.15(b). So far, there is no compression because there were 64 values going in and the same number of values came out. The zero-frequency coefficient is referred to as the DC coefficient; it is the average of all the pixels in the block. The remaining coefficients are for finite spatial frequencies in the block. The beauty of this transform is that it defines most of the block's information in the lower spatial frequencies and many of the higher-frequency coefficients come out to be zero. This is because adjacent pixel values are most often similar. Although the transform itself does not provide any compression, the transformed values are amenable to substantial compression by subsequent processes as described below.

In most algorithms, the DCT coefficients are quantized according to an adaptive table and then they are ordered in ascending spatial frequency (see Section 3.12.1), which causes most of the zero values to be grouped. This facilitates further compression by run-length encoding or statistical techniques.

Because the DCT equations involve transcendental functions, which can only be approximated in a real system, this transform is inherently lossy. However, it lends itself to making the losses invisible in most images and is therefore a highly effective compression technique.

3.11.7 Statistical techniques

Statistical techniques can be used for bandwidth reduction by making use of the fact that some image values are less frequently encountered than others. Shorter words are used to define the more frequently encountered values with a reduction in the total number of bits. This technique, called *Huffman coding* [8], requires the transmission of a code table so that the length of each word can be identified at the receiving end.

3.11.8 Cascaded compression

To achieve a desired degree of compression it is usually necessary to combine more than one technique and to compress the signal in a number of successive steps. For example, a compression of 32 to 1 could be achieved by 4:1, 4:1, and 2:1 compressions performed successively.

3.12 JPEG and MPEG Compression

The *International Electrotechnical Commission* (IEC) some years ago recognized the need for digital video and audio compression standards and formed the *Motion Picture Expert Group* (MPEG) to develop standards in this important technology area. The work of MPEG has led to the MPEG-1 and MPEG-2 standards, which are becoming the basis for most widely distributed forms of compressed video. Additional MPEG standards are under development.

MPEG was actually preceded by another working group of the IEC, the *Joint Photographic Expert Group* (JPEG) for standardization of still-image compression techniques. This was important because still-image compression is widely applicable to the fields of desktop publishing, graphic arts, facsimile, medical imaging, and others. JPEG developed a standard that is configurable and applicable to this wide range of end uses [9]. The standard supports a variety of algorithms and defines how these are combined to address different applications.

The idea of configurability is significant because it does not lock the industry into a single algorithm; in fact, it even allows for new algorithms to be added in the future. Combining this with the flexibility of software-controlled processing allows continued growth of compression system performance. The JPEG still-image compression concepts form the basis of MPEG motion video compression.

3.12.1 JPEG Still-Image Compression

The JPEG standard supports both lossless and lossy compression of images of practically any size and bits per pixel structure. It is parameterizable so the user can make tradeoffs of degree of compression vs. quality, and it has four modes of operation:

1. Sequential—encode in the order that the image was scanned
2. Progressive—multipass encoding so that a coarse image is transmitted rapidly, followed by repeated passes at progressively higher resolutions
3. Lossless—exact reproduction is guaranteed (all other modes are lossy)
4. Hierarchical—the image is encoded at multiple resolutions

Basic JPEG lossy compression uses the DCT algorithm, encoding 8 × 8 blocks of pixels at a time. The coefficients resulting from the DCT are quantized according to a quantization table that is supplied by the application. This is one place where the algorithm is configurable—the quantization table becomes part of the compressed data stream. The next step, called *zig-zag ordering* is shown by the dotted lines in Figure 3.15(b). This arranges the DCT coefficients serially so that zero-value coefficients tend to be together at the end of each block, as shown in Figure 3.15(c). An *end-of-block code* (EOB) identifies the point beyond which all coefficients are zero. This facilitates the next step, which is statistical coding. Huffman coding is typically used here and again, the coding table is configurable.

With most high-resolution color images, JPEG can compress 10:1 or so with excellent results. For example, a 16- or 24-bpp image can be transmitted at less than 2-bpp and reproduced with no visible loss of resolution. In some cases, compression up to 30:1 is achievable. The amount that can be used depends on the image content and the viewing conditions for the final decompressed image.

The JPEG standard is usable for motion video compression by performing the algorithm separately for each frame of the motion sequence. This is called *motion-JPEG* and has been applied extensively in situations that don't need the maximum compression. However, it fails to exploit the compression opportunities that are inherent in the redundancy between frames of a motion sequence. The MPEG algorithms were designed to do that.

3.12.2 MPEG Architecture

The technique for exploiting frame-to-frame redundancy is called *motion compensation* and it is done to reduce the frame data before it gets coded by techniques similar to JPEG.

3.12.2.1 Motion compensation

The basic idea of motion compensation is to find the parts of each frame that have moved or changed from the preceding frame and code only the changes, which are called *residuals*. Each frame is then built by adding the decoded residuals to the prediction based on the previous frame. This is a form of adaptive prediction, where the previous frame predicts the next frame except for places where something has changed or moved. These frames are called *P-frames*.

The example in Figure 3.16 shows how frames are divided into blocks for the purpose of motion compensation (the number of blocks is reduced to simplify the figure). The frames are taken from a sequence of a crow flying over a city. The camera tracked the crow, so the city background is moving to the left. The previ-

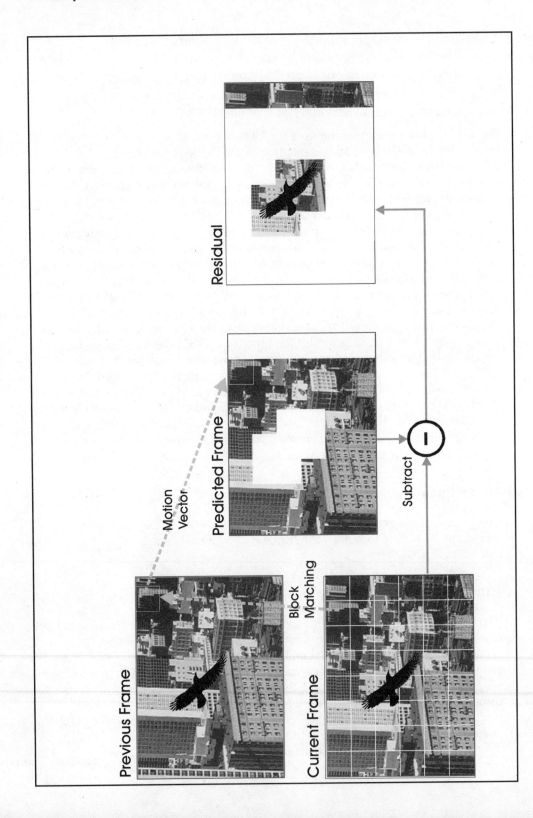

Figure 3.16 Motion compensation encoding.

ous frame is searched for each block of the current frame (this is called block matching). If a match is found, the location of the block (called a *motion vector*) in the previous frame is coded. A trial predicted frame is then built by moving blocks from the previous frame using the motion vectors. This process is shown in the figure for one block. The predicted frame is subtracted from the actual current frame to make the residual image, which is then coded for transmission. At the receiving end, the process is reversed: the predicted frame is built from the previous frame, the residual image is decoded and added to the predicted frame.

Searching the entire previous frame for a block from the new frame requires a fantastic amount of computing. Since the probability of finding the block drops as one searches further away from the block's position in the new frame, the search can be restricted to a small region surrounding each new block. This is defined by an encoder parameter called the *search range*. If the search range is too small, the encoder will be overloaded for even small amounts of motion from one frame to the next. Greater search range allows more motion to be handled, but it makes the encoder more expensive.

Transmitting a long motion sequence by means of frame prediction has two problems:

1. If an error occurs in one frame, it will tend to be propagated to future frames until a change obliterates it. This makes errors more visible.
2. In many cases video must be editable, meaning that it should be possible to cut any group of frames from a video sequence and use them in another sequence. With predictive coding, a video sequence must always begin at the first frame.

Both problems are solved by periodically sending a full frame that is not a prediction of the previous one (similar to a JPEG image). If this is done at a moderate rate (say every 10th frame), it doesn't add much data to the stream. Thus, in a 30-frames-per-second system, edits could be started at any of three points in every second. This is not frame-accurate editing, but it is sufficient for many purposes. Such full frames are called intracoded (*I-frames*).

There is a third strategy for frame coding in MPEG, which uses bidirectional interpolation between two frames. These are called *B-frames* and they provide the most compression, but they also require the most processing during compression and the most complex decompression strategy. Figure 3.17 shows how the three frame types might be combined in a data stream and how they depend on each other. The exact frame sequence is configurable.

3.12.3 MPEG-1 and MPEG-2

The first MPEG standard was MPEG-1 [10], which is designed for data rates up to 1.5 Mb/s. This supports the single-speed CD-ROM rate used in computers and the T-1 data communication channel. MPEG-1 performance is often described as similar to a VHS videotape recording. It is satisfactory for many entertainment applications or for partial-screen display on computers, but it is not as good as NTSC or PAL broadcast television.

MPEG-2 is designed for higher data rates, up to 15 Mb/s, and higher-quality

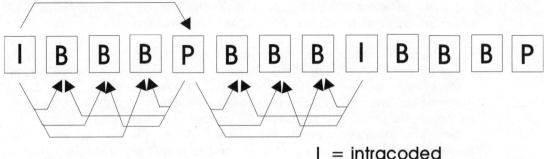

Figure 3.17 A typical MPEG frame sequence.

performance, including HDTV (see Section 7.5). It supports full-screen display on both computers and television receivers.

MPEG-4 and MPEG-7 are under development. They will extend the application of MPEG to multimedia applications including interactivity and all forms of graphics, text, animation, as well as audio and video.

3.13 Summary

It is safe to say that digital video technology has reached the point where it is the preferred approach for all new system designs. In most cases, digital video performance vs. price has surpassed analog systems. Satellite broadcasting is already digital, and with DTV, terrestial broadcasting and cable systems will be digital. Digital video technology allows integration with computer data networks and it will be an essential ingredient to the developing Information Superhighway.

3.14 References

1. A. J. Viterbi, "Error Bounds for Convolutional Codes and an Asymptotically Optimum Decoding Algorithm," *IEEE Trans. Information Theory*, Vol. IT-13, pp. 260-269, April 1967.

2. R. W. Hamming, "Error Detecting and Error Detecting Codes," *Bell Syst. Tech. J.*, 26:147-160, 1950.

3. I. S. Reed and G. Solomon, "Polynomial Codes Over Certain Finite Fields," *J. Soc. Ind. Appl. Math*, 8:300-304, 1960.

4. R. V. L. Hartley, "Transmission of Information," *Bell Syst. Tech. J.*, Vol. 7, July 1928.

5. C. E. Shannon, "A Mathematical Theory of Communication," *Bell Syst. Tech. J.*, Vol. 27, July-October 1948.

6. Arch Luther, *Digital Video in the PC Environment*, McGraw-Hill, New York, 1991.

7. R. Kishimoto, N. Sakuri, and A. Ishikura, "Bit-Rate Reduction in the Transmission of High-Definition Signals," *J. SMPTE*, 96(2), February 1987.

8. D. A. Huffman, "A Method for the Construction of Minimum Redundancy Codes," *Proc. IRE*, 40, pp. 1098-1101, 1952.

9. ISO/IEC 10918-1. *Digital Compression and Coding of Continuous-Tone Still Images*, 1993.

10. ISO/IEC 11172. *Coding of Moving Pictures and Associated Audio for Digital Storage Media at Up to About 1.5 Mbits/s*, 1993.

Chapter 4
Elements of Image Quality

4.1 Overview

The technical performance of a video system is measured by the quality of the images it produces. An image is judged to be of high quality if it is a near-perfect replication of the original scene, which may be a natural scene viewed by a camera or an artificial image created by an artist or a computer. Deliberate distortions of the image may be introduced for esthetic purposes—for example, by defocusing the background areas of the scene to better emphasize the foreground—but the basic technical objective is to duplicate the appearance of the original as closely as possible.

The use of the word "appearance" in the preceding sentence indicates that the evaluation of video images is ultimately subjective. The video engineer, however, cannot be satisfied with exclusively subjective criteria for image quality. The design and operation of video systems require that they be defined in objective and measurable terms. An important engineering task, then, is to develop objective picture quality yardsticks that coincide with the subjective perceptions.

Video systems are designed to optimize the presentation of natural scenes while computer displays are designed to best present artificial images, graphics, or text. However, there is more and more need to show both image classes on all systems, a situation that is sometimes called "convergence" between TV and computers. The basic criteria for computer display of "artificial" images are covered in Chapters 3 and 8.

This chapter will concentrate on the criteria for display of natural images, whether on TV or computer. It begins with a brief summary of the most important quality criteria for monochrome video and the luminance component of color television. That is followed by the application of these criteria to practical operating systems, both analog and digital, that use NTSC broadcast and high-definition

TV (HDTV) production standards (SMPTE 240M). The implications of these criteria to the display of natural images on computers is also treated. The criteria for the chrominance components of color TV are described in Chapter 2.

4.1.1 Television and film

The complexity of establishing objective quality criteria for video images is increased by the necessity of comparing television and photographic systems. Photography preceded video, and film quality has historically provided a basis for defining video image quality. A frequently stated goal for the design of broadcast TV systems is to try to duplicate the quality of 16-mm motion picture film. Similarly, the goal of HDTV systems is often described as the duplication of the quality of wide-screen theatrical film.

Comparison of the quality of TV and film images is difficult and cannot be rigorous. There are inherent differences between the video and photographic processes that make it almost impossible to duplicate precisely the appearance of a film image with a video system or vice versa. The video engineer nevertheless must establish practical image quality standards that are equivalent to those used for film, even though the comparison cannot be exact.

The challenge of establishing objective image quality criteria for video and of comparing the quality of inherently dissimilar film and video systems has stimulated an enormous amount of research by the photographic and video industries over the past decades. Practical definitions of the most significant measures of video image quality, together with their specifications and measurement procedures, have emerged from this research. A comprehensive listing of the criteria for evaluating the quality of video images, together with performance standards for both color and monochrome, is contained in a joint publication of the Electronic Industries Association (EIA), and the Telecommunications Association (TIA): "EIA/TIA STANDARD—Electrical Performance for Television Transmission Systems, EIA/TIA-250-C." Although this document is intended to set standards for transmission systems alone, it provides a useful basis for evaluating complete system performance. Since performance degradations may be cumulative, the standards for the transmission system are more rigorous than for the complete system.

4.1.2 Production, transmission, and display standards

Video systems have three components—production, transmission, and display. Each of these has a different set of constraints affecting image quality.

The NTSC television system was designed to operate within the constraints of a 4-MHz bandwidth and to have compatibility with monochrome video standards. It is fundamentally a transmission standard (see the EIA/TIA standards noted above), but of necessity it became a de facto production standard as well. This results in reduced performance in applications such as studio program production where limited bandwidth and monochrome compatibility are not required. This problem has been recognized and efforts are now underway to develop production standards for high-performance systems that use NTSC scanning standards but that are not limited by transmission system constraints or a requirement for compatibility. Component systems are an example (see Section 7.11).

In HDTV, the differing requirements of production and transmission were recognized early and separate standards have been proposed or established for each.

SMPTE Standard 240M is a widely used HDTV production standard. It originally specified an analog signal, but a digital version has now been developed (SMPTE 260M). The 240M and 260M standards are used to define the performance of HDTV systems in this chapter.

A transmission standard for HDTV broadcasting has been adopted by the FCC. This specifies a digital system with the parameters listed in Table 7.1. Note that with digital compression it is possible to transmit the HDTV signal in the same 6-MHz channel used for NTSC broadcasting.

4.2 Basic Image Quality Criteria

Image quality criteria can be divided into two categories—basic criteria that apply to all imaging systems, and image defects, some of which are unique to television images. Basic criteria include image definition, limiting resolution, gray scale, signal-to-noise ratio, and colorimetry.

4.2.1 Image definition

Definition is the degree of "in-focus" appearance of the image. It is a subjective parameter, but it is primarily determined by the shape of the transitions between dark and light areas of the picture. Although definition is related to the resolution of fine detail (see Section 4.2.2) in the image and is sometimes confused with it, image definition describes a different aspect of picture quality.

4.2.2 Limiting resolution

Limiting resolution indicates the maximum number of alternate light and dark lines that can be distinguished per unit distance in the image. This is a measure of reproduction of fine detail.

4.2.3 Gray scale

Gray scale describes the image brightness as a function of scene brightness. It is sometimes called the *transfer characteristic*. It can be specified simply by three parameters: highlight brightness, the brightness of the brightest areas of the image; *contrast ratio*, the ratio of the brightness of the brightest areas of the image to the darkest; and *gamma*, the slope of the transfer characteristic of image brightness as a function of scene brightness, both expressed logarithmically.

The three-parameter specification of gray scale is an approximation, especially in specifying the slope of the transfer characteristic by a single number for gamma.

Other factors, such as highlight compression or clipping, may upset the slope at certain levels. In such cases, a complete curve for the transfer characteristic must be specified.

4.2.4 Signal-to-noise ratio

In the broadest sense, "noise" is any unwanted signal, but in video systems it usually means random or thermal noise of a type that produces a moving speckled effect in the picture, called *snow*. This is specified by giving the ratio between the maximum signal level and the noise level—*signal-to-noise ratio* (SNR).

4.2.5 Summary

Definition, gray scale, and SNR are fundamental criteria that apply to all imaging systems. They are also interrelated. For example, the perceived definition of an image is strongly influenced by its gray scale. For analytic purposes, however, it is usual to consider them separately. These basic quality criteria are described in more detail in Sections 4.5 to 4.16.

4.3 Image Defects

In addition to the basic criteria described above, *image defects* must be considered in evaluating video picture quality. Image defects can be divided into three categories:

1. Defects that are inherent in the system such as flicker and aliasing.
2. Defects such as lag and smear that result from deficiencies in the production of the video signal.
3. Spurious signals that are introduced by recording or transmission systems such as hum and other forms of electrical interference, cochannel and adjacent channel interference, receiver-generated interference, and "ghosts."

The following sections summarize the nature of these defects. Section 4.16 contains more detailed descriptions.

4.3.1 Flicker

Flicker (see Section 4.16.4) results when the frame repetition rate is not high enough to cause the eye to perceive it as a continuous image. In an interlaced system (see Section 1.6.2), the visibility of large area flicker is determined by the field rate and of small area flicker by the frame rate.

4.3.2 Aliasing

Aliasing (see Section 4.8) is the production of spurious signals as the result of sampling in space or time.

4.3.3 Lag

Lag is a measure of the rate of change of the video signal at a fixed point on the raster when the scene at the corresponding point changes. Ideally, the signal voltage at any point would change instantaneously with a change in scene; if it does not, the result is a loss of resolution of moving objects or in extreme cases a trailing tail or smear.

4.3.4 Geometric distortion

Distortion of image geometry, i.e., of the shape or size of objects, results from defects in the camera's optical system or nonlinearities in the scanning process of the camera or display device.

4.3.5 Hum

The term "hum" is borrowed from audio engineering and it describes interference caused by spurious power source voltages that have been introduced in the system. In its most common form it results in a horizontal *hum bar* that moves vertically through the picture at a rate equal to the difference between the field rate and power source frequency.

4.3.6 Cochannel interference

Cochannel interference occurs in TV broadcasting when two stations operating on the same frequency are received at the same location. The most common effect is "venetian blinds," alternate black and white horizontal bars across the picture that result from the "beat" between the two carriers.

4.3.7 Receiver-generated interference

A number of spurious signals can be generated by TV receiving systems. Two of these, local oscillator radiation and intermediate-frequency (IF) images, result from the use of the superheterodyne principle in receivers. A third, intermodulation interference, may result when strong signals are received from two stations having carriers separated by the i-f frequency of the receiver. These problems are particularly troublesome for UHF stations, but they are mitigated by the FCC's policy of channel assignments that reduces the possibility of receiver-generated interference from the use of the superheterodyne principle. Another potential source of interference is intrachannel beat frequencies between the sound and picture carriers.

4.3.8 Ghosts

A *ghost* is a duplicate of the main image, slightly displaced from it, usually to the right, and much fainter. In broadcasting, it is caused by multipath transmission, i.e., the signal radiated from the transmitter arrives at the receiver over two or more paths, usually a direct path and a reflection from a building or other large object. In spite of the enormous velocity of radio waves, the path-length differences between direct and reflected waves that are often encountered in practice are sufficient to produce noticeable displacements of the ghost image. Radio waves travel 63,000 feet during the time interval of one scanning line, and a 1000-foot path difference would produce a displacement of 1/63 of the scanning line length or slightly more than 1/3 inch on a display with a 20-inch width.

4.4 Characteristics of Human Vision

The human eye is the final link in the perception of video images, and a consideration of its capabilities is critical in the design of video systems and the establishment of objective performance criteria. To state the obvious, there is no advantage in designing a system with image quality that exceeds the ability of the eye to perceive it.

The characteristics of human vision are exceedingly complex, and a detailed discussion of this subject is beyond the scope of this volume. The most significant characteristics for television systems design are the perceptions of brightness and contrast, the visibility of sharpness and detail, and flicker. These properties are described in this chapter. The perception of color was described in Chapter 2.

4.4.1 Brightness and contrast perception

Brightness is a subjective term that indicates the magnitude of the visual sensation produced by a source of light. *Contrast* is the ratio of brightnesses between the darkest and lightest sections of the image. For example, if the darkest area of an image has a brightness of one one-hundredth of the brightest area, the contrast ratio is 100:1.

Luminance is the objective equivalent of the subjective sensation of brightness and is a measure of the *luminous intensity* of the surface of an extended light source. The luminous intensity is equal to the *luminous flux* or *lumens* emitted per unit solid angle. It is calculated by multiplying the energy or radiated flux expressed as a function of wavelength by a weighting factor, the *luminosity function*, that expresses the sensitivity of the eye to radiation of different wavelengths. Luminance is specified in *lamberts*, or *foot-lamberts*. Figure 4.1 shows the luminosity function, i.e, the response of the eye with normal levels of brightness, as standardized by the CIE (*Commission Internationale de l'Eclairage*) in 1924.

Care should be exercised to avoid confusing luminance and *illuminance*. The former is a measure of the brightness of an area of an image. The latter is a measure of the illumination of a scene, for example, a television studio set, by an external light source. Illuminance is specified in *foot-candles*.

The visibility of an object in an image depends on the contrast between different areas. In common with most physical sensations, the magnitude of this perception tends to be proportional to the brightness ratio rather than to the abso-

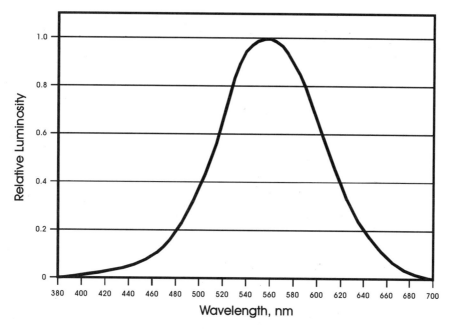

Figure 4.1 The luminosity function.

lute brightness difference. This is expressed in *Weber's law*:

The increase in stimulus necessary to produce an increase in sensation of any of our senses is not an absolute quantity but depends on the proportion that the increase bears to the immediately preceding stimulus.

The perception of contrast between two areas also depends on the sharpness of the boundary. If the boundary is sharp, a much smaller brightness difference can be perceived.

4.4.2 Detail and image sharpness perception

One of the most significant characteristics of the eye for establishing the design parameters of a television system is its ability to discern detail and the sharpness of edges in an image. This attribute determines the eye's perception of two of the basic quality criteria for video images—definition and SNR. It also has a direct effect on the bandwidth requirements for the system.

The ability to discern detail is known as *visual acuity*, which affects the perception of limiting resolution. The eye's visual acuity can be described by specifying the smallest angular separation at which individual lines can be distinguished. Visual acuity varies widely, as much as from 0.5′ to 5′ (minutes of arc), depending on the contrast ratio and the keenness of the individual's vision. An acuity of 1.7′ is often assumed in the design of video systems.

The angular separation between lines can be converted to television lines after specifying the *viewing ratio*, which is the ratio of viewing distance to picture height, and this is a more common method for specifying this parameter.

With 480 active scanning lines (as in 525-line broadcast systems) and a visual acuity of 1.7′, individual scanning lines can be distinguished at a viewing ratio as great as 4.4. At a viewing ratio of 4 (the minimum usually assumed for broadcast systems), individual scanning lines can be faintly distinguished by a typical viewer. A viewing ratio of 3 or less is usually assumed in designing HDTV. Desktop computer displays operate at viewing ratios as low as 1:1.

The difference in the eye's acuity for detail and edge sharpness at the viewing ratios assumed for HDTV, broadcast television systems, and computer displays establishes some of the most important differences in their design requirements. HDTV pictures, observed at a low viewing ratio, must have higher definition to give the same perception of sharpness. Broadcast-quality pictures seen from a low viewing ratio will appear fuzzy. On the other hand, an HDTV picture will appear no sharper than a broadcast-quality picture at greater viewing ratios. The practical result is that HDTV requires a larger screen at normal viewing distances to take advantage of its superior definition and SNR.

4.5 Image Definition

The sharpness of an image is expressed by its image definition parameter.

4.5.1 Image definition defined

The dictionary meaning of image definition, the distinctness of the outlines in the image, is also satisfactory as a technical definition. A less precise definition is the degree to which the image appears to be "in focus." Stated in still other terms, the definition of an image is a measure of the sharpness of the transitions or edges between its dark and light areas. In a high-definition system, these edges must be very sharp.

While the perceived definition of an image is determined primarily by the width of the edges between light and dark areas, there are second-order effects that depend on the shape of the edge transition. As compared with film, the slope of the transition curve for a TV system is steeper in the center of the brightness scale, and there is often an overshoot. This tends to give the TV image a "harder" appearance that is sometimes described pejoratively by the film industry as "the television look." The perceived definition of an image is also affected by the gray scale and the SNR. These factors are treated with greater detail in later sections.

4.5.2 Criteria of image definition

Although the perceived definition of an image is determined by the sharpness of its edges, edge width is not a satisfactory working criterion for image definition. It

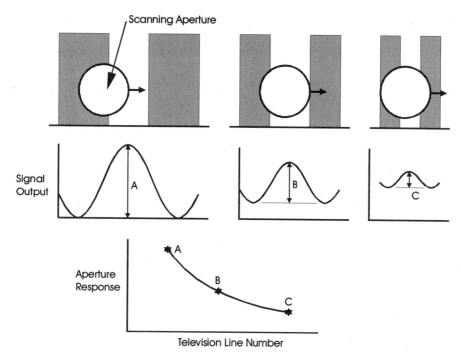

Figure 4.2 Definition of aperture response.

is difficult to measure and equally difficult to handle analytically. As a result, television image definition is usually specified by one of two indirect criteria—aperture response and limiting resolution. The measurement of aperture response is more complex, but it is a more complete criterion and is indispensable for system analysis. Limiting resolution, which is the highest resolution that can be seen, is easy to measure and is particularly useful for routine maintenance.

4.6 Aperture Response

Aperture response is a universal criterion for specifying picture definition and other aspects of imaging system performance. It can be used for film images, camera lenses, television camera imagers, video amplifiers and other bandwidth limiting components, the scanning process, display devices, and for the human eye. It is used frequently in this volume for describing the performance of video system components.

4.6.1 Definition of aperture response

The term "aperture response" is defined in Figure 4.2. Assume that a pattern of black and white lines of varying widths is scanned by a narrow light beam (the aperture), and the peak-to-peak variation in the reflected light from the black and white lines is measured. On lines that are much wider than the diameter of the spot, these variations will be of full amplitude. As the width of the lines is decreased so that the scanning spot always overlaps a portion of a black and white line, the amplitude of the variations will decrease. When the width of the lines is twice the diameter of the spot, the variations disappear.

The width of these lines is specified by its reciprocal, the number of alternate black and white lines (counting both black and white lines) that can be fitted into the vertical dimension of the picture. This parameter is known as *TV lines of resolution* (TVL).

The aperture response of a component or system is a graph of the peak-to-peak amplitude of its response, or of the variations in reflected or transmitted light, as a function of the television line number.

Aperture response can be specified either by the response to a square-wave pattern, i.e., alternate dark and light bars, known as the *contrast transfer function* (CTF), or of the response to a mathematically defined pattern in which the cross-sectional darkness of the bars varies sinusoidally, known as the *modulation transfer function* (MTF). The CTF is physically measurable, but the MTF is more useful for analytic purposes. The conversion can be made by means of Eq. (4.1):

$$\mathrm{MTF}n = \frac{\pi}{4}\left(\frac{\mathrm{CTF}3n}{3} - \frac{\mathrm{CTF}5n}{5} + \frac{\mathrm{CTF}7n}{7} \dots\right) \tag{4.1}$$

where $\mathrm{MTF}n$ is the modulation transfer function (aperture response to a sinusoidal pattern) at line number n, and $\mathrm{CTF}n$ is the contrast transfer function (aperture response to a square wave pattern) at line number n. Aperture response curves, $\mathrm{MTF}n$ and $\mathrm{CTF}n$, for a representative charge-coupled device imager are shown in Figure 4.3.

4.6.2 Aperture response of components in series

The aperture response of a system having several components in series can be calculated by multiplying the MTFs of the individual components, just as with the frequency response of cascaded electrical components. The ability to calculate the cascaded aperture response of a number of components in this manner is a very useful characteristic of aperture response as a criterion of performance.

4.6.3 Limiting resolution and aperture response

Limiting resolution is determined by visual acuity (see Section 4.4.2) and also by the line number at which the aperture response curve disappears into the noise (see Figure 4.3).

4.6.4 Aperture response of the human eye

Figure 4.4 shows the aperture response, MTF, of a typical human eye at viewing ratios of 3, 4, and 8. This response function should be considered as nominal rather than precise because the visual acuity of individuals varies widely.

Elements of Image Quality

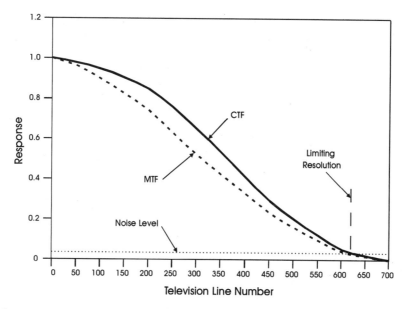

Figure 4.3 Aperture response, showing MTF and CTF curves.

4.6.5 Aperture response and image definition

The correlation of aperture response with definition or edge sharpness is not intuitively obvious, but it can be demonstrated by calculating the output of a system of known aperture response with a step function input, i.e., to an input signal that

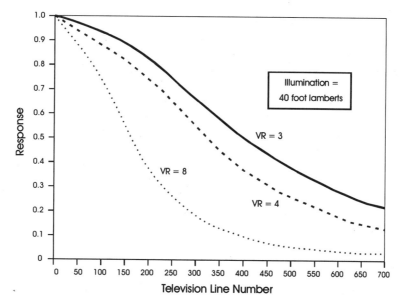

Figure 4.4 Aperture response of the human eye.

makes an infinitely steep transition between two levels. The shape of this transition as a function of the system aperture response is calculated by an *inverse Fourier transform*:

$$e(x) = \int g(n)e^{jn}dx \qquad (4.2)$$

where $e(x)$ is the *transient response* of the component or system as a function of its position on the scanning line, i.e., along the x axis of the raster, and $g(n)$ is the aperture response of the system as a function of the television line number, $n = 2\omega x$. With the aid of this equation, the correlation between aperture response and definition can be verified.

4.6.6 Equivalent line number, N_e

An aperture response curve is the most complete indication of the definition of a system, but it is not as convenient or easily compared as a single number. In an effort to define a single number that would specify the definition of a system to a reasonable degree of approximation, Schade [1, 2] developed the concept of equivalent line number or N_e. Although N_e is not widely used as a criterion of picture quality, it is very useful for system analysis.

N_e is defined in Figure 4.5. It is the line number that defines a rectangle having the same area as the area under the aperture response squared curve. (It is equivalent to the noise bandwidth in the electrical analog.)

The validity of N_e as a measure of picture definition can be demonstrated by calculating the edge configurations of systems having the same N_e but differently shaped response curves. Figures 4.6 and 4.7 compare two imaging systems with the same N_e but with aperture response curves typical of television and film system components. Figure 4.6 shows the square of their aperture responses while Figure 4.7 shows their responses to a step function input. Note that the limiting resolution of the film system is much higher but its MTF response is lower at lower line numbers. This causes minor differences in the shape of the edges, but the edge width is approximately the same and the pictures appear equally sharp, thus confirming the validity of N_e as a measure of image definition.

The edge width is given by Eq. (4.3):

$$\text{Edge width} = \frac{\text{Picture height}}{N_e} \qquad (4.3)$$

Within reasonable limits of inequality, the approximate perception of definition of images that have unequal horizontal and vertical definition is given by Eq. (4.4):

$$N_e = \sqrt{N_e(\text{vert})\, N_e(\text{horiz})} \qquad (4.4)$$

Because of limitations imposed by the scanning process and the improved performance of modern cameras, the inequality in H and V definitions often exceeds "reasonable limits" and the horizontal definition is noticeably greater.

Elements of Image Quality 99

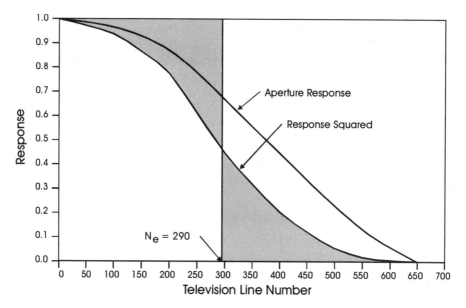

Figure 4.5 Definition of N_e.

4.7 Sampling and Aliasing

Sampling is a fundamental technique that is pervasive in video systems. It was introduced in Section 1.3. Applications of sampling in analog television include the following:

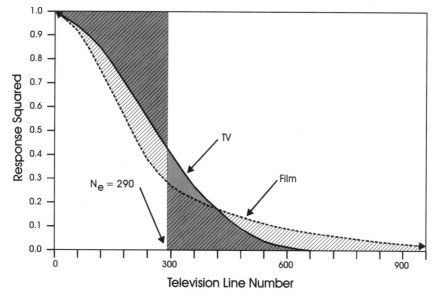

Figure 4.6 TV and film MTF compared.

100 Chapter Four

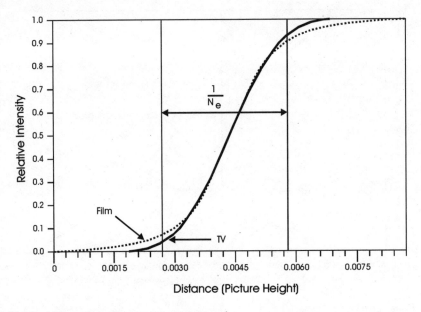

Figure 4.7 TV and film edge sharpness compared.

Sampling process	Parameter
Scanning lines	Vertical detail
Frames	Image motion
Shadow mask kinescope	Horizontal detail
CCD imagers	Horizontal detail

Sampling is an essential ingredient of digital video systems (see Chapter 3) and in these systems the image is sampled horizontally as well as vertically. In a related process, the amplitude of the signal is quantized, i.e., it is represented by discrete levels rather than by a continuous function. Thus digital systems sample the image in both dimensions and quantize the signal amplitude.

4.7.1 Sampling aperture response (MTF)

Sampling a signal potentially gives rise to two distortions. The first is aliasing, as described above, for sampling rates below the Nyquist criterion. The second is a reduction in the MTF for higher frequency components of a spatial or temporal signal. The $MTF_{(H,V)}$ of the sampling process for the horizontal (H) or vertical (V) components of a spatial signal can be calculated by means of Eq. (4.5). This equation is strictly valid only for image frequency components below the Nyquist limit; higher frequencies are subject to aliasing (see Sections 4.7.4 to 4.7.7 below).

$$MTF_{(H,V)} = \frac{\sin x}{x} \qquad (4.5)$$

$$x = \pi \frac{W_{(H,V)} f_{(H,V)}}{F_{(H,V)}}$$

where $F_{(H,V)}$ is spatial sampling frequency in the horizontal or vertical dimension, $W_{(H,V)}$ is (sample width)/(sample spacing), e.g., (scanning line width)/(scanning line spacing), and $f_{(H,V)}$ = spatial frequency component of sampled image. Examination of Eq. (4.5) shows that the MTF is determined by the sample width and is independent of the sampling frequency provided that frequency exceeds the Nyquist frequency.

4.7.2 Scanning line aliasing

Vertical brightness variations are sampled by the horizontal scanning lines, and if the dimensions of the vertical detail are less than the scanning line spacing, aliasing will occur. The result is a third and spurious set of lines having a line number equal to the difference between the number of scanning lines in the raster and the line number of the vertical detail. With typical subject matter, this produces an objectionable moiré pattern in the image.

This form of aliasing is particularly visible with subject matter that has a strong geometric pattern, e.g., a herringbone suit. Motion in the picture also enhances the effect, and a geometric pattern may appear to twinkle. It is most noticeable in sharp high-definition images, and it was not a serious problem in early television pictures because early cameras had limited resolution and their fine detail did not have sufficient amplitude to interact visibly with the scanning lines. With the improved definition of current broadcast systems, it has become a significant problem, and it is a major consideration in HDTV system design.

Figure 4.8 shows the MTF of the vertical scanning sampling process (without vertical aperture correction) and its relation to aliasing as calculated by Eq. (4.5) for an NTSC and representative HDTV signal that meets the SMPTE HDTV production Standard 240M.

Aliasing for the NTSC signal begins at 240 TV lines, one-half the number of active scannning lines. The MTF is high for spatial components above 240 TV lines, and, if scanning were the only element determining the MTF, aliasing would be severe. In practice, the system MTF at these higher line numbers is substantially reduced by the aperture responses of an optical filter, the camera lens, the camera imager, and the display device. The vertical MTF for a complete NTSC system, which includes these elements (as described in Section 4.9), is shown in Figure 4.15. Aliasing is present but at a much lower amplitude than would be produced by the scanning pattern alone because of the attenuation of the other elements. System vertical aliasing can be seen as twinkling in the horizontally-oriented line patterns of the EIA resolution chart, Figure 4.9.

Vertical aliasing for the HDTV system begins at 474 TV lines, one-half the number of active scanning lines. The increase in the beginning of aliasing from 240 TV lines in NTSC systems to 474 is one of the major factors in the improved picture quality of HDTV systems.

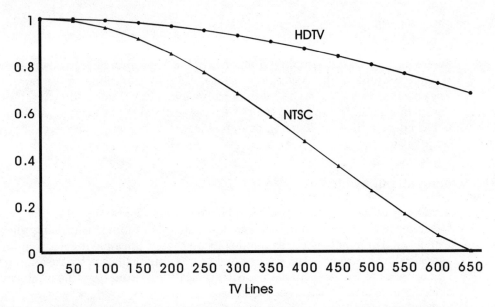

Figure 4.8 Scanning MTF curves for NTSC TV and HDTV. Parameters: NTSC—525 scanning lines, 480 active lines, $W_v = 0.9$; HDTV—1080 active lines, $W_v = 0.90$.

4.7.3 Frame aliasing

Frame aliasing occurs when a cycle of a repetitive horizontal pattern moves vertically more than one scanning line during the frame interval. For example, if a fine-grained repetitive pattern moves upward 1.5 lines during the frame interval, the visual effect will be the same as though it had moved downward 0.5 lines. This is the cause of the motion picture phenomenon in which wagon wheel spokes appear to be turning backward.

Frame aliasing is an inherent sampling defect, and can only be reduced or eliminated by increasing the frame repetition rate.

4.7.4 CCD aliasing

Signals from CCD imagers are generated by rows and columns of discrete photosensitive elements and hence have been sampled in both the horizontal and vertical dimensions. This causes aliasing in both dimensions as described more fully in Chapter 9.

4.8 Limiting Resolution

Although it is not as scientific as an MTF curve, limiting resolution is widely used as a measure of image definition.

4.8.1 Limiting resolution as a measure of system performance

Limiting resolution has been used historically to specify the performance of optical instruments and film systems. It indicates the maximum number of alternate light and dark lines (or points of light) that can be distinguished per unit distance in the image. Following the lead of the film industry, limiting resolution was initially used to specify the performance of television systems. It has, however, two serious weaknesses as a performance criterion for television systems that reduce its value for this purpose.

The first difficulty is that limiting resolution defines only one point on the aperture response curve—the line number at which the curve falls below the system noise level. This is not a problem when comparing two systems with aperture response curves of very nearly the same shape—as for two types of film—but it can be very misleading if the shapes are different—as with film and television. Image definition, as measured by edge sharpness, is determined by the aperture response over the whole range of line numbers, not just at a single point. Figure 4.6 showed examples of film and television systems with almost equivalent definition, although the limiting resolution of the film is much higher.

A second problem arises when limiting resolution is used to evaluate the performance of systems that include horizontal sampling (see Section 4.7). The relationship of image detail to the sampling points is random, and the limiting resolution is less than the number of sampling points. More seriously, the Nyquist criterion predicts that aliasing will occur for image detail components that have line numbers more than one-half the sampling rate. The limiting resolution may be an ambiguous and irrelevant figure if the Nyquist number is less than the limiting resolution, and significant aliasing occurs before the limit of resolution is reached.

In spite of its flaws as a criterion of image definition, however, limiting resolution plays a useful role in the design and maintenance of television systems, and it continues to be widely used. It is easy to measure, and the result is a single, easily understood number. The Electronic Industries Association (EIA) standard resolution test chart, Figure 4.9, has been widely used for evaluating system performance.

4.8.2 Vertical limiting resolution and the Kell factor

Because of the random relationship of image detail and scanning lines, the vertical limiting resolution is less than the number of scanning lines. The ratio between the observed limiting resolution and the number of scanning lines is known as the *Kell factor*. This must be evaluated empirically rather than analytically, and various values have been determined by different observers in the range from 0.53 to 0.85, with 0.70 being a commonly accepted value. All of these values are above the Nyquist limit of 0.5 and are, therefore, in the region where aliasing occurs.

In television's early years, the aperture response of cameras and other system components at line numbers above the Nyquist limit of the scanning process was

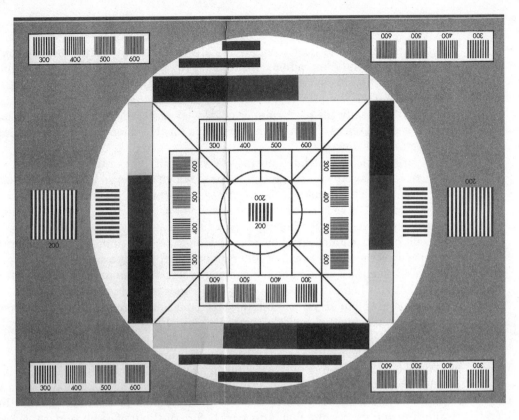

Figure 4.9 The EIA resolution chart.

low, and vertical aliasing was small enough to be of minor significance. Limiting resolution under these conditions was an important specification, and it was frequently quoted.

With cameras of recent design, CCD imagers have delivered a major increase in the vertical aperture response in the region above the Nyquist limit, and aliasing is now a significant factor in system design. This is illustrated in Figure 4.15, which shows the locations of limiting resolution calculated with a Kell factor of 0.7 and the onset of aliasing in the vertical MTF of a representative NTSC television system. The limiting resolution of 336 lines is well above the Nyquist limit, and it is relatively insignificant as a measure of system performance since aliasing begins at lower line numbers. In consideration of this fact, the vertical resolution of cameras is seldom specified.

4.8.3 Horizontal limiting resolution and bandwidth

Aliasing in the horizontal dimension occurs with CCD imagers because the pixels cause sampling along each scanning line. The number of effective horizontal samples per line is much higher than the number of scanning lines, however, typically over 1100 lines with spatial offset imagers (see Section 9.4.3) and the Nyquist limit is then over 550 lines. At these line numbers, the aperture response of the camera optics and other system components is typically low enough that the aliasing amplitude is relatively insignificant. Under these conditions, the horizontal limiting resolution of the system components rather than the Nyquist limit may determine system performance.

The horizontal limiting resolution of an analog video system—the number of black and white vertical lines that can be distinguished in a dimension equal to the picture height—may be determined by its electrical bandwidth or by the aperture response of the system components. In cases where the bandwidth is limiting, each cycle reproduces a black and white line-pair, i.e., two TVL. This leads to Eq. (4.6) for the horizontal limiting resolution of analog systems:

$$R_H = 2 \frac{C_H \, BW}{AR \, NL \, FR} \qquad (4.6)$$

where R_H is the horizontal limiting resolution in TVL, C_H is the fraction of time in each scanning line devoted to the transmission of picture information after subtracting the time required for horizontal blanking, BW is the system bandwidth in Hz, AR is the aspect ratio, NL is the total number of scanning lines per frame, and FR is the frame rate per second.

The horizontal limiting resolution, R_H, of NTSC broadcast systems (U.S. and Japan), PAL broadcast systems (Europe), and the analog HDTV system defined by the SMPTE Standard 240M are compared in Table 4.1.

High definition requires a high limiting resolution, and the 4.2-MHz bandwidth limitation of standard broadcast channels is a fundamental problem for analog HDTV. A tremendous amount of engineering research has been devoted to the development of systems to acheve high resolution and definition with a limited bandwidth. These efforts were only marginally successful with analog technology. Thus, the emerging HDTV standards all use digital formats, which can use compression techniques that can equal the performance of wide-band analog systems in much narrower bandwidths. The bandwidth requirements for digital HDTV systems are described in Section 3.7.2.2.

Table 4.1 Horizontal limiting resolution comparison

	NTSC	PAL	HDTV (SMPTE 240M)
BW (MHz)	4.2	5.0	30
NL	525	625	1125
C_H	0.85	0.80	0.87
AR	1.33 (4/3)	1.33 (4/3)	1.78 (16/9)
FR	29.97	25	30
R_H (TVL)	352	409	869

4.8.4 Picture elements (pixels)

A video raster can be described as a grid of picture elements or pixels. A pixel is defined basically as the smallest element of a picture whose properties can be described independently of adjacent elements. The height of each pixel is usually considered to be the spacing between scanning lines (or alternatively the spacing divided by the Kell factor) and its width is the dimension of the narrowest object that can be reproduced. For analog signals, that is approximately one-half cycle of the highest video frequency in the transmission bandwidth. The number of pixels in a frame is a rough indication of its information content.

The number of pixels in a frame is:

$$NP = 2\, \frac{C\,KF\,BW}{FR} \tag{4.7}$$

where C is the fraction of time occupied by active scanning (after subtracting vertical and horizontal blanking periods), KF is the Kell factor (optional), BW is the bandwidth of the system in hertz, and FR is the frame rate. In the broadcast and HDTV systems described in the preceding sections, the broadcast system has 152,000 equivalent pixels while the HDTV system has 1,162,000.

With CCD imagers, the effective H and V pixel counts are determined by the number of elements in the CCD array.

4.9 NTSC Television System Aperture Response

This section describes the aperture response of the functional components of a representative NTSC television system and their cascaded response in a complete system.

The components can be divided into two categories: the electro-optical components that generate and display the images, and the processing and transmission components that generate and process the video signal and transmit it from camera to receiver. The principal system components are:

Electro-optical components
  Camera lens
  Optical filter
  Camera imager
  Display device
Processing and transmission components
  Video bandwidth
  Aperture correction
  Image enhancement
  Vertical scanning
  Vertical aperture correction (optional)

Figure 4.10 shows the relationship of these components in the complete system.

Elements of Image Quality

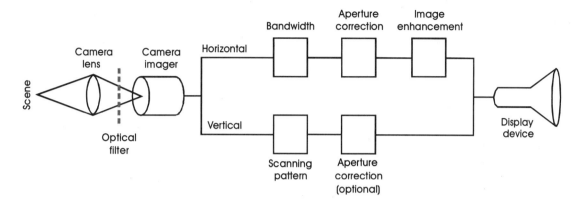

Figure 4.10 Aperture response-determining components.

4.9.1 Electro-optical components

Aperture response curves for typical broadcast-quality electro-optical components and their overall response when cascaded are shown in Figure 4.11. The performance of commercial products varies widely, and these curves are representative and not intended to describe specific models.

4.9.1.1 Camera lens

A high-quality fixed-focus lens such as that used as an example in Figure 4.11 has a very high response out to 300 lines, the approximate upper limit of NTSC systems, and it has only a minor effect on the performance of these systems. A zoom lens makes some compromises in aperture response in order to provide the variable focus feature, and the performance of these lenses must be considered when used in HDTV systems.

4.9.1.2 Optical filter

CCD imagers have aliasing, an inherent result of sampling. Once introduced, aliasing cannot be eliminated from a signal by electrical filtering. It can, in theory, be removed by filtering the spatial frequency components above the Nyquist limit from the optical image before CCD sampling. This could be done in a brute force fashion by defocusing the lens, but that would produce an unacceptable sacrifice of picture definition. The solution would be to use an optical prefilter with unity aperture response below the Nyquist limit of the scanning pattern and zero response above it. Such a filter cannot be made, but the necessary response can be very roughly approximated by practical filters.

The response of a typical optical filter is shown in Figure 4.11; it is a characteristic $\sin x/x$ curve with a zero ($x = \pi$) at 1200 TV lines (14.8 MHz).

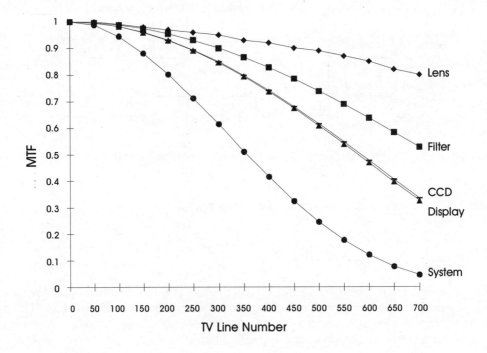

Figure 4.11 Electrooptical components MTF—NTSC.

4.9.1.3 Camera imager

A wide range of camera imager aperture responses is available from commercial cameras as described in Chapter 9. Figure 4.11 shows the responses for a camera using 768(h) × 494(v) CCDs with spatial offset.

4.9.1.4 Display device

The display device aperture response curve shown in Figure 4.11 is for a shadow-mask color CRT (see Chapter 6) with 763 apertures per picture width, giving a horizontal limiting TVL of 574 (763/1.33). Aliasing, in this case the interference pattern between the apertures and horizontal detail, begins to appear at about 287 (574/2) lines. The horizontal limiting resolution, unless it is limited by other factors, is 400 lines (574 lines times a Kell factor of 0.7), and aliasing rather than resolution limits the performance of the device.

4.9.1.5 Cumulative response, electro-optical components

The cumulative horizontal aperture response of the electro-optical components shown in Figure 4.11 is approximately 0.5 at 340 TVL, which is the same as the upper limit established for the U.S. broadcast system by the video passband. This response can be improved further by the use of image enhancement techniques (see Section 4.9.2) to give better results within the bandwidth, aliasing, and scanning line limitations of the system.

4.9.2 Horizontal processing

This section discusses processing of high frequencies..

4.9.2.1 Video channel bandwidth

The video bandwidth of an analog system establishes a fundamental limit on its resolution and definition. The resolution is given by Eq. (4.6). The aperture response of the video channel, which is also its amplitude response, is one of the factors that determines the system's definition.

In analog systems, the amplitude response drops to zero at TVL numbers above the limit set by the video bandwidth as given by Eq. (4.6).

4.9.2.2 Video channel aperture response

The amplitude response of the video channel can be converted to aperture response by translating electrical frequencies to TVL numbers in accordance with Eq. (4.8):

$$\text{TVL} = 2 \frac{C_H f}{AR\ NL\ FR} \qquad (4.8)$$

where C_H is the fraction of time in each scanning line devoted to the transmission of picture information (after subtracting the time required for horizontal blanking), f is the frequency, AR is the aspect ratio, NL is the total number of scanning lines per frame, and FR is the frame rate per second.

Controlling the amplitude response of the video channel (it is important that this be accomplished without introducing phase distortion) is an effective way of increasing the aperture response of the system and hence improving its definition. A number of techniques with the generic title of *image enhancement* have been developed to accomplish this.

Aperture correction is a commonly used method. It boosts the amplitude response of the video channel to the higher-frequency components in the signal (taking care that this does not introduce unwanted phase shifts) to compensate for the fall-off in the response of the electro-optical components.

Another method (once called "contours out of green") is to create a waveform by passing the green signal through a linear-phase high-pass circuit that produces spikes at each rapid change in the signal. These spikes are then combined with the luminance signal in a manner that sharpens its edges.

Figure 4.12 shows the aperture response of a video channel with a flat response out to a bandwidth limit of 4.2 MHz and with a response enhanced to compensate for the reduced MTF of the electro-optical components at higher frequencies.

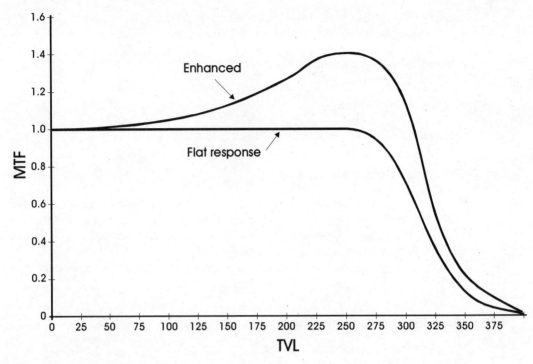

Figure 4.12 Video amplitude response—NTSC.

4.9.3 Vertical processing

This section discusses processing of the vertical detail components.

4.9.3.1 Scanning

The aperture response of the scanning pattern, which affects the sharpness of horizontal edges and the reproduction of vertical detail, is the complex result of the size of the scanning aperture and the sampling process. The major criteria of scanning are aperture response or MTF, limiting resolution, and aliasing. These were described in Sections 4.7.3 and 4.7.4.

4.9.3.2 Vertical image enhancement

Image enhancement by aperture correction or equivalent techniques is more difficult in the vertical dimension than in the horizontal because the correction cannot be accomplished along a single line. Analog vertical image enhancement circuits use delay circuits of one or more lines delay that make it possible to compare the video signals on adjacent lines. The image enhancer accentuates the differences in signal levels between corresponding points on adjacent lines as would occur on a horizontal edge, thus sharpening its appearance. (The same technique

Elements of Image Quality

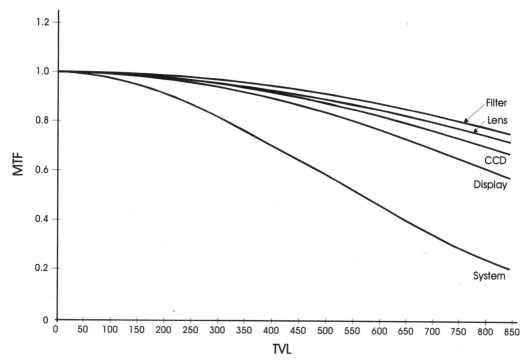

Figure 4.13 Electrooptical components MTF—HDTV.

can be used for horizontal image enhancement by the use of delay lines that delay the signal by one or more picture elements.)

The necessary delays for vertical enhancement are easily achieved in digital circuits. In fact, the process can be done more rigorously with digital techniques, because signals from multiple adjacent lines can be read from appropriate memory locations and combined to achieve a better enhancement.

4.10 HDTV System Aperture Response

The higher demands on picture quality in an HDTV system require even more attention to the aperture response.

4.10.1 Electro-optical components

Figure 4.13 shows the MTF curves for HDTV electro-optical components.

4.10.1.1 Lens

The aperture response of a high-quality fixed-focus lens is adequate for use in HDTV systems. The response of zoom lenses may be compromised somewhat to achieve the variable-focus feature; this must be considered in HDTV systems.

4.10.1.2 Optical filter

The optical filter for HDTV cameras should be designed to have a higher cutoff frequency than for NTSC cameras. Aliasing begins at higher line numbers; this permits a higher cutoff, and a higher cutoff is needed to avoid needless attenuation of the higher-frequency components in the optical image. The filter response in Figure 4.13 has its first zero at 2000 lines (25 MHz) as compared with 1200 lines in the filter assumed for the NTSC example in Figure 4.11.

4.10.1.3 Camera imager

The aperture response of the camera imager used in an HDTV system must be substantially greater than for an NTSC broadcast system, and it must be designed for the wider aspect ratio of 16:9. If the imager is a CCD, the size of the photosensitve elements must be reduced to improve the aperture response and their number increased to reduce aliasing. A typical HDTV CCD has a 1920(h) × 1036(v) element matrix as compared with 768(h) × 494(v) for NTSC. Figure 4.13 shows the aperture response of a 1920 × 1036 CCD.

4.10.1.4 Display device

To obtain the full visual benefit of HDTV, it is usually necessary to use larger display devices. Otherwise, the viewers would have to be too close to the screen to see all the detail that is present. HDTV displays may be either direct view or projection. However, the maximum practical size for direct view devices using CRTs is 35 to 40 inches. Direct-view HDTV display tubes may use either the shadow-mask or color-stripe design. One of the latter type, a *trinitron* design, has a picture diagonal of 38.5 inches with 1861 vertical stripe triads. The MTF of this tube is shown in Figure 4.13.

Projection displays can be larger, but most do not equal direct view in brightness or aperture response and they are expensive. They are also physically large, especially in depth. The real answer for HDTV displays is flat panels (see Section 14.6); intensive development effort is being applied on several approaches for this.

4.10.1.5 Cumulative response, electro-optical components

The cumulative response of all the electro-optical components of the HDTV system described above is also shown in Figure 4.13.

4.10.2 Video channel

HDTV systems generally use digital processing and transmission of the signals. Analog signals are generated by the imagers, but are converted to digital early in the processing. The sampling rate must be at least twice the highest-frequency component of the analog signal, the Nyquist frequency, to avoid aliasing. The SMPTE 240M Standard calls for a luminance sampling rate of 74.25 MHz. HDTV systems use video compression (see Chapter 7), which makes it possible to transmit the signal at a substantially lower rate. The effects of compression on picture quality are not considered here.

The aperture response of a sampling process is affected by the ratio of the width of the samples to their spacing. If the width is one-half the spacing, the aperture response exceeds 0.9 at frequencies just below the Nyquist limit. Thus, the aperture response of the sampling process can be maintained almost at unity out to the Nyquist limit.

Image enhancement is introduced by generating a detail signal by digital signal processing. The detail signal is then added to the main signal for enhancement. Figure 4.16 shows the aperture response of a video channel with image enhancement sufficient to maintain an overall system response at unity out to 400 lines.

4.10.3 HDTV scanning

The higher number of scanning lines in HDTV systems provide improved definition and also move the onset of aliasing to higher TVL. The high aperture response of HDTV electro-optical components would cause severe aliasing above the Nyquist limit, which makes the use of an optical low-pass filter important.

SMPTE Standard 240M for HDTV has 1035 active scanning lines, which means aliasing occurs for TVL greater than 517. The aperture response for a 517 scanning-line pattern was shown in Figure 4.8.

4.11 Image Definition—NTSC Systems

The next two sections describe the image definition as measured by the modulation transfer function or MTF (also the aperture response) of NTSC broadcast and HDTV systems. N_e is used to specify the definition by a single number.

4.11.1 Horizontal definition—NTSC

The system horizontal definition is calculated by cascading the aperture response of the electro-optical components (see Figure 4.13) with the response of the video channel. The results of this operation are shown in Figure 4.14 for the NTSC system.

The N_e of the system is 206 TVL with a flat video response and 286 TVL with the enhanced response. The N_e of the enhanced system could be increased further by greater enhancement but a high N_e in combination with a sharp frequency cutoff requires very careful control of the phase/frequency response to minimize overshoots on the picture edges.

The maximum resolution is limited by the 4.2-MHz video bandwidth to 355 TV lines.

The CCD imager assumed in the system has 640 columns of photosensitive elements, and aliasing of the horizontal signal will begin at 240 (640/1.33/2) TVL. If spatial offset is used with 3 CCDs, the onset of aliasing would be doubled to 480.

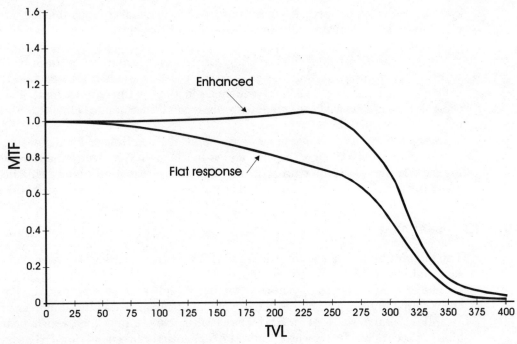

Figure 4.14 Horizontal definition—NTSC.

4.11.2 Vertical definition—NTSC

The vertical definition is calculated by combining the aperture response of the electro-optical components (see Figure 4.13) with the response of the scanning pattern (see Figure 4.8). The results of this calculation for the system in this example are shown in Figure 4.15.

The vertical N_e is 194. This could be increased to more than 225 TVL by the use of vertical aperture correction as described in Section 4.9.3.2. Aliasing begins at one-half the number of active scanning lines or 240 TVL. The limiting resolution as calculated with a Kell factor of 0.7 is 336 TVL, but because of aliasing it has little significance.

4.12 Image Definition—HDTV Systems

The determination of the image definition of HDTV systems is complicated because they include digital compression to reduce the bandwidth requirements while retaining the perceived definition of conventional wider-band analog systems.

To simplify the analysis here, an analog HDTV system based on SMPTE Standard 240M is assumed for the purpose of comparing the performance of NTSC and HDTV systems. The effects of video compression are not considered.

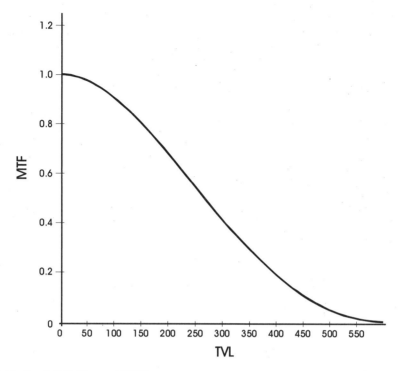

Figure 4.15 Vertical definition—NTSC.

4.12.1 Horizontal definition—HDTV

The cascaded aperture response of the HDTV electro-optical components and the video channel is shown in Figure 4.16. The responses with two video channels are shown, one with flat response out to the limit established by a 30-MHz bandwidth and one with aperture correction or other image enhancement that maintains a unity aperture response for the system out to 400 TVL.

The N_e of the system with flat video response is 375 TVL and of the enhanced system, 539 TVL. The limiting resolution as established by the 30-MHz bandwidth is 808 lines.

4.12.2 Vertical definition—HDTV

In the absence of vertical aperture correction, the vertical aperture response would be that of the electro-optical components (see Figure 4.13) cascaded with the response of the scanning pattern (see Figure 4.8), which in the SMPTE 240M standards has 1035 active scanning lines. This is shown in Figure 4.17.

The vertical N_e of this system would be about 350 lines and this could be increased to more than 400 lines with vertical aperture correction. Now that inexpensive memory circuits, a prerequisite for vertical aperture correction, are avail-

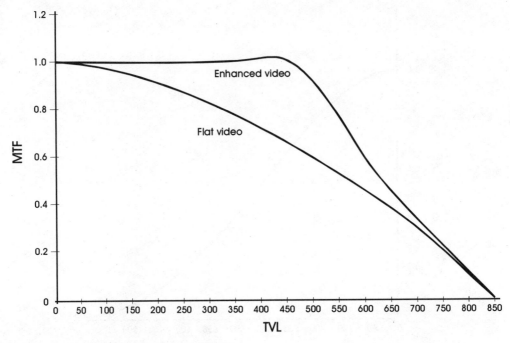

Figure 4.16 Horizontal definition—HDTV.

able, it is expected that the use of vertical aperture correction will become common in future HDTV systems.

The high aperture response of the components at 400 TVL and above indicates the importance of increasing the number of scanning lines in HDTV systems to limit aliasing.

4.13 Visual Perception of Broadcast and HDTV Images

Figure 4.18 shows the combined aperture responses of the eye and of NTSC and HDTV television systems at viewing ratios of 3 and 8. They are calculated by cascading the response of a normal eye as shown in Figure 4.4 with horizontal aperture response of NTSC and HDTV systems as shown in Figures 4.14 and 4.16.

Figure 4.18 illustrates the following:

1. At a viewing ratio of 8, there is very little difference in the perception or appearance of NTSC and HDTV systems. A smaller viewing ratio is necessary to take advantage of the higher definition of HDTV images.

2. As the viewing ratio is reduced below 8, the difference between HDTV and NTSC images begins to become apparent. The difference is striking at a viewing ratio of 3.

3. At viewing ratios of 8 and above, the eye rather than the television system becomes the major limitation in the perceived definition of the image.

Elements of Image Quality 117

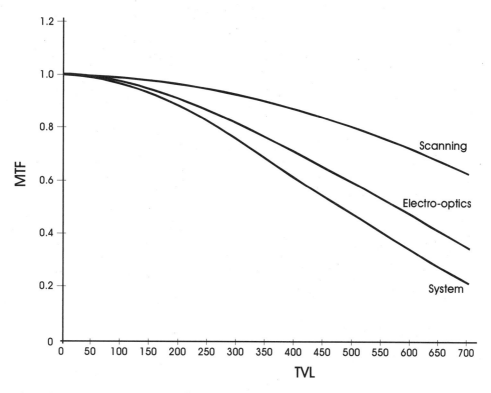

Figure 4.17 Vertical definition—HDTV.

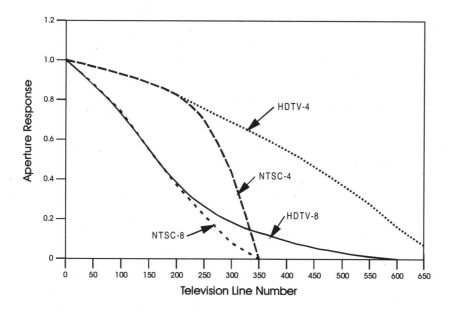

Figure 4.18 Visual perception of NTSC and HDTV images.

4.14 Image Gray Scale

The importance of gray scale as a measure of image quality can be demonstrated visually by turning on the room lights while photographic slides are being projected on a screen. With the lights out, the images appear crisp and in focus. When the room lights are turned on, thus increasing the ambient lighting on the screen, the images appear washed out and fuzzy. The addition of a fixed amount of ambient illumination to the slide image over a large area of the screen not only distorts the image's gray scale but also its apparent definition.

In color systems, as described in Chapter 2, careful control of the transfer function (the parameter that specifies gray scale, see below), of the three primary color signals is required to avoid color distortion.

Unlike image definition, improved gray scale reproduction requires no changes in bandwidth, numbers of scanning lines, or other transmission standards. It is, nevertheless, as important to the production of high-quality HDTV pictures as improved definition. A better gray scale will also improve the quality of standard broadcast pictures with no change in transmission standards.

4.14.1 Gray scale transfer function

The gray scale of an imaging system is specified by its *transfer function*, a plot of image brightness as a function of scene brightness. Figure 4.19 is the transfer function of a typical system, including camera, video channel, and display device. It is the television equivalent of the *Hurter and Driffield* (H & D) curves that specify film density as a function of exposure for photographic negatives. Because the eye perceives brightness differences in terms of ratios rather than numerical differences, a logarithmic scale is used for both scene and image brightness.

The transfer function can also be used to specify the input-output performance of video system components, including camera imageries, video amplifiers, and display devices. The system gray scale function is obtained by cascading the transfer functions of its components.

As with most imaging systems, the television transfer function is an S-shaped curve with a linear section in the middle that slopes off at both ends. The upper limit of the curve is called the *white field brightness*, and is usually established by the limitations of the display device, ordinarily a kinescope. The lower limit is called the *unexcited field brightness* and is the brightness of the kinescope when it is turned off, primarily due to reflection of ambient light.

It is often convenient to specify the gray scale of a television image by numerical parameters rather than graphically. The three most commonly used parameters, as shown in Figure 4.19, are as follows:

1. *Highlight brightness*: The white field brightness, i.e., the brightness of the brightest areas of the image.
2. Contrast ratio: The brightness ratio of the brightest and darkest areas of the image. For an imaging system, it is the ratio of the white field and unexcited field brightnesses.

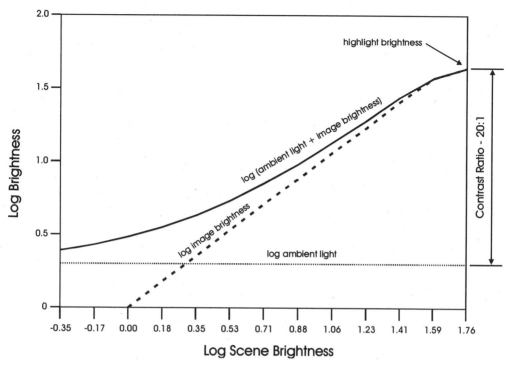

Figure 4.19 TV transfer function.

3. Gamma: The slope of the transfer function curve. Gamma is a term borrowed from photography where it is used to specify the slope of the H & D curve at a single point in its linear portion where the slope is the greatest. Its usage is slightly different in television where it can specify the slope of the transfer function at any point.

4.14.2 Brightness and contrast ratio

The brightness and contrast ratio of television images are determined primarily by the characteristics of the kinescope (or other viewing device) and the ambient lighting. Great progress has been made in recent years in increasing both the brightness and contrast ratio by increasing the white field brightness and reducing the unexcited field brightness of kinescopes.

Between 1962 and 1982 the white field brightness of typical color tubes was increased from 10 to 65 foot-lamberts. This was achieved by utilizing metalized backing on the faceplate, improved phosphors, higher beam voltage, and improved electron gun design. The greater brightness of the phosphors was achieved by deviating somewhat from the NTSC primary color specifications (see Section 2.9). This has not been a serious concern for broadcast systems, but it has been questioned for HDTV systems that strive for perfection.

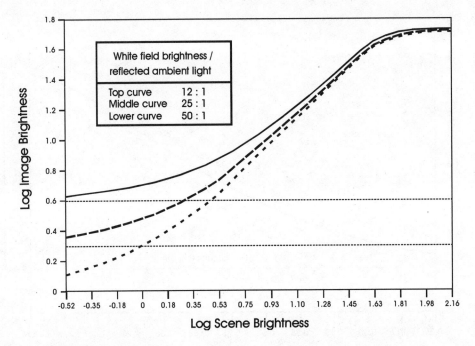

Figure 4.20 Effect of ambient light.

Most of the unexcited or dark field brightness results from reflection of ambient light from the picture tube face. Its effect was originally reduced by placing a neutral density filter in front of the kinescope. Light from the kinescope passed through the filter once, while light reflected from the kinescope passed through it twice. If the filter transmitted one-half the light, the white field brightness would be reduced by one-half but the unexcited field brightness would be reduced by three-quarters and the contrast ratio doubled.

The unexcited field brightness of modern color kinescopes has been reduced by the *black matrix/black surround* and other techniques that substantially reduce the light reflected by the phosphors. The results can be seen by observing the kinescope of a color receiver of recent design when the set is turned off. By contrast with older sets in which the kinescope was whitish gray, new sets appear black.

A high contrast ratio is a technical ideal and, with some exceptions, an aesthetic one as well. Improved kinescopes have made it possible to meet the desired contrast ratio standards for high-quality broadcast TV, provided the ambient lighting is not excessive. Optimum results for HDTV require control of ambient lighting by creating a theaterlike environment for home viewing.

Because of the eye's logarithmic response to brightness differences, the reflected ambient light from the kinescope screen, which establishes the unexcited field brightness, can have a greater effect on the contrast ratio than the highlight or white field brightness. A relatively small numerical difference in reflected light can have a major effect on the contrast ratio.

Table 4.2 System brightness and contrast ratio (brightness values in foot-lamberts)

System	Desired Brightness	Typical Brightness	Desired Contrast	Typical Contrast
Motion picture theater	25	10–12	150:1	50–100:1
HDTV (studio)	150	60	300:1	100–150:1
HDTV (home display)	60	40	150:1	10–30:1
Broadcast TV (studio)	150	100	200:1	100–150:1
Broadcast TV (525 home)	60	40	80:1	10–30:1

This effect is illustrated in Figure 4.20, which shows the transfer function under three different conditions. All assume a white field brightness of 50 foot-lamberts, but with unexcited field brightnesses of 1, 2, and 4 foot-lamberts as the result of differences in ambient lighting. The resulting contrast ratios are 50, 25, and 12.5. The gamma is also reduced as the ambient light is increased. The gamma of the image with low ambient lighting is 0.94, and it decreases to 0.76 as the lighting is increased.

The combination of a lower contrast ratio and a reduced gamma with high reflected light results in a washed-out, less sharp picture, the opposite of the effect desired for HDTV and undesirable for broadcast television.

Table 4.2 shows the approximate ranges of contrast ratios that are typical of or desired for operating systems. In practice, it depends to a great extent on the ambient lighting on the screen, a parameter over which the system designer often has no control.

4.14.3 Gamma and picture quality

An image gamma of unity would indicate that it is precisely reproducing the gray scale of the scene. This is the technical ideal, although deliberate distortions may be introduced for aesthetic reasons.

If gamma is greater than unity, the image will appear sharper but the scene contrast range that can be reproduced is reduced. If gamma is increased sufficiently, the image will be a silhouette with only blacks and whites. (This was a problem with many early kinescope recordings, which had such a high gamma that most of the image areas were either black or white with very little shades of gray between.)

Reducing the gamma has a tendency to make the image appear soft and washed-out.

Selecting the video picture gamma that is most pleasing is an aesthetic judgment. For color images, a gamma of 1.2 to 1.5 is generally accepted as optimum.

4.14.4 Gammas of video images and components

The gamma of the image produced by a complete television system can be calculated by multiplying the gammas of its individual series components, specifically the camera imager, the display device, and any nonlinear video amplifiers that are added to the system for gamma correction.

With the exception of vidicons, which have a gamma of about 0.7, most camera imagers have a gamma of unity. CRTs, which are almost universally used as display devices, have gammas from 2.2 to 2.4.

Gamma correction circuits usually have a gamma of less than unity to compensate for the high gammas of kinescopes. For example, if it is desired to have a system gamma of 1.2 with a unity-gamma camera imager and a kinescope gamma of 2.2, the gamma correction circuit should have a gamma of 1.2/2.2 or 0.55.

The FCC Rules for NTSC Color Transmissions specify gamma by saying: "gamma-corrected red, green, and blue voltages suitable for a color picture tube having a transfer gradient (gamma exponent) of 2.2 associated with each primary color." "Suitable for" is not a precise specification, and no tolerance was set on the value of gamma.

The SMPTE Standard 240M for HDTV production also assumes a display device gamma of 2.2, but it specifies the signal gamma more explicitly and in greater detail; the specified transfer function is linear at low brightness levels up to 2.28 percent of reference white and logarithmic with a gamma of 0.45 thereafter. The specification is:

$$V_C = 4 L_C \text{ for } L_C < 0.0228 \tag{4.9}$$

$$V_C = 1.1115 L_C^{(0.45)} - 0.1115 \text{ for } L_C > 0.0228 \tag{4.10}$$

where L_C is the light input, and V_C is the camera output voltage (both are normalized to equal unity at reference white). This more explicit definition was adopted in anticipation of the availability of technology for achieving compatibility between the transfer functions of a variety of camera and display devices.

4.14.5 Gamma correction and system noise

Gamma correction circuits with a gamma of less than unity "stretch" the black portion of the signal, which causes the noise in the black areas of the picture to be amplified more than noise in the white. The effect on the perceived SNR in the image will depend on the location of the gamma correction circuit with respect to the noise source.

Noise generated in the dark areas of the picture ahead of the gamma corrector, e.g., in the imager, will be amplified more than the picture signal. Depending on the picture content, this may reduce the overall SNR by several decibels. The adverse effect of this phenomenon is exacerbated by the fact that noise is more visible in the dark areas.

Conversely, dark area noise generated after gamma correction, as in the transmission link, will be amplified no more than the signal, and its amplitude will be reduced by the high gamma of the kinescope.

4.15 Signal-to-Noise Ratio

Noise exists in all video systems. This section discusses some of the considerations of noise.

4.15.1 Sources and properties of noise in video systems

Random electrical disturbances, which are usually thermal in origin, produce "snow" in the picture, and are the most common type of noise in video signals. The relationship between temperature and noise is close and *noise temperature* (see Section 18.6.4) is a frequently used specification for system components.

The most significant sources of random noise are components where low signal levels are encountered. They include camera imagers, camera preamplifiers, videotape recorders, cable circuits, microwave and satellite circuits, and broadcast receivers. The noise generated by these elements establishes basic limitations on system performance and is fundamental in establishing its design parameters.

Most random noise is "white," i.e., the noise energy is uniformly distributed through the frequency spectrum as with white light. An important exception is the noise generated in microwave and satellite links that employ frequency modulation for analog signals. These system components have the characteristic that the noise power generated per hertz increases as the square of the frequency (see Chapter 18).

The effect of noise on digital systems is to introduce errors into the bit stream. These only show in the picture if the error protection system is unable to correct them (see Section 3.8).

The nature and amplitude of the noise generated by CCD imagers is described in Chapter 9.

4.15.2 Definition of signal-to-noise ratio

The *signal-to-noise ratio* (SNR) is the basic specification for random noise. It is the power ratio, usually expressed in decibels, of the peak-to-peak signal voltage or current, from black level to reference white but not including the sync pulses, to the rms value of the noise.

4.15.3 Unweighted and weighted noise

Noise power can be specified either as *unweighted* or *weighted*. Unweighted noise is measured by an instrument with uniform frequency response. Weighted noise, a measure of its visibility, is measured with an instrument having a frequency response, the weighting function, that simulates the cascaded aperture response of the eye (see Figure 4.4) and the system components following the noise measurement point. This accounts for the reduced visibility of higher-frequency (fine-grained) noise.

The variables in the calculation of the weighting function include the viewing ratio, the assumed response of the eye, the screen brightness, the width of the

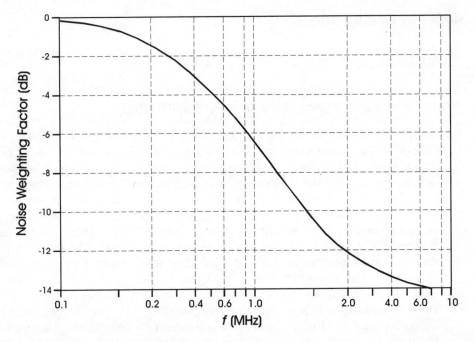

Figure 4.21 EIA noise weighting factor.

scanning lines, and the aperture response of the kinescope and other system components following the point of measurement. There are differences between the visibility of noise in monochrome and composite color images because of interaction of the high-frequency color subcarrier and noise components in this frequency range. The weighting factor, therefore, cannot be both universal and precise.

The industry has found it important, however, to establish a standard weighting function that approximates practical situations so that systems can be specified and compared on a uniform basis. The Electronic Industries Association has published a widely used standard for SDTV broadcast transmission systems, which is part of EIA/TIA-250-C and is shown in Figure 4.21. It is based on a somewhat more pessimistic assumption as to the aperture response of the eye than is indicated by Figure 4.4.

The difference between the weighted and unweighted signal-to-noise ratio is called the *weighting factor*. Using the weighting function shown in Figure 4.21, it is about 8 dB for white noise and 12 dB for FM noise. It is larger for FM noise because most of the noise is concentrated at the upper end of the transmission spectrum where its visibility is less and the weighting function is lowest.

At present, a standard industry weighting factor for HDTV does not exist, but it has been variously estimated as being 3 to 5 dB less than that for broadcast service. This would indicate that the unweighted SNR for HDTV should be 3 to 5 dB higher for the same weighted value. The weighting factor is smaller because a lower viewing ratio is assumed and fine-grained high-frequency noise is more visible.

Table 4.3 Signal-to-noise ratio standards (dB)

Service	Weighted	Unweighted (white)	Unweighted (FM)
HDTV	55	50	47
Broadcast (good)	53	45	42
Broadcast (fair)	43	35	32
Broadcast (poor)	37	33	26

4.15.4 Signal-to-noise ratio in digital systems

Noise power is additive in analog systems, and the SNR for the complete system is inevitably lower than that of the worst stage. Similarly, with multiple-generation recordings, the SNR of the final generation is lower than any individual generation.

With digital transmission or recording, noise causes bit errors, and so long as the bit error rate is low enough, correction circuits can restore the original bit stream and there is virtually no degradation of SNR. If the noise power is increased sufficiently, however, the bit error rate becomes so large that error correction is not effective and the signal is lost.

In summary, an analog signal suffers *graceful degradation* as it passes through successive stages or recording generations. Its quality gets steadily worse, but it remains useful as noise is added. By contrast, the SNR in digital systems suffers little reduction as the channel noise increases to a critical level, but at that level it suddenly becomes unusable—the cliff effect (see Section 3.8).

4.15.5 Signal-to-noise ratio standards

A number of industry and governmental standards have been established for video SNR. Some are for a complete system from camera to receiver while others cover only a portion of the system, e.g., a microwave link. Some are for white noise and others for FM noise. Some are for weighted and others for unweighted noise. When using published standards, it is important to determine exactly what is being specified and the basis of the specification.

Table 4.3 shows one commonly accepted set of SNR standards for broadcast systems. It also shows possible goals for HDTV system standards. The HDTV goals are very demanding, possibly not always obtainable, but would result in superb noise-free pictures. The broadcast standards are based on a study by the Television Allocations Study Organization (TASO) for the FCC. Although the report is a number of years old, its conclusions are still reasonably valid.

Since noise is additive for analog transmission, a practical standard must take the number of system stages into account. As an example, the EIA weighted SNR standards for analog transmission circuits are:

Circuit	SNR (dB)
Short haul	67
Medium haul	60
Satellite	56
Long haul	54
End-to-end system	54

4.16 Image Defects (continued from Section 4.3)

The remainder of this chapter expands the description of video image defects that were defined in Section 4.3. The magnitude of these defects, together with system performance with respect to the basic criteria of definition, gray scale reproduction, and SNR determine the quality of the images produced.

4.16.1 Hum

Hum bars were a major problem in the design of early television receivers. To alleviate the problem, field rates equal to the prevailing power frequency (60 Hz in the United States, 50 Hz in Europe) were chosen. Spurious patterns resulting from hum were then stationary and less noticeable. Small deviations from this frequency became necessary as the industry developed. Networking brought signals into areas where the frequency of the power source was not necessarily synchronized with that at the program source. And NTSC color standards specified the use of a field rate, 59.94 Hz, a submultiple of the subcarrier frequency, that was slightly below the nominal 60-Hz power source frequency.

The effect of these small deviations, which would cause the hum pattern to move slowly up or down on the raster, were eliminated by better receiver designs that greatly reduced the crosstalk of the power source into the video signal. The design of modern receivers has now been improved to the extent that hum should not be a problem even with large differences between power source and field frequencies.

4.16.2 Cochannel interference

The FCC originally specified a minimum desired-to-undesired carrier ratio of 20:1 (26 dB) to reduce cochannel interference to an unobjectionable level. Subsequently it was discovered that the visibility of the interference (venetian blinds) could be greatly reduced by an offset carrier technique in which the carrier frequencies of adjacent cochannel stations were offset by $\pm 1/2$ the line rate or $\pm 7,625$ Hz, for a total difference of 15,250 Hz. The interference bars then have a width equal to that of the scanning lines and are less visible by about 20 dB than the wide bars that are produced when the carriers are only slightly separated in frequency.

This offset was applicable to two stations, but if a third station was added, two of the three would be on the same frequency. A compromise was adopted in which the carriers are separated by ±10 kHz. This results in an improvement of about 12 dB in the SNR and offset carriers are now specified by the FCC.

The use of digital transmission also shows great promise as a means of reducing the minimum acceptable desired-to-undesired signal ratio. Field tests have shown improved cochannel and adjacent-channel interference susceptibility between DTV and NTSC transmissions.

4.16.3 Geometric distortion

Geometric distortion of the final image may originate either in the camera or the display device. The distortion produced by the display device is usually much greater, in part because it is a mass-produced product with wider tolerances and in part because the very wide deflection angle required by most picture tubes makes scanning linearity more difficult.

The geometric distortion produced by a camera chain can be measured with the EIA linearity test chart shown in Figure 4.22, sometimes called a ball chart. The inner diameter of each circular ring is 1% of picture height, while the outer diameter is 2%. Linearity is measured by superimposing an image of the chart as produced by the camera with an electronically generated grating pattern in which the intersections of the vertical and horizontal lines should coincide with the centers of the circles. A well-designed camera will maintain the grating intersections within the 1% circles (±0.5%) in a circle at the center of the raster having a diameter equal to picture height and within the 2% circles (±1.0%) in the remainder of the raster. This was a difficult specification in the days of tube imagers, but it is not much of a problem with CCDs.

Before the advent of color, competitive cost pressures caused manufacturers to design receivers with marginal scanning linearity and geometric distortion. Color, however, imposed a requirement for the registration of the three colors that could not be compromised. Eliminating registration errors also led to improved linearity and reduced geometric distortion. Typical geometric distortion specifications for production receivers are as follows:

Aspect ratio	5%
Overscan	5-10%
V and H linearity	5-8%
Parallelogram	90±1° at each corner

The vertical and horizontal linearity of 5 to 8% is to be compared with the typical 1% tolerance for cameras (see above).

4.16.4 Flicker

Flicker is an inherent defect that results from the frame rate used in the display of video signals. It can be minimized or eliminated by the use of a sufficiently high field or frame repetition rate. Both video and motion pictures depend on the retentivity of the eye to merge a rapid sequence of images into a single continuous

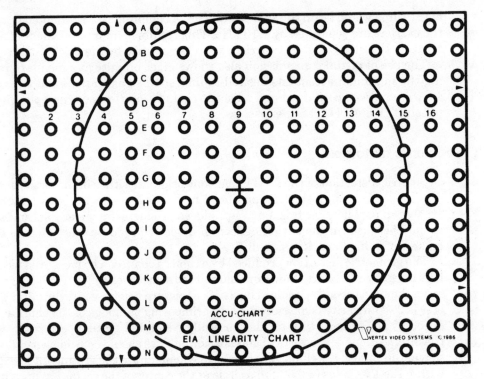

Figure 4.22 EIA linearity chart.

one. If the repetition rate of the images is too low, the eye will fail to merge them, and flicker results. The lowest repetition rate at which a continuous image is perceived is called the *critical fusion frequency*.

Two types of flicker are encountered in video systems using interlaced scanning. Large area flicker occurs at the field rate, 60 fields per second in the NTSC system and 50 fields per second for PAL. Small area flicker, described in Section 1.6.1, affects only vertical detail. It occurs at the frame rate, 30 per second for NTSC and 25 for PAL.

Large area flicker involves the entire image area and is the most troublesome. The magnitude of the flicker effect depends on two parameters, the field repetition rate and the illumination of the retina by the image. Flicker can be reduced and ultimately made unnoticeable by increasing the field repetition rate or reducing the retinal illumination. At the 60 field-per-second rate employed by NTSC systems it is seldom a problem, but it can be a problem at the 50 field-per-second rate. One study showed that the retinal illumination can be as much as three times as great at 60 fields as compared with 50 fields before flicker is observed.

The retinal illumination is determined by the image brightness and the size of the pupil's aperture. The latter, in turn, depends on the brightness of the area around the image, the *surround brightness*, as well as the image itself. In a darkened motion picture theater, the surround brightness is low, the pupil aperture is relatively large as the eye automatically adjusts to the low light level, and the

Table 4.4 Flicker-brightness tolerance

	Field/s	Relative brightness (%)
NTSC TV	60	100
PAL TV	50	29
Theater movies	48	20
Home movies	48	20

retinal illumination is high in relation to the screen brightness. In a typical home environment, the surround brightness is higher, the pupil aperture is smaller, and the retinal illumination is lower in relation to the image brightness. As a result, the image brightness can be greater without causing visible flicker in a home TV set than with motion pictures in a theater. The difference is increased still further by the use of the lower field rate of 48 per second in motion picture systems.

The image brightnesses that can be tolerated without flicker under typical viewing conditions, relative to a 60 field-per-second NTSC system, have been estimated to be as shown in Table 4.4.

Large area flicker has not been a problem with 60 field-per-second NTSC systems with the image and surround brightnesses that are normally encountered. Looking ahead to future HDTV systems, it is a potential problem. In an effort to increase the contrast ratio, viewing conditions with a lower surround brightness may be chosen and the white field brightness increased. Both of these conditions would increase the vulnerability to flicker. Also, the use of a wide screen and lower viewing ratio places the edges of the screen on the periphery of vision where flicker is more noticeable (presumably one of nature's natural protective functions to make the individual aware of threatening movements out of the line of vision).

Small area flicker affects only vertical detail, i.e., horizontal edges, and it occurs at a rate of 30 or 25 Hz in systems using interlaced scanning. It has not been a problem with broadcast systems because the vertical definition is not great enough to produce sufficiently sharp edges. The higher numbers of scanning lines used in HDTV systems significantly reduce this problem, and the use of progressive scanning would eliminate it.

The situation with computer displays is quite different. They are viewed at close distances and the subject matter typically is small alphanumeric characters. Under these conditions, small area flicker can be extremely annoying. Accordingly, most computer displays employ sequential scanning, and frame rates of 72 to 85 Hz are common.

4.16.5 Camera defects

In addition to the image defects described above that are the result of system imperfections, there are others including lag, smear, blooming, and comet tails that are characteristic of camera performance (see Chapter 9).

4.17 Comparison of Film and Television Picture Quality

Film and TV systems both serve the purpose of delivering moving pictures and sound to viewers, and it is only natural that picture quality should be compared between the two media.

4.17.1 Overview

A comparison of the picture quality of photographic film and TV images is complex and controversial, and it has been the subject of an extensive literature. The complexity of the comparison is illustrated by two recent papers [3, 4]. Reference 3 is written from the point of view of the film industry and the more recent reference 4 is written from the perspective of the television industry.

There are a number of reasons for this complexity:

1. The inherent dissimilarity of photographic and television images as described earlier in this chapter, with the result that images produced by the two media will never appear exactly the same. (A somewhat analogous situation arises in comparing pipe organs and electronic organs. Although it is possible in theory to duplicate the sound of a pipe organ by electronic synthesis, the duplication is never exact. On the other hand, there are sounds and effects produced by an electronic organ that cannot be made by a pipe organ. The two instruments simply are different, but this does not imply that one is "better" than the other.)
2. The number of criteria that must be considered in the comparison and the problem of weighing their relative importance.
3. The wide range in the specifications of different film types and electronic imagers so that no single comparison is valid for all of them.
4. The rapid changes in the technologies of both media so that a comparison at any particular point in time is a moving target. The changes have been particularly significant in recent years, and there have been major improvements in the performance of both film and TV cameras since the first edition of this volume was written.
5. The basis of the ultimate comparison must be aesthetic rather than technical. This means that it will be subjective and subject to individual preferences.

In spite of these problems, it is useful to make this comparison, although one must recognize its inevitable limitations. The remainder of this section discusses the quantifiable differences.

4.17.2 Production and display systems

It is necessary to begin the comparison by specifying whether it is production or display systems that are being compared. Four combinations are possible:

Production	Display	Designation
Film	Film	F-F
Film	Television	F-T
Television	Television	T-T
Television	Film	T-F

The difficulties of comparing film and television display systems are the greatest and probably of the least importance. This comparison would have to consider not only the properties of the medium but also the ambiance of the display area. Film is almost invariably displayed in a darkened theater or room—in fact this is necessary to achieve a satisfactory contrast ratio. By comparison, satisfactory results can be achieved with television under conditions of ordinary room lighting. Film displayed in a darkened room may have a better gray scale than television in a well-lit room, but this is not an indication of the relative performance of the media but rather of the manner in which they are used.

This is one reason for limiting the discussion in this volume to systems with television displays, i.e., F-T and T-T. They are the most important in the television industry, and this limitation greatly simplifies the comparison.

4.17.3 Image-quality determining components

Table 4.5 lists the components that determine the image quality in each of the four systems. Subsequent sections compare their individual and cascaded performance in the F-T and T-T systems.

4.17.4 Component performance

Representative performances of the components of the F-T and T-T systems listed in Figure 4.4 are summarized in Table 4.6 and described below.

4.17.4.1 Definition (MTF)

The F-T system requires the use of a *telecine camera* (see Section 9.12) to reproduce film on TV. The telecine reproduces an image of the scene from negative film rather than the scene itself. This would appear to be an advantage for the T-T system's MTF because there are fewer aperture-limiting components in T-T. This advantage has been minimized by significant recent improvements in film performance and can be offset by using a telecine with a higher MTF than a live camera. The higher telecine performance is possible because it can be designed for maximum MTF without the necessity of tradeoffs for sensitivity, variable focal length lenses, and other constraints that limit the MTF of live cameras.

The MTF of the F-T systems at 400 and 600 lines tabulated in Table 4.6 are as reported in [1]. To demonstrate the potential of the film medium, the telecine employed an experimental high-definition storage tube and optimized optics with

Table 4.5 Film and television system components

F-F	F-T	T-T	T-F

Production components
Camera lens
Film negative
Film print
	Camera lens
	Film negative
	Telecine
	  Lens
	  Imager
	  Scanning
	  Sig. processing
	  Bandwidth limit
	  VTR
		TV camera
		  Lens
		  Imager
		  Scanning
		  Sig. processing
		  Bandwidth limit
		  VTR
			TV camera
			  Lens
			  Imager
			  Scanning
			  Sig. processing
			TV/film transfer
			Film print

Display components Projection lens	Kinescope	Kinescope	Projection lens
Viewing ambience Darkened room	Room lighting	Room lighting	Darkened room

a fixed-focus lens. Similar results could probably be obtained with high-quality CCD imagers (see Chapter 8). The MTF of the TV systems was as described in Section 4.9 for NTSC and 4.10 for HDTV (for HDTV it was very close to the performance of the HDTV camera reported in [1]).

The performance of the NTSC TV system was compared with two film formats, 16-mm and wide-screen Super 16. The film type was Eastman EXR 7245 (the sensitivity of the faster type ECN 7292 was also shown).

The performance of the HDTV system was compared with 35-mm type EXR 5245.

This comparison of T-T and F-T systems on the basis of selected points on their MTF curves shows an advantage for film, which is small if the ultimate display medium is television. This advantage can easily be overcome by the use of aperture correction.

There is a subtle but important difference when the complete MTF curve is considered. The film MTF has a more gentle cutoff as the limiting resolution is approached, and this produces a different edge shape as shown in Figure 4.7. This difference is reduced but not eliminated when the film is reproduced by a television system, and it probably accounts for much of the different visual perceptions described as the "film look" and the "television look."

Although a comparison with F-F systems is beyond the scope of this study, it should be noted that the F-F system has at least two apertures in series after the negative film (positive film and projection lens), and a projected image may have an MTF that is no higher or possibly even lower than that of an F-T system.

Table 4.6 A comparison of TV and film image quality

Parameter	16mm	NTSC Standards Film (TV playback) S-16	TV (CCD imager) (818h × 513v)
Definition (MTF)			
400 lines	0.60	0.69	0.59[2]
600 lines	0.30	0.40	0.28[2]
Resolution (TVL)			
Horizontal	650	675	700
4.2-MHz bandwidth	352	352	352
Vertical	315[3]	315[3]	315[3]
Aliasing region			
Frames	> 24 fps	> 24 fps	> 30 fps
Scanning (vertical)	> 226 lines	> 226 lines	> 226 lines
Pixels (horizontal)	None	None	> 560 lines[1]
SNR (unweighted)	42 dB[4]	44 dB[4]	50 dB[5]
Sensitivity (see Chapter 8)			
Exposure index	50–320	50–320	575

1 Includes effect of spatial offset.
2 Before aperture correction.
3 Based on Kell factor of 0.7. Value is ambiguous because of aliasing.
4 Noise equivalent for film granularity.
5 Based on 51 dB VTR SNR.

4.17.4.2 Resolution

The horizontal limiting resolution in each case is determined by the bandwidth of the system, 4.2 MHz for NTSC and 30 MHz for HDTV, and neither medium has an advantage.

The ambiguities in the use of vertical resolution as a measure of performance because the scanning lines are a form of sampling were described in Section 4.8.2.

4.17.4.3 Aliasing

Frame aliasing, e.g., wagon wheels appearing to turn backward, is determined by the frame rate—or more precisely, by one-half the field rate. The telecine image is derived from film that was photographed at a frame rate of 24. The TV image has a frame rate of 30 and is superior in this respect (although audiences have become accustomed to this effect and seem to expect it).

Scanning line aliasing begins at one-half the number of active lines, which is the same for both the F-T and T-T systems.

Table 4.6 A comparison of TV and film image quality (continued)

Parameter	HDTV 240M Standards	
	Film (35mm film) (TV playback)	TV (CCD imager) (1920h × 1036v)[1]
Definition (horizontal MTF)		
400 lines	0.80	0.78[2]
600 lines	0.65	0.59[2]
Resolution (TV)		
Horizontal	> 1000	> 1000
30-MHz bandwidth	869	869
Vertical	724[3]	724[3]
Aliasing region		
Frames	> 24 fps	> 30 fps
Scanning (vertical)	> 517 lines	> 517 lines
CCD (horizontal)	None[4]	> 490 lines
SNR (unweighted)	50 dB[5]	54 dB
Sensitivity (see Chapter 8)		
Exposure index	50–320	575

1 Includes HDTV telecine.
2 Before aperture correction.
3 Based on Kell factor of 0.7. Value is ambiguous because of aliasing.
4 No scanning processes: film camera uses storage tube telecine.
5 Noise equivalent for film granularity.

CCD aliasing basically begins at one-half the number of horizontal sensors per picture height. By the use of spatial offset, however (see Section 9.4.3), the line number at which aliasing begins can be almost doubled. A factor of 1.9 was used in Table 4.6.

4.17.4.4 Signal-to noise ratio

The high SNR of the latest CCD imagers has made film granularity a major noise source in systems that use film. This accounts for the better SNR performance for the T-T systems.

4.17.4.5 Sensitivity

The latest CCDs have a sensitivity approximately equal to that of very fast film. If the better performance of slower film is desired, CCDs would have the advantage in sensitivity.

4.17.5 Picture quality summary

For TV display, recent improvements in imager performance give television production (T-T) a slight edge over film production (F-T) in the tangible criteria for picture quality, although each has pluses and minuses. Film has a small advantage in MTF and television production systems have a small advantage in frame aliasing, sensitivity, and SNR. Film production systems that employ storage tubes in the telecine do not produce aliasing of a type characteristic of CCDs.

These comparisons do not take into account subtle and difficult-to-quantify differences described as the "film look" and "television look." As a result of these differences, film and TV-produced programs will never look exactly alike, but neither is inherently better than the other. A comparative evaluation must be made by personal preference rather than tangible, objective measurements.

Further improvements can be expected in the picture quality of both television and film production systems for television display, but the quality produced by both is now approaching the limit imposed by the human eye. The picture quality available from both systems is so nearly alike that other factors will probably be preeminent in the choice.

4.18 References

1. Otto H. Schade, Jr., "Electro-Optical Characteristics of Television Systems," *The RCA Review*, Vol. 9, March, June, September, December 1948.

2. Otto H. Schade, Jr., *Image Quality, A Comparison of Photographic and Television Systems*, Princeton, N.J., RCA Laboratories, 1975.

3. Glenn Kennel, LeRoy DeMarsh, and John Norris, "A Comparison of Color Negative Films and HDTV Cameras for Television Program Production," *SMPTE J.*, May 1991.

4. L. Thorpe, F. Nagumo, and K. Ishikawa, "A Comparison Between HD Hyper-HAD Cameras and Color Film for Television Program Production," *SMPTE J.*, June 1994.

Chapter 5
Audio Technology for Video

5.1 Significance of Audio to Video

Video is incomplete without audio. In fact, much of the information content of a visual presentation is contained in the audio tracks. Consider a news broadcast without audio—it would be difficult to figure out the story from the video alone. From the viewpoint of information content, one might even say that video is subordinate to sound. Thus, sound reproduction is an important element in most video systems. This chapter discusses some of the considerations for sound reproduction used with video.

5.2 Natural Sound

Sounds that we hear consist of pressure waves in the atmosphere, in the frequency range from about 20 to over 15,000 Hz. Sound behaves according to the laws of all wave phenomena, where the wavelength λ is determined by the velocity of propagation v and the frequency f according to

$$\lambda = \frac{v}{f} \tag{5.1}$$

The velocity of sound in air at sea level and room temperature is 344 m/s.

Sound waves readily reflect from hard surfaces, which is a property that is both useful and detrimental under different conditions. The study of sound wave phenomena is known as *acoustics*.

5.2.1 Properties of human hearing

As with vision, human hearing is remarkably sensitive to sound waves over a wide range of frequencies and amplitudes. That is accomplished by the combination of the ears and the brain.

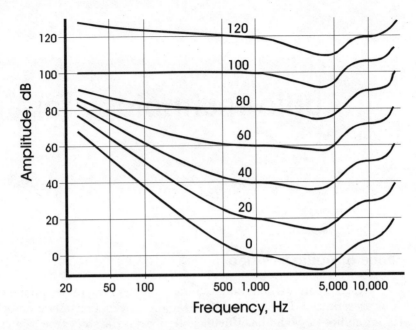

Figure 5.1 The ear's relative sensitivity to sound frequency at different amplitude levels. Reproduced with permission from [1].

5.2.1.1 Sound frequency range

The ability to hear higher frequencies diminishes with age; older people may not hear much beyond 10,000 Hz. However, the ear's sensitivity to sound frequency is not uniform over the range; it falls off at both high and low ends, more so at low sound levels. This may be characterized by curves of frequency response vs. sound level, shown in Figure 5.1.

5.2.1.2 Sound amplitude range

The curves show sound response over a 120-dB range—an amplitude ratio of 1,000,000 to 1. No electronic sound reproduction system has yet achieved such a range. Smaller-range reproduction is successful because the ear cannot hear simultaneous sounds covering the entire range—loud sounds tend to mask quieter sounds.

The perception of sound amplitude is known as *loudness*.

5.2.1.3 Ambience

Natural sound environments generally involve much more than the source and listener. Sound waves reflect from surfaces in the environment, and background sound sources, such as wind, street noises, other voices, and so on, contribute to what the listener hears. The accumulation of all these additional sounds is known

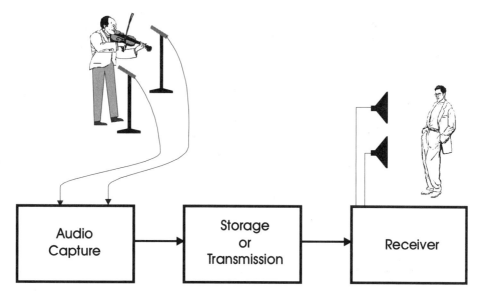

Figure 5.2 Elements of an audio system. Reproduced with permission from [1].

as the *ambience*, which is a significant element in natural sounds.

In an enclosed space, such as a meeting room or concert hall, reflection of sound occurs from all the walls, causing a complex group of reflections that we call *reverberation*. Depending on the size of the room, reverberation may range from an effect that simply interferes with high sound frequencies, to an actual echo effect. Also, sound tends to bounce repeatedly around the room, slowly decaying as a little energy is lost on each bounce. The decaying of the reverberation effect is specified by *reverberation time*. A certain amount of reverberation is desirable to give a room a "live" feeling. A room that absorbs all sound and has no reflections, is called "dead," and it is very uncomfortable to most people.

5.2.1.4 Direction perception

By processing the signals received from the two ears, our brain is able to give us a perception of sound direction. This is the *binaural* effect, and it provides the sensation of spaciousness of sound, as well as the direction to the source(s).

5.3 Audio Systems

A system for the electronic reproduction of sound is an *audio system*, and the signals present in that system are *audio signals*. A typical audio system contains the elements shown in Figure 5.2. The system shown uses two channels of reproduction to simulate the use of two ears—such a system is called a *stereophonic* (sometimes, just *stereo*) audio system.

5.3.1 Electronic capture of natural sounds

A *microphone* is the most common sound sensor—it converts atmospheric pressure variations into an analog electrical signal. Signals from a microphone must be amplified before they can be recorded, transmitted, or sent to a loudspeaker for a listener. The intervening amplifiers and other circuits must have highly linear amplitude characteristics or the reproduced sound will be audibly distorted. Noise and other spurious signal generation must also be kept small to achieve good reproduction.

There are two basic types of microphone operation—*pressure-sensitive* and *velocity-sensitive*. Practical microphones may use one or both of these methods combined with various acoustic elements. By these means, various *patterns* of directivity can be obtained, and the microphone's sensitivity and frequency response can be optimized.

The output from a microphone is at a very low level and must be amplified before it can be put to any use. Microphone preamplifiers must be carefully designed to avoid noise generation in the amplifier itself. Most audio signal operations are done at much higher level—high enough that circuit or cable noise is not a consideration. This is sometimes referred to as *line* level.

The setup and use of microphones is practically an art—the quality of pickup is sensitive not only to the type of microphones chosen, but also to their placement and to the recording environment in general.

5.3.2 Audio storage and transmission

Audio signals may be recorded on magnetic tape, optical disc, or magnetic hard disk. They can be edited, just like video, and replicated for distribution to many users. Audio transmission can be done over short distances using baseband signals on coaxial cables, but for longer distances, analog or digital modulation methods must be used. Similarly, modulation is required for radio-frequency (RF) transmission of audio signals.

5.3.3 Audio reproduction

For the final listener, audio signals must be converted back to sound pressure waves. This is done with a *loudspeaker*, which is an electromechanical actuator that moves a mechanical structure, usually a cone shape, to generate pressure waves that replicate the audio signal variations. To obtain faithful reproduction, the loudspeaker and its housing must be carefully designed according to acoustic principles.

Depending on the loudness required and the size of the listening area, speakers may require considerable audio power delivered to them. This is the job of a power amplifier, and this is another area of audio design art. Large venues may require hundreds or even thousands of watts of audio power, driving large arrays of speakers to cover the area.

5.4 Analog Audio Signals

The properties of analog audio signals are very different from video signals, as explained in this section.

5.4.1 Signal properties

Analog audio signals are typically a complex mixture of many frequency components and their harmonics. The signal is bipolar, and has no dc component. There is no structure such as is produced by the scanning of a video signal. In many respects, an audio signal appears random. Because of this, audio signals are difficult to compress.

The frequency spectra of natural audio sources typically fall off at high frequencies, which offers the opportunity to use preemphasis-deemphasis to improve SNR of audio transmission.

5.4.2 Signal specifications

Audio equipment is typically specified for frequency response, distortion, signal-to-noise ratio (SNR), and wow and flutter where applicable. Because of the accumulation of impairments in cascaded analog equipment, units intended for use in large systems have very tight specifications. Consumer equipment, which may not be cascaded, can have less stringent specifications, although there is an aura about audio specifications, and many units have more demanding specifications than might actually be required simply because this sells equipment.

5.4.2.1 Frequency response

Audio frequency response requirements differ depending on the application. For full high-fidelity audio, a range of 20 to 20,000 Hz is required, calling for a flat response over this range with a tolerance that is often as tight as ±0.1 dB relative to some midfrequency, such as 1,000 Hz. No one can hear a variation in audio response of 0.1 dB, but such tolerances allow for cascading.

Some applications can accept a much looser specification; for example, telephone-quality audio (intended just to reproduce speech) may use a bandwidth of only 400 to 4,000 Hz, with a tolerance of ±1 dB. Other applications, such as AM radio, has a bandwidth from 50 to 5,000 Hz, which is limited by the broadcast channel width of 10 kHz. Similarly, FM broadcasting has a bandwidth of 30 to 15,000 Hz. These different frequency ranges are important because they relate to the cost of equipment—there's no reason to pay for tighter specifications if the rest of the system cannot support them anyway.

Frequency response is measured by applying different-frequency sine waves to the system and measuring the output level with a meter. More sophisticated measuring equipment applies a continuously varying frequency that sweeps across the entire range. The output level during the sweep can be stored in a computer, and can later be plotted on a graph or displayed in other ways. Because of the relatively low frequencies involved, and the need to allow the system to stabilize its response for each frequency, an audio sweep measurement may take a few minutes to perform.

Because the signal is delayed in going through a piece of equipment, there is a phase shift that is proportional to frequency. To the extent that the delay is not constant with frequency, the phase shift vs. frequency curve may depart from linearity. This is *phase distortion*, which is not very important for audio, but it is very important in video, where it is called *envelope delay distortion*. The degree of importance of phase to audio is somewhat controversial, and some equipment specifies this anyway.

5.4.2.2 Harmonic distortion

Amplitude nonlinearity in an audio channel causes distortion in the form of unwanted signal harmonics, which can become audible if they are too large. This is tested by applying pure sine waves to the system and at the output, trapping out the fundamental frequency of the input, leaving only any harmonics generated by amplitude nonlinearity. They can be measured relative to the amplitude of the desired signal; the result is expressed as a percentage and is called *total harmonic distortion* (THD). This can be measured at any input frequency, but the measurement is usually done at 1,000 Hz. A THD of 1%, is good for a system specification—cascading requires much tighter specifications in individual components.

Another related specification is *intermodulation distortion*, which is the amount of amplitude modulation of a high-frequency signal in the presence of a low-frequency signal. This test is seldom used except with variable-density film sound tracks (in most equipment, the THD specification is sufficient to cover amplitude nonlinearities.) In that case, a test specified by the SMPTE [2] uses a low frequency of 60 Hz and a high frequency of 2,000 Hz. The amount of amplitude modulation on the 2,000-Hz signal caused by the presence of the 60-Hz signal is measured. This is similar to the differeitnal gain test in analog composite video systems (see Section 6.10.3).

5.4.2.3 Signal-to-noise ratio

Random noise in an audio system is perceived as a hissing sound in the background. Of course, other forms of spurious interference could also be called "noise," but they can usually be designed out, leaving the random component the dominant noise source. SNR is specified as the ratio between the root-mean-square (rms) signal level and the rms noise level, measured when no signal is present.

The ear has nonuniform sensitivity to different frequencies, as shown in Figure 5.1. This also applies to noise—our ears are less sensitive to high-frequency noise. Thus, a noise-weighting function that attenuates high frequencies the same as the ear's response will give a more pertinent noise measurement. Noise-weighting curves for audio have been developed and standardized for this purpose [3].

5.4.2.4 Wow and flutter

Analog audiotape recorders introduce a degree of frequency modulation caused by speed instabilities occurring in the mechanical motion of the tape deck between recording and replay. This is similar to the time base instability in analog video recorders (see Section 11.2.5). The result is a wavering of the pitch of a continuous sound, called *wow*, or a garbling of the sound caused by higher-frequency FM, called *flutter*. Generally, these two are measured together and a specification for the total of the two is given: *wow and flutter*.

To measure wow and flutter, a steady frequency is recorded. On playback, the signal is passed through a FM discriminator and the output, representing any FM on the signal, is measured as a percentage of the input frequency. A good wow and flutter specification is less than 0.1%—most professional recorders meet this number. Home recorders may have higher numbers, but the effect becomes quite objectionable over about 0.25%. Note that this problem can be totally eliminated in a digital tape recorder.

5.5 Digital Audio Signals

Much of the audio business has converted to digital technology for the same reasons that it is happening in video systems. This has been led by the audio Compact Disc (CD) that was introduced in 1982 for distribution of prerecorded music. The audio CD provides nearly transparent reproduction of sound, virtually free of the noise, distortion, and time base instabilities of previous analog sound recording devices.

5.5.1 Audio ADC

Since audio from natural sources originates as an analog signal from a microphone, the first step of its processing is to convert to a digital format using an ADC. An audio ADC is similar in concept to the ones used for video, but the numbers are vastly different. Table 5.1 shows the differences. Audio has much lower sampling rates and bandwidths, but a higher number of bits per sample. Even so, the data rates are also much lower than video. This is fortunate, because audio data cannot be compressed nearly as much as video data.

Table 5.1 Audio and video sampling compared

System	Sample rate	Bits/sample	Bandwidth	Data rate*
Component SDTV	13.5 MHz	8 or 10	5 MHz	108 Mb/s
Audio CD	44.1 kHz	16	20 kHz	706 kb/s
Speech audio	8 kHz	8	3.5 kHz	64 kb/s
Production quality	48 kHz	20 or 24	20 kHz	960 1152 kb/s

* Per channel

5.5.2 Audio data compression

Considering that video data may be compressed as much as 50 to 100:1, compressed video data rates can be similar to the uncompressed data rate of a high-quality audio channel. Since there are usually two audio channels for stereo and more in some cases, there is a need for audio compression so as to not add too much to the total data rate of a video program.

As already mentioned, an audio signal has none of the structure or redundancy of a video signal that can be exploited to compress the data, the opportunities for audio compression are different, and fewer.

5.5.2.1 ADPCM

On of the compression opportunities in audio arises from the already-stated falloff of high-frequency energy in the typical natural sound. This means that the differences between adjacent samples in an audio stream will likely be smaller than the samples themselves. Thus, if the differences are coded instead of the samples, fewer bits per sample should be sufficient. This is known as *differential pulse code modulation* (DPCM).

Audio samples having 16 bits can be coded, for example, with 4-bit DPCM. However, there will be occasional situations where the 4-bit signal amplitude range will not be enough to handle the differences, causing the problem of slope overload. This can be alleviated by making the difference coding adapt to the amplitude range required by the current sample differences, which is *adaptive differential PCM* (ADPCM).

As with any difference coding scheme, DPCM and ADPCM have the problem that an error in the bit stream will result in a signal error that continues until the coding is restarted in some way. The coding must take this into account to prevent error sensitivity from being too great. A second problem with ADPCM is that the adaption information must somehow be included in the bit stream so the receiver knows how to adapt. Both these problems are dealt with in the standards for audio ADPCM. The typical performance of ADPCM is 4:1 compression with only slight reduction of audio quality.

5.5.2.2 Dolby Digital

More audio compression is achieved with a much more sophisticated approach, using a technique that is somewhat comparable to the transform coding used in video (see Section 3.11.6). *Dolby Digital* (formerly called *AC-3*) audio compression [4], which is used in the ATSC DTV standards, relies on a *filter bank*, which is a kind of transform that takes a block of audio data and convertes it to a frequency-band representation.

Figure 5.3 is a block diagram of Dolby Digital encode processing. The filter bank processes the signal into coefficients for the signal spectrum of each block of 256 audio samples. The coefficients are expressed in a special floating-point format; the exponents and mantissas are processed differently, as shown. The

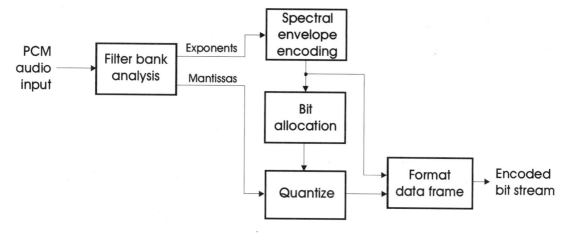

Figure 5.3 Block diagram of Dolby Digital encoding.

results from six audio blocks are combined into an audio frame, which goes to merge with the video stream. Compression is achieved by appropriate quantizing of the data according to the sensitivity of the ear to the frequency blocks combined with the masking effect where loud sounds mask quiet ones. The result is that 5.1 channels of high-quality surround-sound audio are compressed into 384 kb/s. The number 5.1 refers to five full-range channels (left, right, center, left surround, right surround) and one narrow-band subwoofer channel (the 0.1). This represents more than 10:1 compression.

5.5.3 The audio CD

The hardware technology of the CD is described in Section 10.20.1. This section describes the record format is more detail. The two stereo channels of audio are each digitized at 44.1 kHz with 16 bits/sample and linear quantization. This produces a continuous incoming bit rate of 1.41 Mb/s.

A continuous data stream such as this must be partitioned into blocks before recording for error protection, synchronization, and other purposes. The CD block is called a frame and contains the data from 6 stereo samples or 192 bits (6 × 32). Additionally, 8 bits of subcode data (nonaudio information), 64 bits of error-protection code, are added per frame. The error-protection coding is *cross-interleaved Reed-Solomon* coding (CIRC), which applies R-S twice to the data grouped as 8-bit symbols, with data interleaving between the two codings. This is to prevent burst errors from overloading the error protection.

The output from the CIRC coder is then modulated with a version of *eight-to-fourteen modulation* (EFM). EFM maps each 8-bit symbol to 14 bits, choosing only those 14-bit words that meet specific conditions for run-length of bits. It is further necessary to add 3 merging bits between each 14-bit word to essure that the run-length condition remains met during concatenation of the 14-bit words.

Thus, the data size is expanded by 17/8 times in the EFM process. Adding 27 bits of synchronization (includes 3 merging bits) to each frame, the frame containing 192 bits of raw audio samples has grown to 588 bits to be recorded, a total data rate of 4.3218 Mb/s. This seems like a lot of overhead, but it provides a very robust and reliable system—proven by hundreds of millions of players in use.

5.6 Audio Signal Processing

Audio signals often need to be processed to adjust frequency response or levels, to add effects, or to combine (mix) signals. These functions can be done in analog or digital systems, although digital systems have somewhat more flexibility in applying them.

5.6.1 Equalization

In spite of flat-response specifications everywhere, it is often necessary to adjust the frequency response of an audio channel to achieve the desired subjective reproduction of the original sounds. This is done with an *equalizer*, which usually takes the form of a multiplicity of adjustments to control the response in small bands of frequency. When slider controls are used and they are placed side by side in ascending frequency order, the control handles trace out a frequency response curve, hence this is called a *graphic equalizer*.

5.6.2 Audio noise reduction

Because of the masking effect where a loud sound masks a simultaneous quiet sound, noise is not audible when the sound level is high. However, when the sound becomes quiet, the system noise may be heard. This can be exploited by making the system adapt to the incoming audio level to always keep the signal large enough to fill the system's dynamic range. This process is called *companding*. It operates by having a variable-gain amplifier at the input of the system, which is adjusted automatically by a feedback loop that senses the level going into the system, as shown in Figure 5.4.

At the output of the system, the gain changes must be reversed to restore the dynamic range of the signal. The trick of this is to make the output changes track what happened at the input. There are many ways to do that, and numerous systems are in use, both analog and digital.

5.6.3 Electronic reverberation

The effect of sound bouncing around in a concert hall (reverberation) can be simulated electronically with the use of delay elements. This is an application where digital technology is most suitable and "reverb" units are available for both analog and digital audio systems, where the amount of reverberation or echo, reverb

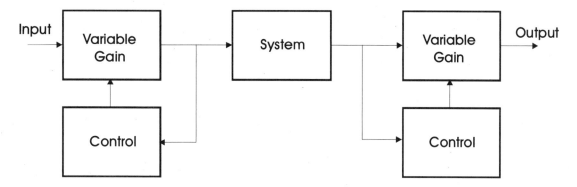

Figure 5.4 Companding. Reproduced with permission from [1].

time, and hall characteristics can be selected. These are widely used in the music industry, and are standard fixtures in audio postproduction studios.

5.6.4 Audio mixing

The process of combining multiple audio signals into one track is *mixing*. Most mixing is simply linear addition of the signals, with control of amplitude for each of the input signals. More advanced mixing is *stereo panning*, where the proportion of each signal that goes to either of the stereo channels is adjusted to position that sound in the stereo field. Mixing is done in both analog and digital systems.

5.7 Music

For thousands of years, people have made music using acoustic devices (instruments) to create pleasing sounds. Only in the last 20 or so years, has electronics become a factor in the music industry. Analog musical instruments were first, but more recently, they have all become digital and the extreme flexibility and control of this technology has taken over the electronic music business. At the same time, digital musical instruments have spurred much research into digital techniques that have advanced digital video as well. Many musical concepts come up in the digital audio and video fields, too, so some of them are covered here.

5.7.1 Pitch

A single musical note is generally a tone of a recognizable frequency—its *pitch*. Combinations of notes together and in time sequence is the basis of most music. Many notes contain multiple frequencies; the pitch is the dominant frequency that most listeners will recognize. In most music, there is a *scale*, which are specific points along the pitch continuum that produce a pleasant sequence. The most common scale in western music is the *equally tempered scale*, where successive notes are in the ratio of $\sqrt[12]{2}$ (1.05946) to each other. Thus, every 12th note is exactly double in frequency. This interval in music is known as an *octave*.

Pitch must be reproduced with reasonable accuracy, because most listeners will recognize pitch errors of 1% or so. In some music, pitch is deliberately varied, either periodically, as in *vibrato*, or in a sweeping way from one note to the next, called *portamento*. Either way, an audio system must correctly reproduce the pitch and its variations.

5.7.2 Timbre

Most musical notes have a lot more character than just frequency; they may vary in amplitude from start to end (the note's envelope); and they have harmonic content, either constant or varying with time. These attributes represent the notes *timbre*, which is what makes different instruments sound differently.

5.7.3 Tempo

The speed of a sequence of musical notes is the *tempo*, which may be constant during a piece or may vary in expressive ways to add to the musical effect. Generally an audio system that reproduces pitch with sufficient accuracy will also reproduce acceptable tempo.

5.7.4 MIDI

Music can be created electronically by a *synthesizer* (see the next section) that knows about pitch, timbre, and tempo and can create sequences of notes under computer control. Formats are necessary to store and communicate this information so that electronic instruments can talk to each other and work together. The music industry addressed this early in the electronic era and standardized in 1984 on a single format, called the *musical instrument digital interface* (MIDI). This standard is managed and distributed today by the MIDI Manufacturers Association [5].

MIDI is actually a bus standard that allows multiple devices to be daisy-chained together for control purposes. It has also been adopted in the PC industry and all PCs have a built-in MIDI synthesizer. Because MIDI transmits only control information for notes and their properties, it is an extremely efficient way to handle music, using as much as 100× less data than sampling live music. Its limitation, of course, is in the synthesizer that must play it.

5.7.5 Music synthesis

An electronic musical instrument is generally called a *synthesizer*. Various methods are used to approach, in some cases, remarkable synthesizing of acoustic instrument sounds. Equally valuable, however, is the synthesizer's ability to create sounds that do not exist in nature, and this has greatly expanded the portfolio of musical capabilities available to the musicians of the world.

There are two general classes of synthesizers: those that *generate* notes from electronic oscillators and modification processes, and those that *sample* natural sounds and reproduce them under note control. The first synthesizers were analog and used the generation method. Analog synthesizers have one or more oscillators to create pitch, combined with various filters and modulators to create timbre. This is very effective and has been widely used. However, most new synthesizers are digital, and various digital forms of the generator-synthesizer are on the market.

When digital technology came along, the sampling type of synthesizer became practical. In this method, a real instrument is sampled at various points along its scale. The samples are stored in a memory and can be called up and played at different sample rates to change pitch so as to fill out the entire scale without needing to sample every note. Digital modification can be applied to the samples as they are played to vary timbre and other features. With a sampling synthesizer, it is also possible to sample nonmusical sounds, for example, a hammer blow, and create a scale of notes that can be played just like any musical instrument. This has also expanded the musical possibilities for those who like to experiment.

5.8 Conclusion

The importance of audio to a good video presentation is well established. Thus, audio technology has been an important part of the video industry from the start. In fact, many of the problems of video, such as processing, storage, recording, dustribution, and so on, were solved first in the audio industry. This is still true as video moves into the digital era.

5.9 References

1. A. C. Luther, *Principles of Digital Audio and Video*, Artech House, Norwood, MA, 1997.

2. SMPTE RP-120.

3. K. Blair Benson, *Audio Engineering Handbook*, McGraw-Hill, New York, 1988, p. 16.11.

4. ATSC Standard A/52 (1995), *Digital Audio Compression Standard*, http://www.atsc.org.

5. MIDI Manufacturers Association, PO Box 3173, La Habra, CA 90632-3173, http://www.midi.org.

Chapter 6

Analog Video Systems

6.1 Introduction

This chapter describes analog video systems (as distinguished from digital systems; see Chapter 6) that are used for the transmission and storage of color video signals. They are based on the colorimetry of television as described in Chapter 2.

Analog systems can be classified as composite or component, depending on the format of their signals. Composite formats multiplex the luminance and chrominance components into a single signal, whereas component formats employ a separate signal for each component.

Composite formats are used for analog TV broadcasting. Component formats are widely used for recording in complex program production systems where the highest quality must be achieved. They are also used for satellite transmission and DTV.

The principal composite analog formats are NTSC, PAL, and SECAM. The configurations and signal formats of NTSC and PAL composite systems and of the principal component systems are described in this chapter.

6.2 Standardization

The objective of most analog video systems is mass distribution of programs and information. This may happen by broadcasting or other means of distribution such as cable transmission or videotape. In any case, the equipment at the receiving end must know the signal format before successful reception is possible. That means the signal format must be standardized—precisely defined so that any manufacturer around the world can build receiving equipment. Because this is a multicompany and multinational problem, various organizations have been formed specifically for this purpose. For example, in the United States, standards for broadcasting are set by the Federal Communications Commission (FCC). However, the FCC does not develop standards, they only endorse them. Other industry bodies are relied upon to propose standards to the FCC. These are organizations such as the *Society of Motion Picture and Television Engineers* (SMPTE), the *Institute of Electrical and Electronic Engineers* (IEEE), the *Electronic Industries Association* (EIA), and special ad-hoc groups that are formed from time to time. One of the latter class is the *National Television Systems Committee* (NTSC), which was formed in the 1950s to develop color television standards that were proposed and later adopted by the FCC. Similar approaches have been taken by other countries.

However, the television industry is worldwide and it would not be practical for every country to have a different standard. That would require receiver manufacturers to have different products for every country and that would increase the cost of equipment for everyone. Therefore, another class of standards groups exists with a worldwide purview. These include the *International Standards Organization* (ISO), the *European Broadcasting Union* (EBU), the *Consultative Committee on International Radio* (CCIR), and the *Consultative Committee on Telephone and Telegraph* (CCITT). Like the FCC, these organizations usually do not develop standards but rather they rely on others to make proposals. However, they will in some cases form special committees to work in specific areas, such as the JPEG and MPEG groups of the ISO (see Setion. 3.12).

The standards resulting from these activities are voluntary; the standardizing organizations do not have authority to enforce their standards. However, most manufacturers comply with them because of the important commercial advantages of worldwide standards.

6.3 Video Signal Formats for Color

All video signal formats are derived from the primary signals E_R, E_G, and E_B or their gamma-corrected derivatives E'_R, E'_G, and E'_B, which are generated by the camera chain. The camera signals are in analog form and have amplitudes proportional to the spectral energy for the primary colors in the scene. At the receiver, the inputs to the receiver kinescope or other display device are modified replicas of these signals. Between the camera chain and the kinescope input, however, the signal is usually transformed to another format—analog or digital—that is better suited to the functions of signal processing, recording, transmission, and broadcasting.

For all formats, a luminance signal E'_Y is generated by matrix circuits in accordance with Eq. (6.1):

$$E'_Y = 0.299\, E'_R + 0.587\, E'_G + 0.114\, E'_B \qquad (6.1)$$

The coefficients of the components E'_R, E'_G, and E'_B are based on the relative sensitivity of the human eye to the primary colors. The magnitude of E'_Y is then proportional to the perceived brightness of the scene.

Composite color formats can be either *compatible* or *noncompatible* with the monochrome format. Compatible color transmissions can be received on monochrome receivers (displayed in monochrome, of course) and monochrome transmissions can be received on color receivers. All formats currently used for broadcasting are compatible.

6.3.1 Composite formats

The FCC's experience with incompatible sequential color demonstrated that a successful format for color broadcasting would have to be compatible with the existing monochrome format. To be compatible for broadcasting and related services, a format had to meet three requirements:

1. The luminance and chrominance information had to be contained in a single channel (hence it is *composite*).
2. The luminance component had to be sufficiently similar to the monochrome signal so that color broadcasts could be received on monochrome receivers and vice versa.
3. The bandwidth had to be no greater than required for monochrome.

The television industry devoted enormous efforts to the development of a compatible system, and the result was proposed to the FCC by the NTSC in July 1953. It was approved by the FCC for color broadcasting in the United States in December 1953.

The NTSC system has been remarkably durable—in a fast-changing industry it is still in use 40 years later. It is also the approved system for color broadcasting in Japan and in most western hemisphere countries.

Two additional analog color systems, PAL and SECAM, are used in other countries—PAL in Europe (except in France and Russia) and in the countries of the British empire (except Canada), and SECAM in France and Russia. These systems are similar in principle to NTSC in that the color information is transmitted by one or two high-frequency subcarrier(s) (3.58 MHz for NTSC, 4.43 MHz for PAL) superimposed on the analog luminance signal.

6.3.2 Component formats

A major disadvantage of subcarrier composite systems is the impossibility of completely eliminating crosstalk between the luminance and chrominance components as the result of nonlinearities in the system and overlaps of the sideband spectra of the subcarriers. The nonlinearities cause *differential gain* and *differential phase* distortion of the chroma component—variation in the amplitude and phase of the subcarrier with brightness—and visibility of the subcarrier in the luminance component. The spectrum overlaps cause *cross-color* and *cross-luminance distortion*.

Differential gain and phase distortions are described in Section 6.10. Cross color results when the luminance signal contains frequency components near the subcarrier frequency, and spurious colors are generated in the image. Cross luminance occurs when the signal representing a given point changes from field to field or on the edges of objects so that the subcarrier cancellation is not complete. A dot pattern then occurs in the luminance channel.

Another disadvantage of subcarrier composite systems results from the fact that the noise in transmission systems that use frequency modulation, e.g., microwave and satellites, increases linearly with the modulating frequency. This

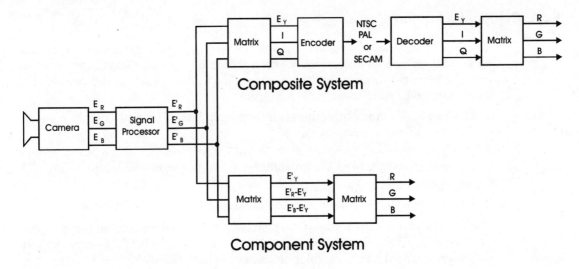

Figure 6.1 Functional diagrams for color TV systems.

puts the subcarrier at the high-noise end of the spectrum and can lead to a poor SNR for the chrominance signal.

Signals used for editing, point-to-point transmission, and recording are subject to none of the constraints of broadcasting, and it is possible to choose a component format that eliminates the use of a subcarrier with its crosstalk and noise problems. Component formats have no cross-luminance or cross-color distortions. The most common component format employs three components—luminance or E_Y, and the color-difference components, $E_R - E_Y$ and $E_B - E_Y$. They may be transmitted on separate channels or time-multiplexed on a single channel using the *multiplexed analog component* MAC system (see Section 6.13).

6.4 Analog Color TV System Configurations

Figure 6.1 is a simplified functional diagram that shows the configurations of composite and component color television systems. Both begin with a camera chain that includes a transducer for each primary color, which may be a tube or solid-state device, and signal-processing circuitry that generates the primary signals, E'_R, E'_G, and E'_B. The processing functions include white and black level control, aperture correction and image enhancement, color correction by matrixing, gamma correction, and sync and blanking insertion.

In composite systems, the camera chain is followed by an encoder that produces the composite signal by multiplexing the primary signals. This composite signal is transmitted to the receiver, which decodes the signal to recover the original E'_R, E'_G, and E'_B primary signals that are the inputs to the picture tube.

In component systems, the luminance and color-difference signals are generated by matrixing the primary signals, i.e., by mixing them in controlled ratios. After matrixing, the component signals may go to a video recorder; to a micro-

Table 6.1 Color TV system bandwidths

Country	System	Bandwidth (MHz)
United States, Japan	NTSC	4.2
Canada, Mexico	NTSC	4.2
Great Britain	PAL	5.5
Germany, Austria, Italy	PAL	5.0
France	SECAM	6.0
Former USSR	SECAM	6.0

wave, fiber-optic, or satellite transmission system; or to an ADC for conversion to a digital format.

6.5 Composite Signal Components

Composite signals are made up of their own kind of components—luminance and chrominance. These are discussed in this section.

6.5.1 The luminance signal component

The amplitude of the luminance signal component, E'_Y, was given by Eq. (6.1). Its bandwidth varies from country to country (Table 6.1). Most European countries use UHF for color broadcasting; this accounts for the greater bandwidths available for their systems.

6.5.2 NTSC chrominance signal components

It is possible to conserve spectrum space by deriving two chrominance components, E'_I and E'_Q, called the I and Q signals, from the primary components:

$$E'_I = 0.60\, E'_R - 0.28\, E'_G - 0.32\, E'_B \tag{6.2}$$

$$E'_Q = 0.21\, E'_R - 0.51\, E'_G + 0.30\, E'_B \tag{6.3}$$

The chrominance components, E'_I and E'_Q, have the following characteristics:

1. E'_I is at a maximum for orange or cyan hues, while E'_Q is at a maximum for green or magenta.
2. The ability of the eye to discern fine detail in colors is less than for monochrome, and it is less for magenta than for orange. As noted above, this property is utilized in the NTSC for reducing the bandwidths of the chrominance signal components (see Table 6.2).
3. With white light $E'_R = E'_G = E'_B$, and all chrominance (or color-difference) components, E'_I, E'_Q, $(E'_R - E'_Y)$, and $(E'_B - E'_Y)$, disappear.
4. The amplitudes of the modulated chrominance components relative to the

luminance signal are a tradeoff between a large value that would lead to excessive crosstalk with the luminance component and a small value that would lead to an inadequate chroma SNR.

6.5.3 PAL chrominance signal components

With video bandwidths of 5 to 5.5 MHz available, the PAL system does not need to be as frugal as the NTSC system in its usage of spectrum space. Accordingly, there is no need to add to the complexity of PAL systems by creating the I and Q signals, and the color-difference primaries are used directly for producing the color subcarrier.

The two chrominance components in the PAL system are E'_U and E'_V. Their equations are:

$$E'_U = 0.493 \, (E'_B - E'_Y) \tag{6.4}$$
$$E'_V = 0.877 \, (E'_R - E'_Y) \tag{6.5}$$

The bandwidth of the E'_U and E'_V components is 1.3 MHz (to the –3 dB points).

6.6 Subcarriers

The essence of the difference between the NTSC and PAL color TV systems is their choice of subcarrier frequencies, phases, and formats. The subcarrier frequencies in both systems are chosen to be as high as possible while keeping their sidebands within the video passband. Their relationship to the frequencies and phases of the scanning lines and fields must also minimize the crosstalk between the luminance signal and the subcarrier and the visibility of the subcarrier in the composite signal. The solutions to these problems differ for each system.

6.7 The NTSC System Subcarrier

The NTSC subcarrier format was the first to be developed (in the United States in the early 1950s), and it is the simplest.

6.7.1 The NTSC subcarrier format

The NTSC subcarrier combines both amplitude and phase modulation, its amplitude determining the saturation of the image and its phase the hue. The use of phase to transmit hue requires the transmission of a reference phase that is supplied by the color burst, a train of eight cycles of the subcarrier frequency with a fixed phase on the back porch of the blanking pulse. The NTSC burst is shown in Figure 6.2.

The NTSC subcarrier, e_{scn}, can be derived either from the I and Q signals or from the color difference signals, $(E'_R - E'_Y)$ and $(E'_B - E'_Y)$.

If e_{scn} is derived from I and Q, its equation is:

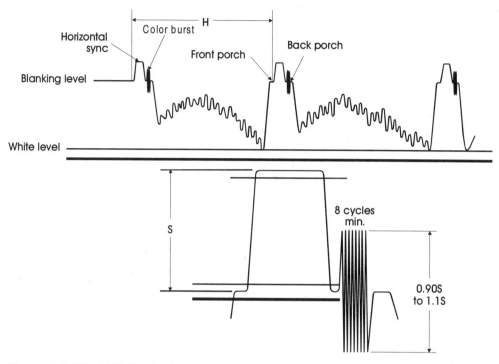

Figure 6.2 The NTSC color burst.

$$e_{scn} = [E'_Q \sin(\omega t + 33°) + E'_I \cos(\omega t + 33°)] \tag{6.6}$$

where $\omega = 2pf_{sc}$ and f_{sc} is the subcarrier frequency.

If e_{scn} is derived from the color-difference signals, the equation is:

$$e_{scn} = 0.877[0.562(E'_B - E'_Y) \sin(\omega t) + (E'_R - E'_Y) \cos(\omega t)] \tag{6.7}$$

Note that the subcarrier vanishes for white light. Also note that its average value is zero, so that in a linear system, the luminance signal is unaffected by the chroma. This is known as the *constant-luminance principle*. Stated otherwise, there is no crosstalk from chrominance to luminance unless there are nonlinearities in the system.

The chrominance subcarrier e_{scn} and its terms can be represented by vectors with the sine and cosine functions in quadrature. Figure 6.3 shows the phase relationship of these functions to the burst and the significance of the 33° angle in the equation.

The NTSC subcarrier format is subject to crosstalk from luminance to chrominance, resulting from nonlinearities in amplifiers, recorders, and transmission systems that cause variations in gain or phase shift with the brightness level of the luminance signal (differential gain and differential phase). Differential gain causes distortions of the image saturation and differential phase its hue.

Differential phase shift is a particularly serious problem because the eye is very sensitive to hue distortions. After the introduction of NTSC, the Europeans, who were considering other systems, jokingly said that NTSC meant "never the same

158 Chapter Six

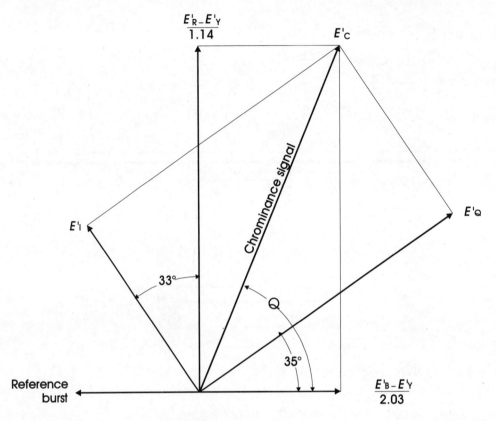

Figure 6.3 NTSC subcarrier vector diagram.

color." With the passage of time, advances in technology have greatly reduced the differential gain and phase of video amplifiers, transmission systems, and recorders, and their nonlinearities no longer need be significant problems. They still must be considered, however, in system design and operation, and the vulnerability of the NTSC format to system nonlinearities continues to be a potential problem.

Other forms of crosstalk are cross color and cross luminance (see Section 6.3.2).

6.7.2 NTSC subcarrier frequency

The NTSC subcarrier frequency is derived from the scanning line frequency, which in turn is an integral submultiple of the 4.5-MHz audio-video carrier spacing used in broadcast systems. This relationship minimizes the beat between the audio and color subcarrier. It is also desirable to have the line frequency as close as possible to the monochrome line frequency of 15,750 Hz. As a result of these considerations, the line frequency is the 286th submultiple of 4.5 MHz:

$$f_{line} = \frac{4.5 \times 10^6}{286} = 15{,}734.26 \text{ Hz} \tag{6.8}$$

The field frequency is related to the line frequency by the number of scanning lines:

$$f_{field} = \frac{f_{line}}{525/2} \tag{6.9}$$

It would be desirable for the field frequency to equal that of the primary power source, 60 Hz (as for monochrome), but it is mathematically impossible to match this frequency precisely while other requirements are being met. The small difference between 59.94 Hz and 60 Hz causes hum bars to move slowly through the image, but this is not a serious problem with modern receivers because of their excellent filtering.

To reduce the visibility of the subcarrier in the luminance signal, its frequency is an odd multiple of one-half the line frequency. With this relationship, the dot pattern resulting from the subcarrier is stationary, and the phase of the subcarrier (for a given hue) at the beginning of each line reverses on successive scans. With the phase reversal, the dots on successive scans have opposite polarity and tend to cancel.

On the basis of these relationships, the equation for the subcarrier frequency, f_{sc}, is:

$$f_{sc} = 455 \frac{f_{line}}{2} = 3.579545 \text{ MHz} \tag{6.10}$$

The relationship between subcarrier frequency and line rate must be maintained very precisely to avoid dot crawl. Usually the line rate is established by counting down from the subcarrier frequency by a series of frequency dividers:

$$f_{line} = \frac{2 f_{sc}}{(5)(7)(13)} = 15{,}734 \text{ Hz}$$

Although the line and field rates are slightly different than those for 525-line monochrome TV, the differences are so small that they cause no problems with monochrome receivers.

6.7.3 NTSC subcarrier reference phase

If 0° and 180° designate the subcarrier reference phases at the beginning of successive lines, their phases are:

Line	Field 1	Field 2	Field 3	Field 4
1	0°		180°	
2		180°		0°
3	180°		0°	
4		0°		180°
5	0°		180°	

Note that four lines are required for the pattern to repeat.

Crosstalk of the chroma signal into the luminance channel, which produces a dot pattern in the image, is minimized by the choice of subcarrier frequency and phase. By relating the subcarrier and line frequencies, the dot patterns are stationary and thus less visible. And the reversal of the phase of the subcarrier on alternate frames causes at least a partial cancellation of the visual effect of the dots.

6.7.4 NTSC signal spectral content

The video bandwidth requirements and the relative amplitude of the luminance and I and Q chrominance components over the video passband for a typical signal are shown in Figure 6.4.

The luminance and chrominance components of the signal can occupy the same portion of the spectrum with a minimum of crosstalk because of their fine-grained spectral content. The content of the luminance component was shown in Figure 1.9. It consists of groups of frequency components, each with a center frequency that is a multiple of the line rate and surrounded with a cluster of sideband-like components separated by the field rate. Unless there is an unusual amount of motion in the picture, there are wide spectral gaps between these groups in which additional signal information can be inserted.

The spectrum of the subcarrier consists of similar groups of sideband components. Because of the integral relationship of the subcarrier frequency and the line rate in the NTSC system, the groups of subcarrier sidebands fit in the gaps in the luminance spectrum.

The interleaving of the luminance and chrominance components in the frequency domain is an indication that the chrominance signal will cause a minimum of crosstalk between the two signals. In the time domain this is evidenced by the fact that the subcarrier pattern is stationary and that it has opposite phases on successive scans.

6.8 The PAL System Subcarrier

The PAL system was developed in Europe after NTSC had been developed in the United States. Thus, it benefited from the early experience with NTSC and some improvements were included.

6.8.1 The PAL subcarrier format

The format of the PAL subcarrier was designed to minimize the effects of differential phase, one of the results of crosstalk from the luminance to the chrominance channel. It is similar to the NTSC format except that the phase of the subcarrier component derived from the color-difference component, $E'_R - E'_Y$, is shifted ±90° at the end of each line with respect to the $E'_B - E'_Y$ component—hence the system's acronym is PAL for *phase-alternating line*.

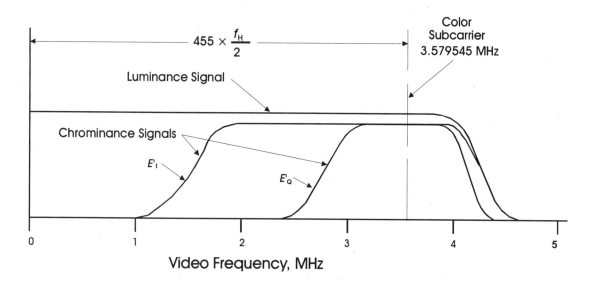

Figure 6.4 NTSC subcarrier frequency spectrum.

Because of the wider frequency band used for most PAL systems, it is not necessary to use the I and Q signals with their greater complexity, and the subcarrier is derived directly from the color-difference signals, $E'_R - E'_Y$ and $E'_B - E'_Y$. The equation for the PAL subcarrier is:

$$e_{scp} = 0.493(E'_B - E'_Y)\sin(\omega t) \pm 0.877(E'_R - E'_Y)\cos(\omega t) \qquad (6.11)$$
$$= E'_U \sin(\omega t) \pm E'_V \cos(\omega t)$$

The ± indicates the phase reversal on successive frames.

The phase relationship between the burst and the $E'_R - E'_Y$ and $E'_B - E'_Y$ subcarrier components is shown in vector form in Figure 6.5. Unlike NTSC, the PAL burst does not maintain a fixed relationship with the subcarrier components but shifts 90° at the end of each line. This is sometimes called a *swinging burst* and it provides some cancellation of errors as explained below.

The dotted vectors show how the phase error due to differential phase is reversed on successive scans of the same line. For example, if differential phase causes a shift of the subcarrier phase and the corresponding image hue toward the $E'_R - E'_Y$ phase on the first scan, it will cause a shift toward the $E'_B - E'_Y$ phase in the next. Visual averaging of the two hues would then minimize the perceived hue distortion. This results in a number of annoying spurious effects, but these can be avoided by electrical delay-line averaging, in which the chrominance signal of each line is added to an electrically delayed signal of the same line from a previous scan. This technique, however, reduces the vertical chrominance resolution by a factor of 2, which is not normally a serious problem.

As with NTSC, the PAL subcarrier vanishes for white light.

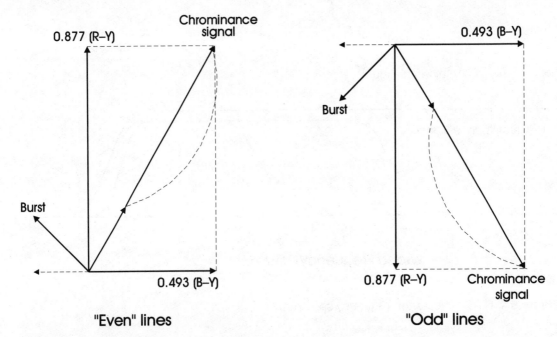

Figure 6.5 Vector diagram of PAL subcarrier.

6.8.2 The PAL subcarrier frequency

In the PAL system the field, frame, and line rates are the integral numbers used for European monochrome systems: The field and frame rates, f_{fld} and f_{fr}, are 50 and 25 Hz, respectively, and the scanning line rate, f_{line}, is 15,625 Hz, 625 times the frame rate.

Since two phase changes are involved in the PAL subcarrier—the phase of the reference carrier at the beginning of the line and the phase of the $E'_R - E'_Y$ component with respect to $E'_B - E'_Y$—the relationship between line rate and subcarrier frequency chosen for NTSC would not be satisfactory for PAL. Instead, the subcarrier frequency is an odd multiple of one-quarter the line rate plus one-half the frame rate:

$$f_{scp} = 1135 \, \frac{f_{line}}{4} + \frac{f_{fr}}{2} = 4.43361875 \text{ MHz} \tag{6.12}$$

The term, $f_{fr}/2$, causes the subcarrier phase on successive scans of the same line to shift by 90°. A rigid relationship between the subcarrier frequency and line rate is also required for PAL systems.

6.8.3 PAL subcarrier reference phase

The relationship between the subcarrier frequency and the line and frame rates results in a 90° phase shift in the subcarrier reference phase at the beginning of each line (as contrasted with 180° for NTSC).

Table 6.2 PAL subcarrier phase relationships

Line	Field 1	2	3	4	5	6	7	8
1	0A		90B		180A		270B	
2		270B		0A		90B		180A
3	90B		180A		270B		0A	
4		0A		90B		180A		270B
5	180A		270B		0A		90B	
6		90B		180A		270B		0A
7	270B		0A		90B		270B	
8		180A		280B		0A		90B

If 0°, 90°, 180°, and 270° indicate the carrier reference phase at the line beginning, and A and B represent the two phase relationships between the color-difference subcarrier components, the phase relationships of the subcarrier and its components are shown in Table 6.2.

To achieve the proper phase of the burst for each line, it is necessary to shift its insertion time by one line at the field rate. This is done with a circuit known as a *meander gate*.

6.8.4 PAL signal spectral content

As a result of the relationship between the PAL subcarrier frequency and the line rate and the phase shift of the $(E'_R - E'_Y)$ components at the end of each line, the PAL chrominance harmonics do not fall neatly between the luminance harmonics as with NTSC. In the time domain, the subcarrier phase shifts only 90° on alternate frames, and complete cancellation owing to phase reversal does not occur. Thus the subcarrier is inherently more visible in PAL than in NTSC, but the effect is mitigated by the higher PAL subcarrier frequency, which makes an uncancelled subcarrier less visible. The relationship between the luminance and chrominance spectral components is shown in Figure 6.6.

6.9 Comparison of NTSC and PAL

PAL systems are usually used in countries where wider video bandwidth is available, often by the use of the UHF band for transmission. This provides a design flexibility that is not available to narrower-band NTSC systems. In addition, the wider bandwidth provides an advantage in signal quality that is independent of the choice of the color system.

PAL systems are sometimes described as being more rugged, meaning that they are less vulnerable to hue distortions resulting from phase errors in the transmission system.

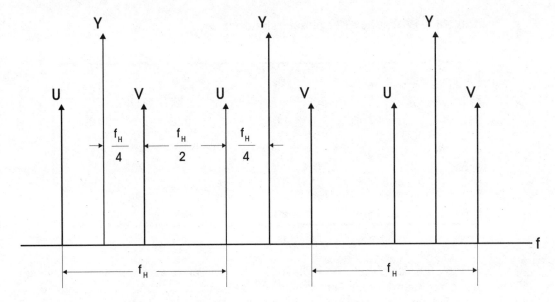

Figure 6.6 PAL frequency components.

On the other hand, NTSC systems are simpler, they do not require quite as much bandwidth, and their subcarriers are less visible. Also, improvements in transmission and recording systems make the greater vulnerability of NTSC to phase errors less significant.

6.10 Color Signal Distortions and Artifacts

Multiplexing the luminance and chrominance components of a color TV signal with a subcarrier is not without problems, and distortions and artifacts (spurious signals produced artificially) result. This section describes some of the most significant impairments that may occur in NTSC and PAL systems. More detailed descriptions of these distortions together with suggested tolerances and methods of measurement have been published by the EIA, FCC, and CCIR [2].

6.10.1 Chrominance-to-luminance gain inequality

The amplitude-frequency response of the luminance channel in a color transmission has the same significance as in monochrome with respect to the definition of the image (see Chapter 4). But since the saturation of the image depends on the subcarrier/luminance signal ratio, it is exceedingly important that this ratio be carefully controlled with special precision in the region around the subcarrier, 4.0 or 5.0 MHz. The end-to-end EIA tolerance for a transmission system in this frequency region is ±7 IRE units (on a signal amplitude scale, an IRE unit is 1% of the range from white level to blanking) or ±0.6 dB.

6.10.2 Envelope delay distortion

Envelope delay is the measure of the time delay of different frequency components as they pass through the system, and is measured in nanoseconds. It is also the variation of the derivative with respect to frequency of the system's phase response. In an ideal system, phase shift is proportional to the frequency, and the delay is constant over the entire spectrum.

As with monochrome TV, envelope delay causes some distortion of the luminance waveform, but in color it can also result in a differential time delay between the subcarrier and luminance signals. The chroma signals are thus displaced from the luminance (the funny paper effect).

The EIA tolerance for end-to-end chroma transmission time in the vicinity of the subcarrier (3 to 4 MHz) relative to 2 kHz to 600 kHz is ±60 ns.

6.10.3 Differential gain

The transmission system for a color television system must maintain its gain as a function of luminance within close tolerances in the vicinity of the subcarrier frequency. Gain variation at the subcarrier frequency as a function of the luminance signal level is known as *differential gain*, a form of cross-luminance distortion that changes color saturation as a function of image brightness.

Tests [1] have shown that differential gain in the range from 10% to 90% of average picture level becomes noticeable to the median observer at ±2 dB and objectionable at ±5 dB. The EIA voltage or current standard for long-haul circuits is ±10% (±0.8 db).

6.10.4 Differential phase

The eye is even more sensitive to differential phase, another cross-luminance distortion, than to differential gain since it affects the hue of the image. In the early days of color television, the tolerance for differential phase distortion was generally considered to be ±10°. This tolerance was established more on the basis of the state-of-the-art capability of the then-current transmission systems than of the visibility of the distortion. Reducing the visibility of the effects of differential phase distortion was perhaps the primary motivation for the development of the PAL system.

With the passage of time and improvements in technology, much tighter tolerances have become practical, and the most recent EIA standard for differential phase of an end-to-end transmission circuit is ±3°.

6.11 Subcarrrier Crosstalk

The presence of the subcarrier on the luminance channel has the potential of creating crosstalk that is visible as an annoying dot patterns on the image. A number of procedures have been developed to minimize the visibility of these patterns.

6.11.1 Relating subcarrier and line frequencies

An important technique that is used in all color subcarrier systems is to relate the subcarrier and line frequencies as described earlier (see Sections 6.7.2 and 6.8.2). This relationship causes the polarity of the dot patterns to reverse or partially reverse on successive scans so that there is a cancellation of their visual effect. It also causes the dot pattern to be stationary.

When a monochrome receiver is used with a color signal, the color subcarrier may be visible as dots in colored areas of the picture because of imperfect subcarrier cancellation. This effect is often mitigated by poor aperture response in inexpensive monochrome receivers.

6.11.2 Low-pass filters

A brute force reduction of dot visibility used in early color receivers was a low-pass filter in the luminance channel with a cutoff below the subcarrier frequency—3.58 MHz in the case of NTSC. This limited the horizontal resolution and definition, but the high-frequency output of the image orthicon camera tubes used in that period was so low that very little information was removed by the filter.

6.11.3 Comb filters

The signals from photoconductive camera tubes and charge-coupled devices (CCDs) (see Chapter 9) have a sufficiently low noise level to permit aperture correction, and the loss of frequency components above 3.58 MHz significantly degrades the picture. To solve this problem, a *comb filter* was introduced in the receiver luminance channel (see Chapter 13). As the name suggests, a comb filter removes alternate bands of frequency components. In the color TV application the comb filter removes the chrominance channel frequency components, which are interleaved with the luminance channel components at harmonics of the line frequency (see Section 6.7.4). A comb filter also has the desirable effect of improving the SNR by nearly 3 dB since it removes the noise as well as the chrominance sidebands from the spaces between the luminance harmonics in the frequency spectrum.

A comb filter is achieved with a delay of one line period; the input and output of the delay circuit are added to create a frequency response that has peaks corresponding to the line harmonics and nulls between them; such a filter is used in the luminance path of a receiver. Subtracting the delayed and undelayed signals has peaks between the line harmonics (where the chrominance sidebands are) and nulls at the locations of the line harmonics, so this filter separates the chrominance components from the luminance.

Although a comb filter is highly effective, it does not completely eliminate the problem of subcarrier visibility. The clusters of chrominance sidebands spread when reproducing edges with sharp color differences, so that they are not completely removed by a comb filter. This creates artifacts in the form of a dot pattern along the edges.

A similar spreading of the chrominance sidebands occurs if there is rapid motion in the picture or if there are major signal changes from line to line. The luminance channel is also affected because the sidebands surrounding the line frequency harmonics spread into the areas attenuated by the filters.

If the dot or line pattern on a color CRT has approximately the same spacing as the subcarrier dot pattern, a heterodyning effect can occur that causes artifacts in the form of a coarse spurious pattern in the picture (called *moiré*).

6.12 Analog Component Formats

Component signals are used in applications where it is especially important to avoid the artifacts and cross-color and cross-luminance distortions that are characteristic of composite systems. These applications include but are not limited to HDTV and complex program production systems where the signals are subjected to repeated recording and rerecording.

Component systems frequently employ a digital format, which further reduces the distortions that occur with analog formats.

Conventional component systems require three channels (see Figure 6.1), the Y or luminance channel, and two color-difference channels. The three channels usually are transmitted simultaneously, but they can also be transmitted sequentially as in the MAC family of formats.

6.13 Multiplexed Analog Component Systems

Multiplexed analog components (MAC) is a term describing a family of time-division multiplex signal formats in which the luminance and chrominance components are time-compressed and transmitted sequentially. In common with other component systems, MAC avoids the problems of crosstalk and sensitivity to nonlinearity described in Sections 6.9 and 6.10 for subcarrier systems. They also have a better SNR because of the absence of strong frequency components in the vicinity of the color subcarrier. Synchronizing information can be sent in digital form, thus eliminating the large-amplitude conventional and digital audio signals inserted for a portion of the sync intervals.

At the same time, MAC shares the feature of composite systems of including the entire signal in a single channel.

Its disadvantages are its lack of compatibility with broadcast systems and somewhat greater bandwidth requirements. Its principal applications, therefore, are in situations where compatibility is not required and where adequate bandwidth is available. It is particularly useful for satellite transmissions.

6.13.1 The principle of MAC transmission

The principle of MAC transmission is illustrated in Figure 6.7, which shows time relationships in one line of B-MAC, a principal member of the MAC family.

The $63.5\,\mu s$ line scan time (it is virtually identical for NTSC and PAL) is divided

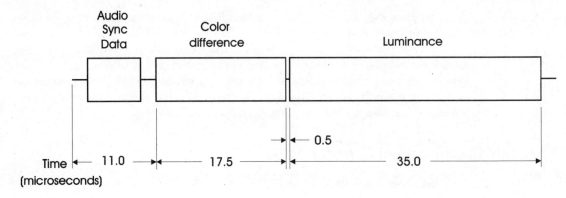

Figure 6.7 MAC signal structure.

into three segments—34.9 μs for the luminance signal, E'_Y, 17.4 μs for the color-difference signals, $E'_R - E'_Y$ or $E'_B - E'_Y$, and the remainder for blanking, burst, audio, and data (optional). (Alternatively, I and Q signals can be used for transmission of the chroma information.) This requires reduction in transmission time for all video signal components—by 2:3 for the luminance signal and 1:3 for the color differences. The $E'_R - E'_Y$ and $E'_B - E'_Y$ components are transmitted on alternate lines. As a result, four fields are required to transmit a complete picture, and the vertical chrominance resolution is reduced by one-half.

To achieve the time-reduced format, the luminance component is sampled at a rate of 15.5 MHz, high enough to avoid aliasing, and the color-difference components are sampled at one-half this rate or 6.75 MHz. The samples are stored in memories and the MAC signal is generated by reading out the samples at a rate of 20.25 MHz in the sequence Y', $R'-Y'$, Y', $B'-Y'$. The luminance samples for one line then occupy 34.9 μs for a time reduction of 3:2 and the color-difference samples 17.5 μs for a time reduction of 3:1. The sequence of the samples for a single line is shown in Figure 6.7. Note that space is reserved for transmission of the burst, audio, and data service.

The process is reversed at the receiver. The samples in the segments of each line corresponding to Y', $R'-Y'$, and $B'-Y'$, are stored in a memory bank and read out in real time to generate the luminance and color difference signals simultaneously. They are then matrixed and processed to recover the original E'_R, E'_G, and E'_B.

The resolution of the chrominance signal is reduced both horizontally and vertically by this process—horizontally because the time reduction is twice as great and vertically because only alternate lines are transmitted.

To achieve resolution of the luminance component that is equivalent to that of the original signal, the bandwidth must be increased by an amount that is inversely proportional to the time compression. In the absence of digital compression (see Section 3.12), a bandwidth of 6.1 MHz (4.2 × 1.5) is required to achieve resolution equivalent to that of the original signal. Equal resolution could be achieved in the same bandwidth, but this would require transforming the signal to a digital format and "MAC" would be a misnomer.

6.13.2 Other MAC formats

A number of MAC formats have been developed to meet the needs of various applications. The complete family includes ACLE, A-MAC, B-MAC, C-MAC, D-MAC, HD-MAC, HD-MAC60, MAC-60, and MUSE. The formats for the video portions of the signals are identical in many cases, and the differences are in the audio and data channels.

B-MAC, described above, is the most common. In its analog form, it requires a bandwidth slightly in excess of 6 MHz (as compared with 4.2 MHz for an NTSC signal).

C-MAC is an occasional variation. It requires a bandwidth of 20 MHz, the additional bandwidth being used for data.

HD-MAC has been proposed for high-definition television in Europe, although it now is superceded by the DVB digital system (see Chapter 7). HD-MAC operates at 50 fields per second, with 1250 scan lines interlaced per frame. The aspect ratio is 16:9.

6.14 Reference

1. D. G. Fink, *Television Engineering Handbook*, McGraw-Hill, New York, 1957.

2. EIA/TIA Standard. "Electrical Performance for Television Systems," EIA/TIA-250-C, Washington, D.C., 1990. FCC Rules, *Part 75*. CCIR, *Recommendations and Reports, CCIR, 15th Plenary Assembly*, Vol. XI, Broadcasting Service (Television), ITU, Geneva, 1982.

Chapter 7
Digital Video Systems—DTV

7.1 Introduction

The *digital television* (DTV) standards that are now being implemented around the world are the result of development efforts that began in Europe, Japan, and the United States in the 1980s or even earlier. These projects all had the objective of improving TV picture and sound quality by utilizing new technologies, recognizing that the existing analog television standards for NTSC, PAL, and SECAM, although they have been extremely successful and are still serving well, are based on the technology of the 1950s, which is now quite obsolete. Because the most obvious TV improvement is increased resolution, this work has been named *high-definition television* (HDTV). Although many systems have been proposed, no country other than Japan had adopted a new system until 1996.

Early HDTV system proposals were based on analog technology and generally required some means to provide more bandwidth for each TV channel. This posed serious problems of spectrum allocation for broadcasting and caused reduction of channel capacity for cable distribution. However, more recent developments using digital technology have shown that HDTV signals can be broadcast or distributed in the same 6-MHz bandwidth used for analog TV. In addition, the flexibility offered by DTV allows many new features that greatly increase the potential to expand the uses of TV.

The history of the development of DTV is long and complex, involving worldwide technical, political, and standardizing activities. This chapter does not attempt to give the history in all its detail; that can be found in other references, such as [1, 2]. This chapter does describe the DTV standards developed in the United States and adopted by the FCC there, and the DVB standards developed and adopted in Europe. Other countries are generally considering these two standards or have already adopted one of them.

7.1.1 DTV Development in the United States

Research in HDTV had been going on in the United States for some time, and in 1987 a group of television broadcasters petitioned the FCC to consider setting broadcast standards for an *advanced television* (ATV) broadcast service. The impetus for this was a demonstration in the United States of the Japanese MUSE (*multiple sub-Nyquist encoding*) system. The FCC response was to set up the *Advisory Committee on Advanced Television Service* (ACATS) to consider system proposals. By the end of 1988, ACATS had received 23 different proposals. At the same time, the EIA established the *Advanced Television Test Center* (ATTC) in Alexandria, Virginia, to test HDTV system proposals. Another testing laboratory, the *Advanced Television Evaluation Laboratory* (ATEL), was established in Ottowa by the Canadian government to also participate. However, consideration of 23 different proposals required more than simply testing them (no small task itself), and in March 1990, the FCC announced some basic ground rules for ATV:

1. ATV should be a separate service broadcasting simultaneously with the NTSC service rather than an add-on to NTSC broadcasting that requires an additional channel. Thus, the ATV service would not have to be compatible with the NTSC service, and there would be a transition period of some years where both services would be broadcast. Further, this meant that ATV must fit into a single 6-MHz channel.

2. Some of the system proposals were for *extended-definition TV* (EDTV), which offered increased definition while maintaining some NTSC compatibility. The FCC said it preferred to first consider HDTV proposals that had increased definition not constrained by any need for NTSC compatibility.

As a result of these guidelines, the system proposers went back to their drawing boards. At this time, General Instrument Corp. made the first all-digital HDTV proposal. This was a surprise, but it caused other proposers to consider digital systems, and by 1991 there were four all-digital system proposals. These and two analog/digital proposals were tested by the ATTC and ATEL and in 1993, the ACATS tried to pick a winner.

The only thing the ACATS could agree on was that the system should be all-digital. However, it endorsed an industry effort already underway to merge the four digital proposals into one—the *Grand Alliance* (GA). This was formally announced on May 24, 1993, and it comprised seven members: AT&T, General Instrument Corp., the Massachusetts Institute of Technology (MIT), Philips Electronics North America, the David Sarnoff Research Center, Thomson Consumer Electronics, Inc., and Zenith Electronics Corp. Over the succeeding months, the GA system details were crystallized and in 1994, system construction and testing began.

The GA published a complete report in December, 1994 [3]. The *Advanced Television Systems Committee* (ATSC) then undertook to actually write the standards documents, which were completed in September, 1995 [4]. The standards were officially adopted by the FCC on December 24, 1996. They are now being implemented throughout the United States.

7.1.2 DTV Development in Europe

In parallel with the activity in the United States, a European initiative called the *Digital Video Broadcasting* (DVB) Project was formed to develop DTV standards for anyone in the world who chooses to use them. At this writing, DVB standards are in use for satellite broadcasting around the world and have been adopted for terrestrial broadcasting in many countries.

7.2 DTV System Objectives

It is not sufficient to develop a new TV system based on advanced technology and call it DTV. The new system must provide meaningful improvements in performance and features that broadcasters, program providers, and consumers will perceive as worth the cost and difficulty of a transition to the new system. The full scope of DTV objectives has evolved over time; the following list gives the objectives that are met by the ATSC standard:

1. High definition was the earliest objective for a DTV system. Analog television provides sufficient definition for viewing ratios down to about 4:1, but this limits the size of displays for use in normal-size rooms to about 35" diagonal. Although larger TV displays are being sold and viewed at lower viewing ratios, the limitations of existing analog TV such as scanning lines, interlace flicker, fuzziness, and color edge effects become clearly visible to anyone. The first step of improvement is to use more scanning lines and scan progressively instead of interlacing. Along with this, the horizontal resolution must be proportionately increased to overcome fuzziness.

2. Most viewers appreciate the advantages of a wider aspect ratio as a result of their experience with it in motion pictures, and that has become a second improvement in all DTV proposals. A wider picture relative to height actually does not require further increase in horizontal resolution because resolution is defined relative to the picture height, but for a given resolution, a wider screen uses more pixels horixontally, and that requires an increase in bandwidth and data rate.

3. Sound quality has been shown to add significantly to the TV viewing experience (see Section 5.1). Multichannel sound and high fidelity are important DTV objectives.

4. Everyone has had the experience of watching noisy TV pictures, maybe because of the use of a rabbit-ear antenna or because of distance from the station. With digital transmission, the possibility of robust reception of noise- and distortion-free pictures right out to the fringe area has become a reality.

5. Digital transmission also allows the incorporation of data channels into the DTV bit stream to support user interactive features to view information in addition to and other than the picture and sound.

6. Because DTV will be broadcast simultaneously with NTSC for some years, the control of interference between the two types of broadcasting is an important objective.

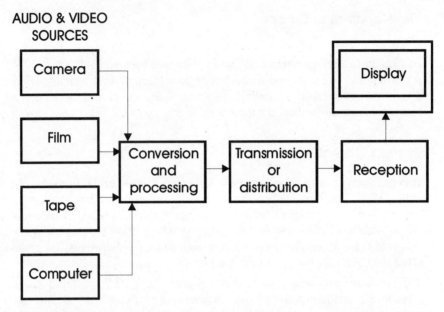

Figure 7.1 Diagram of a video system showing elements that are affected by DTV.

7. Digital video and audio are finding uses in many fields other than television and the interoperability of the DTV signals with these other uses (such as computers, game machines, or video players) must be supported in the standards.
8. The transmission standard must offer the flexibility of either transmitting a single HDTV program, or multiple standard-definition (SDTV) programs. The broadcaster should be able to change the configuration dynamically during the broadcast day.

The resulting DTV standards meet all these objectives, and DTV is an impressively performing system with important new features that are recognizable by any consumer. Meeting the above objectives in a 6-MHz broadcast channel or equivalent distribution bandwidth was a tremendous challenge to the DTV system designers. However, with digital technology, they were able to do it.

7.3 The ATSC DTV System

Figure 7.1 is a diagram of a complete video system showing the elements that must be considered in a DTV system design. The ATSC standard based on the work of the GA defines only the Transmission or Distribution block, which makes it possible for anyone to design receivers. The other blocks are covered by other standards, such as the SMPTE 274M production standard. However, the use of digital technology makes it possible to interface other standards into the system using appropriate conversion hardware or software.

The flexibility of the DTV system is further enhanced by defining the transmission standard in four layers. The layers simply separate the standard into four

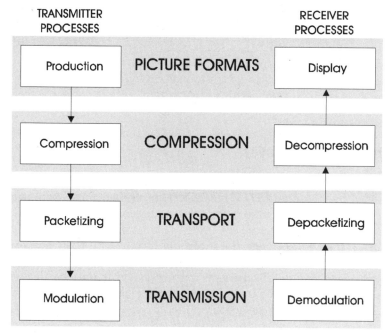

Figure 7.2 Layers of the DTV Transmission standard.

consecutive blocks that have defined interfaces. This not only makes it easier to understand the standard, but it provides means for interoperability with other standards that may have equivalent layers. Figure 7.2 shows the four layers of the ATSC standard.

7.3.1 The ATSC layered architecture

The layers represent the processing steps of the DTV transmission standard. In broadcasting, the layers are traversed top-down, and in a receiver they are processed bottom-up. Thus, the top layer is "Production" for broadcasting and "Display" for receiving. This layer only defines scanning formats that are supported by the transmission standards (there are several formats; see Section 7.4). It does not mean that either the production or display scanning necessarily has to use these formats (in a digital system the conversion between scanning standards is relatively easy), but it does mean that the transmitted signal is defined in terms of one of these formats.

The second layer, Compression, performs audio and video compression in transmission and decompression at the receiver. This layer of the standard defines the video and audio compression techniques that are used; they are based on MPEG-2 for video and Dolby Digital for audio. Details are given in Section 7.5.

The Transport layer during transmission assembles multiple data streams into a single composite stream by means of a packetizing technique (see Section 3.7.1) based on MPEG-2. The transport data stream is demultiplexed in the receiver to recover the individual data streams for video, audio, and auxiliary data. Section 7.6 gives the details of the Transport layer.

In the final layer, Transmission, the transport data stream containing video, audio, and possibly other data is modulated onto a carrier suitable for distribution by terrestrial broadcasting, satellite, or cable. At the receiver, the corresponding process is demodulation, which recovers the packetized data stream. This is covered further in Section 7.7.

7.4 The Picture Formats Layer

The ATSC standard provides for multiple scanning formats, any of which can be decoded by a standard DTV receiver. The receiver must display all the formats, but it may convert them to its own native display format. It is unlikely that receivers will change their actual scanning to match each incoming format they receive. Instead, they will convert the incoming format into their native display format using the frame memory that is required by the video decompression process.

Picture format details are sent to the receiver in data headers that accompany the data. Theoretically, any format numbers could be transmitted by the system, but in practice, receivers are simplified by limiting the number of formats they must accept. Therefore, the GA limited the format choices to variations of only two numbers of active lines: 720 and 1080. For both line numbers, a choice of frame rates is provided and in the case of the 1080-line formats, either interlaced or progressive scanning can be used. The 720-line formats are always progressively scanned. However, in finalizing the standards, the ATSC added another active line choice: 480. This supports SDTV operation and interfacing with computers. Table 7.1 shows the format choices.

7.4.1 Active lines and pixels

The 720-line format always has 1280 active pixels per line, which provides *square pixels* (equal horizontal and vertical resolution) in the 16:9 aspect ratio. Since transmission of blanking intervals is not required in a digital system (all supporting and synchronizing information is contained in headers), numbers for total number of scanning lines or total pixels per line are not relevant for digital systems. Of course, they are important for displays, but that is not part of the ATSC standard. However, typical numbers for total lines and total pixels, such as were used in the GA prototype system, are 750 lines and 1650 pixels.

The 1080-line format provides for nominally 1920 active pixels per line. These were delivered in the GA prototype system by scanning with 1125 total scan lines and 2200 total pixels per line.

Another pixel consideration is whether they are *square*. This means that the horizontal and vertical pixel dimensions are equal, giving equal resolution in each direction. That by itself isn't particularly important because it has been shown that the eye will tolerate unequal resolutions in different directions. However, it is an important consideration to image-processing hardware and software, some-

Table 7.1 Scanning formats allowed by the ATSC standard

Active Lines	Active Pixels/Line	Aspect Ratio	Frame Rate (Hz)*	Scanning Method	Pixel Aspect
480	640	4:3	1,2,4,5,7,8 4,5	progressive interlaced	square square
	704	4:3, 16:9	1,2,4,5,7,8 4,5	progressive interlaced	rect. rect.
720	1280	16:9	1,2,4,5,7,8	progressive	square
1080	1920		1,2,4,5 4,5	progressive interlaced	square square

* Frame rate codes: 1 = 23.98 Hz   4 = 29.97 Hz   7 = 59.94 Hz
                     2 = 24.00 Hz   5 = 30.00 Hz   8 = 60.00 Hz

what simplifying many tasks. Everything else being equal, it is better to have square pixels.

7.4.2 Frame rates and interlacing

Each line format provides for three basic frame rates: 60, 24, or 30 Hz. In the 720-line format, progressive scanning is always used. In the 1080-line format, progressive scanning is used for the 24- and 30-frames/second (f/s) frame rates, but interlaced scanning is used for the 60-Hz vertical-scan option, giving a frame rate of 30 for that choice. This is done because progressive scanning at 60 Hz vertical would make the data rate too high for the transmission channel, or it would be necessary to make unacceptable picture-quality tradeoffs.

The frame rates of 24 and 30 Hz are used when the system input originates on motion picture film at either of those rates. The 60-Hz rate is used when the input is from an HDTV video camera. These numbers are the frame rates implicit in the transmission. For the film rates of 24 or 30, the receiver would scan its display at an appropriately higher rate to avoid flicker.

Because interoperability of HDTV with NTSC signal sources will be important, at least during the transition period, alternative frame rates are also available based on the 59.94-Hz NTSC field rate. This will simplify the display of NTSC-originated material on the HDTV system.

7.4.3 Video sampling rates

Video sampling rates depend on the actual total lines and total pixels used in the originating production equipment. Using the numbers given above for the 720-line format in the GA prototype system, the video sampling rate for the 60 f/s format is:

$750 \times 1600 \times 60 = 74.25$ MHz

Because the 1080-line format uses interlaced scanning (receivers optionally may convert this to progressive scanning for their own displays), the numbers given above for the 1080-line prototype result in the same sampling frequency as the 720-line format:

$$\frac{1125}{2} \times 2200 \times 60 = 74.25 \text{ MHz}$$

Sampling at 74.25 MHz and collecting only the active pixels and lines with 24 bpp gives a raw data rate of:

$$74.25 \times \frac{1280}{1600} \times \frac{720}{750} \times 3 = 171 \text{ MB/s} = 1370 \text{ Mb/s}$$

The Compression layer has to reduce this to a data rate that can be transmitted in a 6 MHz channel!

7.5 The Compression Layer

As will be shown in Section 7.7, state-of-the-art digital transmission systems can handle data rates of approximately 3 bits/Hz of bandwidth with acceptable error rates for video or audio. Thus, a 6-MHz television channel could transmit 18 Mb/s. The objective of the compression layer is to reduce the data rate given above to approximately 18 Mb/s, while also leaving room for audio, headers, and auxiliary data. That calls for a compression factor of about 70.

As explained in Sections 3.10 to 3.12, massive compression of sampled video information is possible because of the redundancy that is inherent in most moving pictures. A combination of techniques is normally used for this and the GA video standard specifies such a combination process that falls within the choices available in the MPEG-2 standard. Compatibility with MPEG-2 draws on the extensive research put into that standard as well as enhancing the possibility of offering the ATSC system as a worldwide HDTV standard.

7.5.1 Compression processing

Using the terminology introduced in Chapter 3, Figure 7.3 shows the video compression processing of the ATSC system.

7.5.1.1 Source-adaptive processing

Since ATSC video encoding can accept various video source formats, the first step must be to conform them all to the same format for further processing. All formats are converted to RGB components at this stage.

7.5.1.2 Color-space processing

The RGB components are digitized (if they are not already digital) and then converted to a luminance and two chrominance component format using a linear

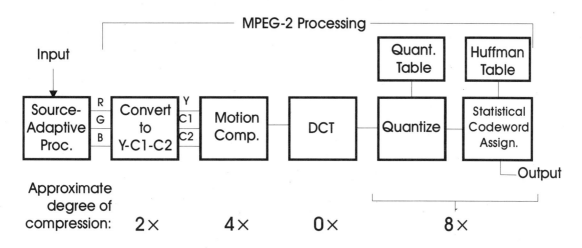

Figure 7.3 Video compression processing.

transformation matrix. The purpose of this is to put the signals into a format where bandwidth can be allocated according to the acuity of the human visual system. Therefore, the resolution of the chrominance components is reduced 2:1 in each direction by filtering and sub-sampling (this is the same as the rec. BT.601-5 4:2:0 format). This accomplishes 2:1 compression.

7.5.1.3 Motion compensation

In most motion video sequences there is considerable redundancy between frames because some parts of the image do not move at all and other parts of the image may be simple translational moves from the previous frame. This is *temporal redundancy* and motion compensation is the technique for dealing with it (see Section 3.12.2.1). In the ATSC system, motion compensation is done by the technique known as *block matching*. The current frame is divided into blocks and each block is compared to a range of adjacent locations in the previous frame. When a near match is found, the block's location in the previous frame (called the translation vector) is stored so that a predicted frame can be built using all the matched blocks moved to their new locations. The predicted frame is then compared to the actual current frame and the difference, which is called the motion-compensation residual, is passed on for further compression.

In most motion sequences, the prediction obtained by block matching is very good and the residual is small and easily compressed. However, this is not always true, especially with high-motion sequences and, of course, always when scene transitions occur. In these cases, the ATSC compression system will give up on motion compensation and the frame(s) will be compressed without dependence on other frames. This is called *intraframe coding* (I-frames). Of course, intraframe coding must be used at the start of a sequence, where there is no previous frame, and it is usually used periodically throughout a sequence to facilitate editing or other actions that require starting decompression in the middle of a sequence.

7.5.1.4 Discrete cosine transform coding

Intra-coded frames or residual frames are compressed using the DCT process followed by adaptive quantization and statistical coding (see Section 3.11 for explanation of these terms). 8 × 8 pixel blocks are used for the DCT processing. Because the same amount of data comes out of the DCT processing as went into it, no real compression is achieved by the DCT process itself, but the data have been transformed into a structure where it is now easy to separate significant from insignificant information as far as the final viewer is concerned. The remaining steps of quantization and statistical coding exploit the new data organization—this is where the real compression of DCT occurs.

Many of the DCT coefficients are typically zero and zigzag ordering as in JPEG (see Section 3.12.1) is used to group most of them together. Run-length encoding is used to send the long runs of zero coefficients that result. The run lengths and the nonzero coefficients are then coded by statistical coding (Huffman—see Section 3.11.7).

7.6 The Transport Layer

The video compression yields a bit stream that contains only the video data along with the header information and table data required to decode it. However, for transmission that stream must be merged with one or more audio streams and possibly an auxiliary data stream. In some applications, there may be more than one stream of each type (such as with multiple-language audio or with multiple programs in one channel). It is the purpose of the Transport layer to merge these streams and add other information needed for efficient, error-free transmission. A packetizing approach is used (see Section 3.7.1).

7.6.1 The transport packet format

The basic packet format is a 188-byte fixed-length block as diagrammed in Figure 7.4. This size was chosen because of several considerations:

1. The packet should be large enough that the overhead due to the transport headers does not represent a significant portion of the channel capacity.
2. The packet size should be consistent with typical block error correction approaches so that packets can be synchronized with error correction blocks. This limits the maximum packet size.
3. The exact number should be chosen to provide several ways to be interoperable with the asynchronous transport mode (ATM) format that is expected to be widely used in other data communication systems.

The packet contains a 4-byte "link" header and a 184-byte data field that may include a variable-length "adaptation" header.

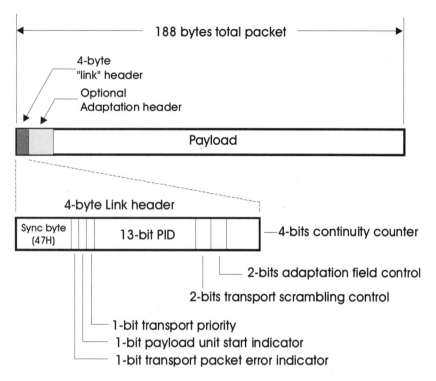

Figure 7.4 The transport packet structure.

7.6.1.1 The link header

The link header supports the following tasks:

1. Packet synchronization—The first byte in every DTV packet is a sync byte that has the hexadecimal value 47H (0100 1111) (see Section 3.2.2).
2. Packet identification—13 bits of the header are assigned as a packet identification (PID) field, which is used for multiplexing and demultiplexing bit streams. This is done by first extracting a packet (PID=0) containing a map of the multiplexing structure that identifies a PID number for each stream in the multiplex structure (see Section 7.6.1.3).
3. Error handling—4 bits are assigned to a continuity counter field. It has a value that cycles uniformly from 0 to 15 at the transmitter for all packets with the same PID value. (Some packets may not have data in them. The continuity counter is not advanced for these packets.) If a discontinuity in the continuity counter value is experienced at the receiver, it normally means that data has been lost and the receiver circuits should take appropriate action.
4. Conditional access—The transport format allows for scrambling of data in the packets to control access at the receiver. This can be done independently for each bit stream in the system. Although the ATSC DTV standard specifies the descrambling approach to be used, it does not specify the descrambling key or

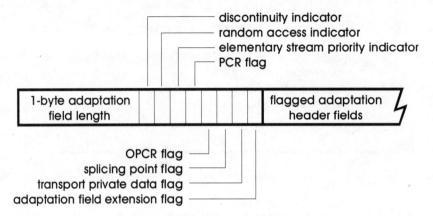

Figure 7.5 The adaptation header of the transport packet structure.

how it is obtained at the decoder.

5. Flags—the rest of the bits in the link header are used as flags for specific purposes as shown in Figure 7.4. For example, a 2-bit field indicates whether the packet contains an adaptation header and/or a data payload.

7.6.1.2 The adaptation header

When indicated by the adaptation field control bits in the link header, an adaptation header follows immediately. If there is an adaptation header, the first 2 bytes are predefined. The first byte gives the total length (bytes) of the adaptation header excluding the first byte. The second byte contains flags for specific purposes as shown in Figure 7.5. Some predetermined uses for the adaptation header are:

1. Synchronization and timing—In the DTV compressed data delivery system, the amount of data delivered for each picture is variable and timing cannot be derived directly from the start of picture data as it could be in a system such as NTSC where the data rates and sizes are precisely known. The digital system has no equivalent to the NTSC sync pulses. Instead, timing information is transmitted in the adaptation headers of selected packets.

    The program clock reference (PCR) flag indicates that this specific adaptation header contains a program clock data field. This is a 6-byte field that contains the value of a time counter running from a 27-MHz clock at the transmitter. The receiver should compare this with its own clock to determine whether the receiver is in sync with the program. Ordinarily, it is used only to adjust the receiver clock rate, but in the case of a program switch, another adaptation header flag (discontinuity indicator) indicates that the receiver counter should be updated to the new value.

2. Random access—A flag in the adaptation header (random access indicator) indicates that this packet is the start of a random access entry point into the bit stream of that packet. Thus, when a receiver is channel switching or searching

for a resynchronization point, it can read packet headers and discard whole packets until it finds the random access indicator flag set. To facilitate rapid channel switching, random access entry points should be included as often as possible.

3. Local program insertion—When it is desired to insert another program into a DTV transport bit stream, such as at a broadcast head end, there are some special considerations. Of course, it must be done at a random access entry point, but that is not enough. The system must also confirm that the insertion will not cause buffer underflow or overflow at the receiver. The encoding system keeps track of the receiver buffer status expected when the bit stream is decoded and identifies appropriate splice-in points. These are identified by setting the splicing point flag in the adaptation header. That flag indicates that this header contains a 1-byte field that gives the number of following packets with the same PID that should be counted until the exact splice-in point occurs. At that point a new program can be switched in. The first packet of the new program must contain an adaption header with PCR flags and fields that will tell the receiver to update its program clock.

4. Other features—additional flags in the adaptation header provide for transmitting "private" (not part of the program data) data to the decoder and for future extensions to the header.

7.6.1.3 Packet multiplexing

As described above, each separate bit stream (audio, video, or auxiliary data) has its own unique PID on all of its packets. The value PID=0 is reserved for packets that tell the receiver how the other PID values are assigned. PID=0 packets contain the *program association table*, which identifies one or more complete programs and what is the PID of packets that contain *program map tables*, which tell the PIDs for each data stream in the program. This is illustrated in Figure 7.6. Special headers are defined for these table packets.

The program tables contain information that will facilitate a receiver selecting a specific program from a multiprogram channel and synchronizing to it in a short time.

There are numerous other features of the Transport layer that are not covered here. The system is highly configurable and expandable for the number of channels or different data types. Characteristics can be changed dynamically, even within one program when necessary. There are also important features for interoperability with other transport systems or other compression systems.

7.7 The Transmission Layer

The packetized bit stream from the transport multiplexer is processed for broadcasting or cable distribution by the Transmission layer. This layer sets up the transport packets in a data frame structure, adds error detection and correction codes, and does vestigial sideband (VSB) modulation onto a carrier for transmis-

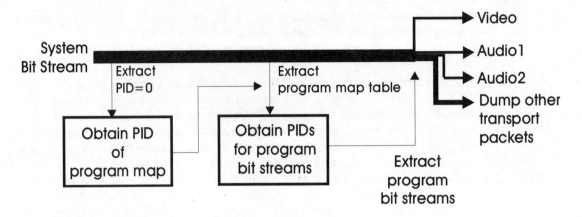

Figure 7.6 Finding a program with the map tables.

sion or distribution. There are two modes of VSB modulation, 8-VSB with trellis coding for terrestrial broadcast and 16-VSB for cable distribution.

7.7.1 The data frame

The transmission layer prepares the data for transmission by converting each packet into a segment by taking the packet contents less the sync byte and processing it for Reed-Solomon (R-S) error protection (see Section 3.8.4). A new sync byte is added to each segment at a later stage in the processing, so the effective segment size as transmitted becomes 208 bytes including 20 R-S parity bytes. This is illustrated in Figure 7.7.

The data are also divided into data fields, which consist of 313 segments. The first segment of each data field contains a Data Field Sync pattern, which has five purposes:

1. indicate the beginning of the data field
2. serve as a training reference signal for the automatic equalizer that corrects for multipath transmission errors in a DTV receiver
3. allow the receiver to decide whether the interference rejection filter should be used
4. used for system diagnostic measurements
5. allow the phase tracker in the receiver to reset its circuitry and determine its loop parameters

7.7.2 Symbols

The VSB modulation is multilevel—for example, 8 amplitude levels for 8-VSB. Thus, each piece of the modulated signal carries more than 1 bit—8-VSB would transmit 3 bits per piece. The multilevel pieces representing multiple bits are

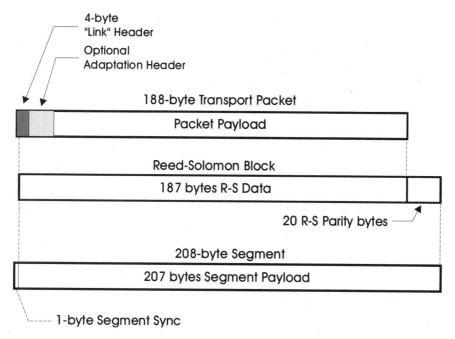

Figure 7.7 From a packet to a segment.

called *symbols*. The GA 8-VSB system actually transmits only 2 bits of data in each symbol. That is because of the trellis-coding that inserts a third bit into the transmission and removes it at the receiver. This is another error-correction strategy. Trellis-coding is explained in Section 3.7.2.

So an 8-VSB symbol actually carries 2 bits of data, meaning that there are 4 symbols per data byte. Therefore, the 208-byte segment described above is modulated in 832 symbols, 4 of which are the segment sync pattern.

The modulation system runs at a fixed symbol rate that is chosen relative to the channel bandwidth. In the ATSC DTV modulation system, the symbol rate is

$$f_S = \frac{4.5}{286} \times 684 = 10.7622 \text{ MHz}$$

The first term (4.5/286) is the NTSC horizontal scan rate in MHz.

The symbol rate must be maintained precisely (±10 ppm) to insure that the spectrum of HDTV transmission relates to NTSC broadcast spectra in a way that minimizes interference between the two services. Considering this symbol rate, the 2 bits per symbol, the 21-byte sync and error protection overhead, and the 1 segment per 313 field sync segment, the net data rate of the 8-VSB channel is:

$$10.7622 \times 2 \times \frac{187}{208} \times \frac{312}{313} = 19.29 \text{ Mb/s}$$

For a 6 MHz channel, that is a transmission speed of 3.2 bits/Hz.

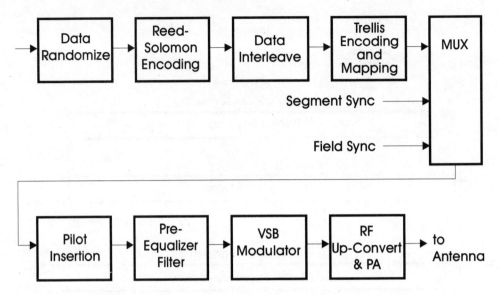

Figure 7.8 Block diagram of a DTV terrestrial broadcast transmitter.

7.7.3 The transmission system

Figure 7.8 is a block diagram of a terrestrial broadcast transmitter. The input to this diagram is the segmented data stream described previously that contains audio, video, and auxiliary data. The features discussed so far are shown.

7.7.3.1 Data randomizing

The first block, Data Randomize, was not discussed above. It ensures that random data is transmitted even when constant data is applied to the input. This might happen, for example, when a data input is disconnected. Keeping the transmitted data random reduces the possibility that this signal could interfere with other services and it also evens out the load on the transmitter. The "randomizing" algorithm is defined and is used at the receiver to derandomize the data. The same pseudo-random bit sequence is simply XORed with the data at both places. (XORing twice restores the original data; see Section 3.5.3.)

The next block in Figure 7.8 (Reed-Solomon Encoding) was covered above.

7.7.3.2 Data interleaving

Further error protection is obtained by interleaving the data after the R-S encoder. This has the effect of spreading the data bytes from the same R-S block over time so that a long burst of noise is necessary to overrun the R-S error protection. The system employs a convolutional byte interleaver that spreads the data over a 52-segment region. Interleaving is performed only on data bytes, not field and segment sync bytes (these are added after the interleaver). Interleaving is syn-

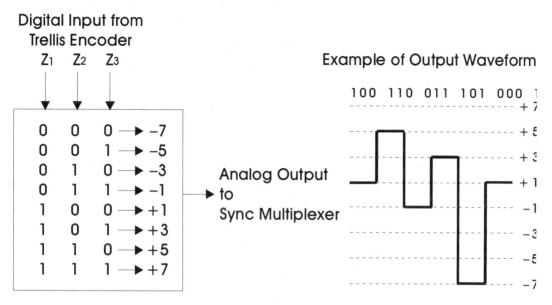

Figure 7.9 Operation of the trellis-code mapper.

chronized to the first data byte of each field. The result is that the system will tolerate an error burst of up to 193 μs.

7.7.3.3 Trellis encoding and mapping

The Trellis Encoding block was covered briefly above, but it should be noted that the trellis coding is also interleaved by having 12 trellis coders that separately process 12 successive symbols. This improves system performance for burst noise. Following the interleaved trellis coders, each 2-bit symbol is represented by 3 bits, which pass to the mapper for conversion to the 8-level analog modulating signal. The operation of the mapper is shown in Figure 7.9.

The insertion of the segment and field sync after encoding is accomplished by multiplexing the sync levels into the 8-level signal from the trellis coder. The sync is transmitted with 1 bit per symbol to improve sync extraction reliability. Modulation levels -5 and $+5$ are used for this.

7.7.3.4 Pilot insertion

The next operation is to provide for inserting a small-amplitude pilot carrier that will assist the receiver in acquiring the signal and maintaining lock in the presence of noise and interference. Since the modulator is balanced to produce a suppressed carrier (zero input produces zero carrier), pilot insertion is accomplished by adding a small DC level to every symbol before the modulator. This level is 1.25, added to the input symbol levels (± 1, ± 3, ± 5, ± 7). The result is a small carrier output—it is 11.3 dB below the data signal power and represents an increase in transmitted power of only 0.3 dB.

7.7.3.5 Pre-equalizer filter

To correct for any in-band ripple or band edge rolloff in the high-power transmitter, an automatic equalizer is used. It responds to a signal sample that is tapped off the transmitter output signal before it goes to the antenna and is sampled by a reference (ideal) demodulator. The reference demodulator contains its own automatic equalizer whose tap weights can be transferred into the transmitter pre-equalizer to provide precorrection of transmitter frequency response distortion.

7.7.3.6 VSB modulation

VSB modulation is amplitude modulation where the lower sideband of the modulation has been suppressed to reduce the required bandwidth by a factor of almost 2. In order to avoid difficult filtering requirements, lower sidebands are not totally removed, but they are rolled off according to a precise (and easily realizable) frequency response curve.

In a typical DTV transmitter, the 10.76 Msymbols/s 8-level trellis-coded composite data signal is amplitude-modulated on two quadrature IF carriers. These carriers are combined to create the vestigial sideband IF signal by sideband cancellation, a technique that is widely used in other forms of single-sideband transmission. For minimal intersymbol interference, the data signal must be properly filtered before transmission in a 6-MHz channel. The proper system (transmitter plus receiver) response is a linear phase raised-cosine Nyquist function, which gives the frequency response shown in Figure 7.10. This system response is achieved in the system by identical filtering at both the transmitter and receiver, so each is actually a root raised-cosine Nyquist filter.

7.7.3.7 Up-conversion

The VSB modulation is on an IF carrier that is the same frequency regardless of the actual transmission channel frequency. In the GA prototype, the IF frequency corresponding to the VSB carrier is 46.69 MHz. The modulated IF signal is up-converted to the final channel frequency and amplified to the power level required by the antenna. The amplifiers, of course, must be linear to preserve the signal properties. Further considerations of DTV transmitters are given in Section 16.16.

7.7.4 Transmission summary

The parameters of the ATSC transmission system are summarized in Table 7.2. The table shows the features for both the 8-VSB with trellis coding format that was described above and the 16-VSB format for cable transmission. The latter format doubles the data rate per channel, which is possible because cable systems can achieve better SNRs and better control of signal levels than broadcast transmission. This means that two HDTV programs could be transmitted simultaneously on one cable channel. The two programs would be multiplexed at the Transport level.

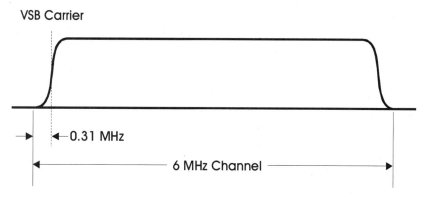

Figure 7.10 DTV transmitter frequency response.

7.8 ATSC HDTV Audio

Sound is just as important as the picture in conveying the message of a video segment. This has not been overlooked in the ATSC system; a state-of-the-art digital audio system is part of the specification. Five-channel surround sound is provided using the Dolby Digital compression system (see Section 5.5.2.2). This system is similar to the one used in the latest motion picture theaters and produces spectacular sound.

Table 7.2 ATSC Transmission system parameters

Parameter	Terrestrial Mode	High Data Rate Cable Mode
Channel Bandwidth	6 MHz	6 MHz
Excess Bandwidth	11.5%	11.5%
Symbol Rate	10.76 MS/s	10.76 MS/s
Bits/Symbol	3	4
Trellis FEC	2/3	None
Reed-Solomon FEC	T=10 (207,187)	T=10 (207,187)
Segment Length	832 Symbols	832 Symbols
Segment Sync	4 Symbols/Segment	4 Symbols/Segment
Frame Sync	1 per 313 Segments	1 per 313 Segments
Payload Data Rate	19.29 Mb/s	38.58 Mb/s
NTSC Cochannel rejection	NTSC rejection filter in receiver	N/A
Pilot Power Added	0.3 dB	0.3 dB
C/N Threshold	14.9 dB	28.3 dB

Numerous options are available and the system is designed to give the best performance possible with whatever audio hardware is available at the receiver. The audio bit stream, which can use up to 384 kb/s of the digital channel capacity for a single audio program, is multiplexed with the video bit stream in the Transport layer using packet interleaving. Additional features allow the system to match the sound level between different programs and to allow end user control of various parameters with the implementation of receiver options. Multiple audio programs, for example, different languages, can be provided by slightly increasing the video compression to make room for more audio data.

7.9 Digital Video Broadcasting Standards

The European DVB standards were being developed concurrently with the activities of the Grand Alliance and the ATSC in the United States. The groups communicated and there are many similarities between the standards; for example, they both use MPEG-2 as the basis for their video compression. However, for good reasons, the transmission portion of the two standards differ; these are described here. For a more complete exposition of DVB, see [2].

7.9.1 Objectives of DVB

During the period of DTV development, Europe has been undergoing the transition to the European Union. Among other things, this has placed European telecommunications entities in a period of transition from government ownership to privatization.

In September 1993, the *European Digital Video Broadcasting Project* (DVB Project) was formed for the purpose of developing specifications for DTV in Europe. Actual standards would be completed by the *European Telecommunications Standards Institute* (ETSI) and published for use anywhere in the world (not just Europe). Some of the objectives of DVB were:

1. To develop specifications for all forms of digital television program distribution (satellite, cable, and terrestrial broadcasting) that will meet the market needs of the European community.
2. Digital broadcasting would utilize existing telecommunications channels and circuits.
3. A primary objective is to direct the data compression capabilities of DTV to achieve more SDTV channels rather than a massive improvement in performance such as intended by HDTV.
4. Performance improvement will not be neglected, however, and the specifications will support the maximum resolution possible within the scanning standards chosen, 16:9 aspect ratio, and multichannel high-quality audio. Further, the error-correction capability of digital transmission will improve the quality and reliability of TV reception, especially with terrestrial broadcasting.
5. Extensive support will be provided for conditional access.

It can be seen that these objectives are somewhat broader than the objectives placed before the GA in the United States, because they address all means of video and audio transport and a more world-based set of market objectives. The most significant difference, however, is that HDTV is downplayed and is left for future addition to the specifications. The first directive meeting these objectives was issued by the DVB Project in September, 1995.

7.9.2 DVB Standards

DVB standards for satellite, cable, and terrestrial broadcasting are discussed in the following sections.

7.9.2.1 Video and Audio Transport Streams

DVB uses MPEG-2 for both video and audio compression. Both SDTV and HDTV formats have been defined. Some of the choices are

Active Pixels (h × v)	Aspect Ratio	P/I
SDTV		
544 × 576	4:3, 16:9	P or I
720 × 576	4:3, 16:9	P or I
352 × 288	4:3, 16:9	P or I
HDTV		
1280 × 720	16:9	P
1920 × 1080	16:9	P or I

DVB broadcast, cable, or satellite channels are capable of carrying multiple SDTV programs; the maximum channel data rates can also support HDTV, so this can easily be implemented in the future as soon as receivers are available in the market.

7.9.2.2 Satellite Broadcasting

The first DVB implementation was for direct-to-home satellite broadcasting in the United States, which was quickly followed by systems elsewhere in the world. This standard is called DVB-S and it uses a single satellite carrier with *quaternary phase-shift keying* (QPSK) modulation. A typical satellite channel can carry up to 18 SDTV programs simultaneously. The DVB satellite transmission system is described in ETSI document EN 300 421 [6].

7.9.2.3 Cable Distribution

The cable inplementation (DVB-C) is based on DVB-S except that it uses *quadrature amplitude modulation* (QAM) instead of QPSK. An 8-MHz cable channel can

Table 7.3 Characteristics of COFDM, as used in DVB terrestrial broadcasting

Parameter	2k mode	8k mode
Maximum carriers	1,705	6,817
Active carriers	1,512	6,048
Pilot carriers	176	701
TPS carriers	17	68
Carrier spacing (Hz)	4,464	1,116
Symbol duration (μs)	224	896
<u>Simulated performance example*</u>		
Guard interval (μs)	56	28
Data bit rate (Mb/s)	26.1	31.7
Max. transmitter spacing (km)	16.8	8.4
Required CNR (dB)	27.9	27.9

** 7/8 code rate, 64-QAM modulation, Rayleigh channel

carry up to 38.5 Mb/s data rates without interference between adjacent channels. This standard is described in ETSI document EN 300 429.

7.9.2.4 Terrestrial Broadcasting

DVB terrestrial broadcasing is defined according to ETSI Document EN 300 744. Because the European terrestrial broadcast environment differs from the United States in that there are many more transmitters covering smaller areas, DVB chose a different method of modulation that is more suited to this environment. It is *coded orthoginal frequency division multiplexing* (COFDM) [7]. This is a multicarrier technique involving a multiplicity of carriers that are generated in a way that they do not interfere with each other (orthogonal). Two systems are specified, the *2k* system has nominally 2,000 carriers, and an *8k* system that has 8,000 carriers.

The carriers are precisely spaced across the entire channel bandwidth and the modulating data are distributed to the carriers such that each carrier conveys a very narrow-band signal—high total data rates are achieved by combining the data from all the carriers at the receiver. Fortunately, the process of modulating and demodulating such a multiplicity of carriers is the same as the *fast Fourier transform* (FFT) process, which can be performed readily in digital ICs.

Since the bandwidth per carrier is low, transmitted symbol lengths are long for each carrier. This means that *multipath interference*, which causes a trail of interference following the transmitted symbols by an amount equal to the maximum path delay difference, can be totally rejected by leaving a *guard interval* longer than the maximum interference delay between the symbols. Since the guard interval reduces the channel capacity, DVB-T provides for a choice of guard intervals depending on each particular broadcasting environment. A choice of modulation methods is also provided: QPSK, 16-QAM, and 64-QAM. The more advanced modulations provide higher data rates but also require higher CNR (carrier-to-noise ratio) at the receiver.

Actually, not all the nominal carriers are used, as shown in Table 7.3, and some carriers are reserved for synchronization (pilot carriers) and transmission parameter signaling (TPS). The example shown in Table 7.3 shows different choices of mode and guard interval, and their effect on data rate, CNR, and maximum transmitter spacing.

7.9.2.5 Single-frequency networks

The ability of COFDM to reject multipath interference up to a certain level, makes possible the concept of the *single-frequency network* (SFN). This is useful in much of the European terrain, where mountains limit the range of individual transmitters. With analog broadcasting, it was necessary to use different frequencies to cover a large area, such as a country, because of problems of interference between same-frequency transmitters in the fringe areas. With SFN, a group of synchronized transmitters on the same frequency can cover a large area without concern for interference as long as the transmitter spacing is kept below the value corresponding to the guard interval chosen. This can eventually make more channels available for use on different programs.

7.10 Summary

This chapter presented DTV systems at a level that exposes most of their features, but it did not go to the full details of hardware and software. Even so, the system appear to be extremely complex compared to their NTSC or PAL predecessors. However, one must remember that actual system implementation uses custom integrated circuits (ICs) and most of the complexity disappears into the ICs. The real details will not be an issue to anyone except the IC designers. Program providers, broadcasters, and cable operators will operate at a level more like this chapter. Consumers will only need to know a small subset of this information to purchase and operate receivers.

DTV systems deliver high definition pictures with high-quality multichannel sound with full digital robustness and does it in the same channel bandwidths that are used for NTSC or PAL broadcasting. It is a superb example of the type of advances that will occur in many other fields as the transition to digital technology takes place.

7.11 References

1. J. Whitaker, *DTV: The Revolution in Electronic Imaging*, McGraw-Hill, New York, 1998, Ch. 1.
2. R. de Bruin and J. Smits, *Digital Video Broadcasting: Technology, Standards, and Regulation*, Artech House, London, 1999.
3. *Grand Alliance HDTV System Specification*, Version 2.0, Grand Alliance, December 1994.
4. Document A/53, "ATSC Digital Television Standard," Advanced Television Systems Committee, Washington, D.C., September 16, 1995. Available for download from http://www.atsc.org.
5. "HDTV and the New Digital Television," *IEEE Spectrum*, April 1995, pp 35-80.
6. ETSI documents are available to download for personal use only at http://www.etsi.fr/download.
7. J. H. Stott, "Explaining some of the magic of COFDM," *Proceedings of 20th International Television Symposium*, Montreux, Switzerland, 13–17 June, 1997. (This paper is available on the BBC Web site: http://www.bbc.co.uk.)

Chapter

8

Digital Video Systems—Computers

8.1 Introduction

Capture, processing, and display of video have become important applications of personal computers. PCs are ideal low-cost platforms on which to implement the features of digital video. This is a recent development, brought about by the combination of several trends:

1. The processing power of PCs has grown approximately 2:1 every 2 years for the past 25 years.
2. Computer video display technology has advanced in resolution and color capability.
3. Mass storage technology continues to improve performance, increase in capacity, and reduce in cost.
4. Digital video compression technology is maturing.

Each of these factors is a story in itself; all are covered in this chapter.

8.2 The Personal Computer

PCs are a development of the late 1970s, when it first became feasible to assemble a microprocessor, digital storage devices, a keyboard, and a video display at a size that would fit on a desktop and at an affordable price. Suddenly, anyone could have computing power that was capable of doing useful tasks right on his or her desktop. Although these early PCs were limited in their processing speed and storage capabilities, they had enough power to attract millions of customers who struggled with all kinds of difficulties to explore the opportunity of having their own computers. Software applications for word processing, spreadsheets, games, and other uses were quickly developed.

However, the remarkable advances of solid-state integrated circuit (IC) tech-

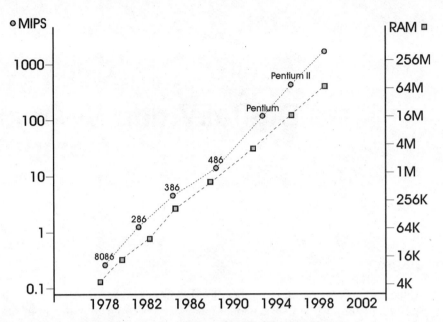

Figure 8.1 Growth of microprocessor speed and RAM chip capacity.

nology have fueled constant growth of the powers of PCs, which still goes on today. As PC power advanced, so did the capabilities of the software, which led to new and more important applications. PC users have become accustomed to the idea that the PC you buy today will be available at a substantially lower price in 6 months and it will be obsolete in 2 years! That is a manifestation of progress in a highly competitive and volatile market.

It is difficult to appreciate the magnitude of progress in ICs that has occurred at such rates and over such a long period of time. Figure 8.1 shows data for two measures of that progress: microprocessor speed (see Section 8.2.1.2) and memory chip capacity—four orders of magnitude over 20 years! Nothing else in the history of industry has shown such growth. The result has been continued advances in circuit power combined with reduction of cost. A $1000 PC today has more power than a multimillion dollar computer of 1970. And the curves are expected to continue going up for many more years.

The same IC features that are utilized in PCs also are driving the conversion of most video equipment to digital technology. Steady cost reduction has brought digital hardware to lower cost than its analog equivalent in most cases.

8.2.1 PC architecture

The purpose of a television receiver is to display the best possible picture from the signal received. The purpose of a personal computer is different—it is to provide the best possible platform for running the software it receives. The PC's display functionality is just one of the ways a PC communicates with its user. For this

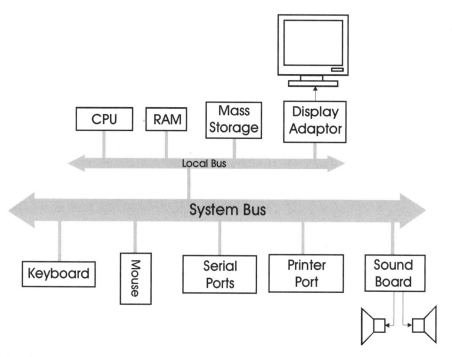

Figure 8.2 Block diagram of a typical personal computer.

reason, early PCs were successful even though their displays were limited to simple monochrome text and they had no ability to output sound beyond beeping. But the architecture of the PC was designed to allow upgrading of video and audio capability when it was necessary and when it made economic sense.

Figure 8.2 is a block diagram of a basic PC. It shows that a PC consists of a main microprocessor (the *central processing unit* or CPU), several types of data storage or memory, and a variety of *input/output* (I/O) devices such as a keyboard, mouse, video display, and sound unit. All are interconnected so that digital data can move freely between them. The CPU is the brain of the system; it performs system control and handles most of the processing functions of the computer.

8.2.1.1 The PC bus

The backbone of a PC is its *bus*, which is a multiwire circuit that makes a parallel connection of all the components of the PC. Having everything in parallel would appear to be awkward until one realizes that each device on the bus is designed to respond only to signals that are designated for it. There are several ways of accomplishing that, but the basic idea of them all is to provide separate sets of wires in the bus for data and address information.

For example, a 32-bit bus has 32 parallel wires for data and another 32 parallel wires for an address. Each device is assigned its own address, and the device watches the address lines and ignores all data unless it first sees its own address. Then it interrogates the data lines to receive data. This type of interface is called memory-mapped, because it is the same access technique used by the computer's main memory.

Important parameters of a PC are its *bus width* and *bus speed*. The width refers to how many bits of parallel data are provided for, and the speed parameter specifies how many cycles of bus activity occur per second. Early PCs had 8-bit data and 16-bit addressing on their buses, and ran at speeds of 1 or 2 MHz. These numbers were consistent with the capabilities of the early microprocessors and other components, but everything in solid-state electronics has progressed since then. Today, 32- and 64-bit buses are common, and bus speeds run from 60 MHz up.

Since (at least) one bus cycle is required for each piece of data equal to the bus width, the bus parameters determine the maximum rate at which data can move through the computer. As the need to move more data has grown, especially to support video display activities, various strategies have been developed to get around the main bus limitation. One of these is the local bus approach, where a separate short bus is used to interconnect only those devices that need the fastest data access. Local buses are wider and faster than the main bus and they are often used to interconnect the CPU, the display adaptor, main memory, or mass storage.

8.2.1.2 The CPU

A CPU is an integrated circuit chip that performs digital operations such as logic, arithmetic, data movement, and I/O. It is controlled by a program that consists of a series of *instructions*. The instructions are digital codes that tell the CPU what kind of operation to do at each step of the program. They are just like other digital data except that they have special meaning to a microprocessor. The program is stored in memory and is read over the bus by the CPU for execution. As a program runs, the CPU may access the bus for other data from memory, mass storage, or I/O.

There are several families of CPUs on the market; they differ in architectural details and in the syntax of their instruction words. Thus, different CPU families require different software. A manufacturer of a CPU family tries to keep new designs in the family *backward compatible* with earlier designs so that existing software will still run on new units. This is important because user's investment in software is often greater than their investment in hardware and they do not want software obsoleted when they upgrade to new hardware. Some current CPU families are the Intel x86/Pentium family (used in IBM-compatible PCs), the Motorola 680x0 family (used in early Apple Macintosh computers), and several *reduced instruction-set computer* (RISC) processors such as the PowerPC, Alpha, R4000, etc.

RISC processors are designed with a simplified instruction set that executes faster per instruction. However, RISC instructions individually are less powerful so it may take more instructions to do a particular task than on a CISC (*complex instruction-set computer*) processor. The end result depends on many factors, but RISC machines are at the high end of today's PC performance curve and CISC processors are even incorporating some RISC features in their new designs (without abandoning compatibility, of course).

The previous discussion indicates that speed is probably the most important parameter of a CPU, but that it may be difficult to compare the speed between different CPU families. However, within a family there are two numbers that are significant: the *clock frequency* and the speed in millions of instructions per second (MIPS).

Clock frequency defines the internal cycle rate of the microprocessor. Depending on the CPU design, instructions may execute in one or more clock cycles. Some CPUs even execute several instructions at once. The resulting instruction execution rate (MIPS) is therefore a complex function of clock frequency, instruction set, and architecture. Today, clock frequencies are in the 400 to 500-MHz range and MIPS ratings are pushing beyond 1,000 MIPS.

8.2.1.3 Memory

Programs being executed and the data they use are stored in the system's *main memory* that can be addressed by the CPU. Main memory is usually solid-state random-access memory, called RAM. RAM is volatile storage—it loses its data when power is removed from the system. Present-day personal computers have 32 MB[1] or more of RAM—the more the better—because when RAM gets filled up, the system must swap data or programs into mass storage, which is much slower. In the fastest systems, the speed of main memory is usually slower than the CPU, so systems may include modest amounts of faster memory, called *cache memory*, to hold the most frequently used data. The latest microprocessor designs also have some on-chip cache memory, which is even faster than off-chip cache.

Most computers have a little built-in software called the *basic I/O system* (BIOS) that provides the first level of interface to the installed hardware. The BIOS is stored in *read-only memory* (ROM), which is nonvolatile solid-state memory that has the data permanently burned into it. A variation of this is called the *flash BIOS*, which has *rewritable* nonvolatile memory. By exercising special procedures, a flash BIOS can be updated with new software.

8.2.1.4 Mass storage

Hard disk, floppy disk, or magnetic tape drives; CD-ROM, and other devices are *mass storage* units that can hold great amounts of data and programs permanently (nonvolatile). Mass storage may be 1,000 times or more slower than RAM or cache, but its capacity and cost per megabyte is much lower. It is essential as a medium of permanent storage. Mass storage cannot be addressed directly by the CPU; data or programs must first be loaded from mass storage to RAM for use by

---

1. In 1995, when we wrote the second edition of this book, a typical PC had 4 MB of RAM, a 100-MB hard disk, and ran at 100 MHz. That's progress!

the CPU. Normally, loading occurs before program execution begins, but in the case of motion video, the data is so large that it cannot all fit into RAM. In that case, data must constantly flow from mass storage to RAM as the video is played. This constant flow of data is called *streaming*.

PCs usually have both nonremovable and removable mass storage. The most common nonremovable storage device is the magnetic hard disk (see Section 9.17), which has a storage capacity from 100 MB to many thousand megabytes; 1,000 MB is a gigabyte (GB). Current desktop PCs have hard disks of 4 GB or greater. Removable storage is used for backup or movement of data between systems. The most common removable mass storage is the *floppy disk*, which has a capacity in the range of 1 or 2 MB. A high-capacity floppy disk, the *Zip disk*, is becoming very popular; its capacity is 100 MB. Other floppy disk products are also coming on the market. For larger removable storage, magnetic tape has been used; it has capacities in the same range as hard disks, but it is much slower for access.

The *CD-ROM* optical disc (see Section 9.18), which is an outgrowth of the audio CD, is a very important medium for audio and video storage. A single CD-ROM disc has a capacity of 650 MB or more and it can be replicated in quantity for less than $1 per unit. However, the process of recording (called mastering) is expensive and the replication of copies takes special facilities. But this is the premier medium for distribution of programs. Thousands of different CD-ROMs are on the market and more are being produced every day.

The optical technology behind the audio CD and the CD-ROM is suitable primarily for program distribution because it is *read-only*. Once a CD has been replicated, its data cannot be changed. This is actually an advantage for distribution use because the distributer is assured that the data cannot be modified by anyone.

CD-recordable (CD-R) drives are also available for PCs. These are somewhat more expensive than hard drives or floppy-disk drives, but the large data capacity and low cost of the medium makes them a good value for backup and small-scale distribution. CD-R blank media now cost less than $2.00 per disc.

A *rewritable* version of the CD (CD-RW) is also available. This is useful when the content of a CD must be updated. However, CD-RW blank medium costs 10 or more times the cost of a CD-R blank. Although the CD-RW capability could be a replacement for a hard disk, the slowness of CD technology so far limits this application (see Section 10.20.1).

Although 650 MB sounds like a lot of data, it represents less than 1 hour of low-quality video in digital compressed form. With MPEG-2 high-quality video or digital HDTV, there is demand for much higher CD-ROM data capacity. This is being addressed by the new DVD-ROM format (see Sections 8.6 and 10.20.2).

8.2.1.5 Input/output

For a computer to do useful work, it must communicate with its user, with printers, and often with other computers via a network. This is done via *input/output* (I/O) hardware interfaces that are connected to the bus and addressed the same way as other bus devices. Video and audio output devices are in this class; these functions have their own architecture that is discussed in detail in Chapter 12.

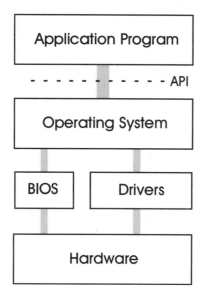

Figure 8.3 Levels of software.

8.3 Software

A PC is useless without software. Software consists of a list of microprocessor instructions that tell the CPU what to do. Although all PCs have some built-in software (the BIOS), that doesn't provide any useful functionality by itself. *Operating system* (OS) software must be installed and executed to provide even the simplest user interface and application software must be added before the computer will actually do any work. These levels of software are shown in Figure 8.3 and are discussed in this section.

8.3.1 The BIOS

The permanent hardware units in a PC (keyboard, mouse, video display, mass storage, printer, etc.) are physically connected to the bus, but that doesn't tell the CPU how to access them. That is the purpose of the BIOS software, which knows the details of the hardware and provides a generic software interface to it. Different PCs may have different designs for their permanent hardware but each PC contains its own customized BIOS to make the different hardware look the same to the next level of software, the OS.

8.3.2 The operating system

The BIOS provides a low-level interface to the built-in hardware of the PC. For example, it provides a raw means to read and write data from and to mass storage devices based on addressing physical locations on the medium. However, much housekeeping is needed to manage mass storage through physical locations and most programs would not want to deal with such detail. One feature of OS software is a *file system*, which allows applications to access mass storage by means of named files and directories. It is the purpose of the OS to take each of the hardware capabilities and create a standardized interface for application programs to use when accessing the hardware. This is called an *application programming interface* (API). All programs are designed to run with a particular OS (this is often stated as running *under* the OS) and the program depends on the existence of that OS's API.

An operating system also provides the first level of user interface, which defines the way a user interacts with the computer. One kind of user interface is text-based (called a *command-line interface*)—the user types commands at a keyboard to control the computer. The video display in a command-line interface shows only text. Another interface is graphical (called a *graphical user interface*—GUI). With a GUI, the user sees pictures of objects (called *icons*) on the screen and he/she interacts with these by use of a *pointing device* such as a mouse, touch screen, or trackball. Of course, when text input is required in a GUI, the user still operates the keyboard. GUIs are discussed further in Chapter 14.

For example, the DOS operating system for IBM-compatible PCs provides a command-line interface. By installing the Windows software package on DOS, a GUI is added. (The Windows 95 or Windows NT versions do not require DOS.) The IBM OS/2 operating system and the Apple Macintosh OS both provide GUIs as their basic interface. In the case of OS/2, a command-line interface is available as an option, but the Macintosh OS does not have this option.

8.3.3 Drivers

The first-level interface to hardware provided by the BIOS only applies to the hardware capabilities installed in the basic computer. Many other types of hardware devices could be added to add other capabilities, or standard units may be replaced by different devices. In either case, the BIOS cannot know the details of the new or additional hardware. This is taken care of by an OS feature called *installable drivers*. Drivers are BIOS-level software specific to a device. The OS provides means to add such software so that the API can include the new or different devices. As shown in Figure 8.3, a driver can (if necessary) bypass the BIOS and go directly to its own hardware.

8.3.4 Application software

Specific computer tasks such as word processing, calculating, or drawing are accomplished with application programs. The operating system includes a launcher that will load and start applications. The user only needs to select the program and tell the OS to launch it. Each application usually has its own user interface, built in the style of the OS.

Application programs are created by programmers who write the program in languages such as C, Pascal, or Basic, which are designed specifically for application programming. The language provides a syntax that describes the application's structure and processes in a format readable by someone who knows the language. However, these languages are not understandable by microprocessors, which require the much lower-level and detailed syntax defined by their instruction set. The translation can be accomplished in two different ways.

Interpretation is a translation from programming language to machine instructions that occurs while the program is running. An interpreter program reads the programming language and generates instructions that are immediately executed by the microprocessor. This is a very convenient mode of operation, but it is inefficient because both the interpreter and the application must share the microprocessor's cycles. The result is that interpreted programs can be slow. For simple applications and fast processors, this is not a problem. However, complex programs, such as image processing or compression/decompression, will be noticeably slowed.

The other type of program creation is *compilation*. In this case, a *compiler* program is used during the programming process to convert the programming language code to a separate executable module that can then run by itself. This produces the fastest programs and most major applications are created this way.

8.3.4.1 Multitasking

There are situations where it is necessary to run more than one program at a time. Some operating systems support *multitasking*, which allows multiple programs to run "concurrently." Of course, if there is only one microprocessor available, only one thing can actually be happening at any instant, but computers are so fast that they can switch between two or more programs (tasks) so rapidly that the user thinks everything is actually concurrent.

A good example of the use of multitasking occurs in an application that requires concurrent video, audio, and other computer functions. A single program might be designed to handle these operations concurrently, but it is much better to have the operating system handle it for the program. If separate programs are written for the purposes, the OS can run all of them and manage the prioritization between them to optimally utilize the computer's resources and keep everything running smoothly. OS/2 and Windows 98/NT are operating systems that support this. They also support another mode, called *multithreading*, which allows a single application to set up its own internal multiple tasks (threads) that the OS will manage concurrently.

8.4 Computer Video Formats

Video in a computer can be packaged as individual still images or as motion sequences.

Table 8.1 Bits per pixel vs. number of colors

Bits/pixel	Number of colors
4	16
8	256
15	32,768
16	65,536
24	16,777,216

8.4.1 Data files

A computer handles its data objects in the form of *files*, which are blocks of sequential data that serve a specific purpose and are stored together. For example, one file might contain a single still image. Such a file includes a header block that identifies the file contents and its format. The header is followed by the image data that lists all the pixel values for the image. If the image file is compressed, the image data does not represent the pixels directly, but rather it is the input to a decompression processor that will generate the actual pixel values.

Files can be stored in a computer's mass storage or they can be transmitted between computers over digital networks or digital communication links. Because most computers include mass storage (some computers on networks don't have their own storage; they rely on the network for it), transmission of data can occur at any time before the file is to be displayed. However, this assumes that the receiving computer has enough storage to accept all the data. With motion video, especially, the data may be too large to store locally. In that case, the system has to display the data as it is received.

8.4.2 Still image formats

The data for uncompressed still images consists of the values for all the pixels in the image. That sounds simple, but there are actually many ways to deliver the information. It becomes even more complicated when compression is included. As a result, there are more than 100 still image formats in use in PCs.

8.4.2.1 Color lookup tables

The bits per pixel parameter (see Section 1.7.2) specifies how many bits are used to carry each pixel value. That has a direct relationship to how many different colors are available to display the image as shown in Table 8.1. When the number of colors is large, as in 15-, 16-, or 24-bpp, the pixel's bits can directly define the color, but for fewer colors it is advantageous to use a *color lookup table* (CLUT). In this approach, the pixel value is an index into a data table that holds the actual color value. The color values in the table can have more bits than the index does, so colors can be selected from a larger palette. For example, 8-bpp images can only use 256 colors at a time, but if the CLUT has 18 bits in each color value position, those 256 colors can each be selected from a palette of 262,144 colors.

When an 8-bpp image is captured, its 256 colors are chosen for the best match to the actual colors in the source image. There are a number of algorithms for that process and the results can be suprisingly good. The most difficult images to reproduce this way are ones that have a lot of smoothly shaded areas such as faces or desert landscapes. These may show contouring, which is the artifact that results from having too few colors to reproduce the shading.

8.4.2.2 Color dithering

A further improvement in CLUT images is available with the use of *color dithering*. This is a process that depends on the viewer not being able to see individual pixels. Adjacent pixels are deliberately selected to have different values that the eye will combine to create the desired color. In the case of smooth shading, dithering will break up the color boundary lines so that contouring is not so easily seen. Dithering is most effective when the viewing conditions are such that the viewer cannot see individual pixels. When individual pixels are visible, the viewer will see the dithering and that spoils the effect.

8.4.2.3 Still image capture

Still images can be captured from live scenes, motion video, or hard copy. Still-image capture from live scenes requires a digital still-picture camera, which delivers a digital image directly to a computer, or a motion video camcorder and a digitizing board in the PC. Hard copy capture can also be done the same way, but a better method is to use an *image scanner*, which is an electromechanical device that optically scans images directly into digital formats and can deliver higher resolution and gray scale rendition than most video cameras.

A block diagram for a video digitizing board that connects to a PC's system bus is shown in Figure 8.4. Such boards often provide both composite and component inputs as shown in the figure. Composite inputs are decoded to RGB and each component is separately digitized. In order to deal with capturing a single frame from live video, the board has its own buffer memory to temporarily hold the captured image. This is necessary because single-frame capture cannot be interrupted once it is started and the PC bus cannot be guaranteed to be free of interruptions.

The user will usually want to view the captured image before it is saved to hard disk, which means that the image in the digitizer's frame memory must be displayed. This is easily accomplished when the digitizing functionality is built on the same board as a display adaptor. In that case, the temporary memory can be the display memory. If digitizing is on a separate board, the temporary image must be moved over the bus to the display adaptor for viewing.

When a captured image is to be saved on the computer's hard disk, it is sent

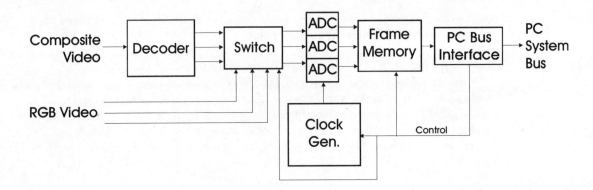

Figure 8.4 Block diagram of PC video digitizing board.

over the PC bus to mass storage along with the appropriate header information for the chosen format.

The sampling clock for the ADCs must be synchronized with the analog video input. Ordinarily, the sampling frequency is chosen so that the proper number of horizontal pixels are generated during the analog video active scanning interval as shown by Eq. (8.1):

$$f_S = \frac{f_H R_H}{C_H} \tag{8.1}$$

where R_H is the desired horizontal resolution (pixels), f_H is the analog horizontal scanning frequency, and C_H is the active horizontal scan fraction. For example, for 640 × 480 resolution and NTSC video, the sampling frequency is:

$$f_S = \frac{15{,}734 \times 640}{0.84} = 11.988 \text{ MHz}$$

Vertical pixel choices are limited by the number of active lines in the incoming video. For interlaced scanning, either one or two fields can be captured. If two fields are captured to get a full frame, there can be artifacts when the source image is moving. Notice also that a moving source image will be blurred because of the electrooptical integration that occurs in the camera imager (see Chapter 9).

8.4.2.4 Pixel formats

CLUT pixels are pure binary numbers that index into the lookup table, but direct-color pixels must be formatted to define the encoding of the RGB values. This is usually done by allocating groups of bits for each of the three colors. For example, 24-bpp pixels use 8 bits for each color and each one is a pure binary number that specifies the intensity of that color. 16-bpp pixels cannot use the same number of bits for each color (16 does not divide evenly by 3), so green is usually given an extra bit because green is the visually brightest color. This gives a bit format of 6-5-5 for G, R, and B, respectively.

Some systems use 15-bit pixels (5-5-5) and the 16th bit is defined as a flag that goes along with each pixel. (Since all computer hardware is oriented in multiples of bytes, it would be too awkward to handle actual 15-bit pixels.) The extra bit in each pixel can define (for example) how that pixel would be combined with the corresponding pixel of another image when two images are mixed. This idea is extended in some systems that actually use 32-bit pixels, which allows 8 bits to be assigned to the control of pixel-mixing operations. These extra 8 bits are called an *alpha-channel*.

8.4.2.5 Compressed still images

The JPEG still image compression standard was described in Section 3.12.1. It is an international standard and is in wide use. However, it is quite compute-intensive and it may take several seconds to decompress a high-resolution image in any except the fastest systems. This can be overcome with dedicated hardware, and some display adaptors have special hardware or a DSP chip to assist JPEG decompression.

Note that the processing time required for still image compression is usually not an issue because compression happens only once and at a time when a user is not waiting to view the image. Decompression, on the other hand, occurs during an application and the user has to wait until it is finished to see the image.

There are still image compression techniques available as options in some of the other image formats. Most of these are simpler than JPEG, they operate more quickly, but they do not offer as much compression at equivalent quality levels. The details of these techniques are beyond the scope of this book.

8.4.3 Motion video formats

Compression must be used for full-screen motion video on personal computers because uncompressed video data rates (see Table 1.2) are too high. However, unlike television, motion video does not have to be full-screen, in fact, most of the time it is not. Computers that have GUI operating systems (now most of them) define rectangular areas of the screen as *windows* and can display different information in each one. Windows can be made any size and, of course, the smaller they are the fewer pixels they contain. One way to get uncompressed motion video to work is to reduce its window in size until the data rate becomes manageable. Unfortunately, a window defined that way becomes pretty small and is often referred to as "postage-stamp video."

Another strategy for simple compression of digital motion video is to reduce the frame rate. A frame rate of 30 per second is considered normal and can satisfactorily reproduce most motion effects. However, if there is not much movement in the scene (such as in a talking-head shot), the frame rate can usually be reduced to 15 frames/second. That is effectively 2:1 compression. Below 15 frames/second, any motion becomes jerky and is not satisfactory.

Motion video is sometimes compressed as a series of still images, generally using JPEG. This is known as motion-JPEG and has the advantage over more advanced motion compression that the frames are independent of one another and the video can be started or edited on any frame. However, more compression is achievable by using a motion compression technique that exploits the redundancy between successive frames.

8.4.3.1 Motion video decompression

When video compression is used, the limit on playback window size is not the data rate but the decompression processing time. Decompression can be done in software alone or with special added hardware. With dedicated hardware, motion video can be full-screen, but with software-only decompression, video is usually limited to a partial-screen window. However, most of the industry focus is on software-only decompression and new algorithms are constantly being introduced and existing ones are being improved. The latest algorithms work well on the latest and fastest machines and, as the curves of future CPU speeds show, software motion video playback will only get better.

One advantage of software-only decompression is that the algorithms can be easily changed since that only requires a change of software, which is loaded for each video playback anyway. When an algorithm is in hardware, it is "cast in silicon" and cannot be improved by updating. The compression field is so dynamic that this is a serious limitation. It can be partially solved by building into the PC system new generic functions that support motion video decompression tasks at a lower level than algorithms. That concept is being aggressively pursued by CPU manufacturers and the display adaptor manufacturers. Current examples of this are the *multimedia extensions* (MMX) in the Intel Pentium CPUs and the *advanced graphics processing* (AGP) functions built into display adaptors. Thus, algorithms can be changed or improved as long as they stick to the generic functions that are contained in the hardware.

8.4.3.2 Capturing motion video

The video digitizing adaptor shown in Figure 8.4 is not satisfactory for motion video capture. That is because it captures into a RAM buffer, which would grow to impractical size to handle any reasonable amount of motion video. Capture of motion video really has to be to hard disk to support video segments longer than a few seconds.

Hard disk capture changes the problem to a data rate one because hard disks are not fast enough for raw video rates. Table 8.2 gives some numbers for different window sizes and frame rates of uncompressed raw video. Considering that current low-cost hard drives cannot achieve sustained data rates higher than about 2 MB/s, it can be seen that only 160 × 120 (postage-stamp) video could be captured. More expensive hard disk solutions can store data rates up to 10 MB/s; these are generally called for in the high-quality nonlinear editing systems that are increasingly using standard PCs (see Section 11.5.3). This is necessary because nonlinear editing should use a little compression as possible to preserve the image quality.

Table 8.2 Motion video data rates for different window sizes (24 bpp—uncompressed)

Window size (pixels)	Frames/s	Data rate (MB/s)
640 × 480	30	27.6
320 × 240	30	6.9
320 × 240	15	3.5
160 × 120	15	0.9

A solution to the storage data rate problem is to perform limited compression on the fly during capture. (Full compression takes far too much processing time unless dedicated hardware is used on the capture board.) However, simpler algorithms can be used, but this is a disadvantage because the simpler algorithms may introduce too much quality degradation. It becomes a tradeoff and different boards have made different choices.

The best choice is to capture to hard disk at the highest quality possible and then perform full compression as a separate step. The captured frames can be read from the hard disk one at a time and the computer can take as much time as needed to compress them before writing them back to disk as the compressed bit stream. The only disadvantage of this is that the compression step may take a long time, such as an hour of processing to compress a minute of video. In most cases, this is not a problem, but if a lot of video is being handled, special equipment is probably indicated to speed up the compression step.

8.5 Computer Video Standards

Motion video algorithms for computers are still being developed at a great rate. The only official standard is MPEG, described in Section 3.12. MPEG is a high-performance standard but it is highly compute-intensive and definitely requires dedicated hardware for real-time compression and requires hardware for decompression except in the latest, fastest systems. However, because of its worldwide standard status and its performance, it is the best choice for applications where the necessary system capability can be made available.

Other algorithms are proprietary to their owners, but this is not a large problem because they are mostly software-based and the software is widely available. It is not necessary to know the contents of a software package in order to use it. Current algorithms such as Intel *Indeo*, Apple *QuickTime*, *CinePak*, *Truemotion S*, etc., are successful for that reason.

Because computers are programmable, they can support many different video formats through software. A fully documented standard for a video format is not required if the appropriate software is widely available. However, this leaves the industry at the mercy of video format distributors who may not always support the software they distribute. In that sense, true standards allow anyone to develop and distribute software (or hardware) that meets the standard. That is a step toward a more mature industry.

8.6 DVD on Computers

The PC industry is rapidly embracing the *Digital Versatile Disc* (DVD) standard (see Section 10.20.2) for removeable mass storage and DVD-ROM drives are appearing on many new PCs. These drives can also play CD-ROMs, audio CDs, and DVD-video discs. In the latter case, a MPEG-2 decoder is required to actually view the video. This can be either in the form of a plug-in board or as software. Software decoding requires the fastest computers to operate successfully. One advantage to the board approach is that an output can be provided to connect the DVD video to a large-screen TV for display.

The DVD-ROM data capacity is 4.7 MB per side (see Section 10.20.2), about 7 times that of CD-ROM. This will allow even larger software packages, games, and data bases to be distributed on single discs.

A DVD-RAM product is also being introduced that provides record and rerecord capabilities similar to the CD-RW drives. At this writing, this is moving slowly because all manufacturers have not yet agreed on standards, but that is probably just a matter of time.

8.7 Digital HDTV on Computers

With the development and expected broad proliferation of DTV in the form of HDTV, there is sure to be an impact on the computer market. The pressure this is putting on the CD-ROM field has already been mentioned in Section 8.2.1.4. HDTV will also up the ante for motion video quality on PCs, which will lead to demand for higher bus data rates, larger mass storage capacities, and higher CPU speeds. These are ongoing trends in the PC business anyway—HDTV is just one of the ways such higher powers will be used.

It seems reasonable that the computer industry should embrace the HDTV standards for higher-resolution computer screens. Hardware will be available in the mass market and it would cost far more to develop different, noncompatible standards. This is a scenario that will play out in the next few years as HDTV becomes a factor in the TV marketplace.

Chapter 9
Video Cameras

9.1 Introduction

The camera is the most basic element of a video system, and its most important component is the *imager*, a device that converts the optical image of a scene into an electrical signal. Cameras are designed around the characteristics of their imagers, and a description of cameras must begin with imager technology. It has changed enormously during the past 50 years, and its development has included three breakthroughs, one of which—the shift from storage tubes to solid-state devices or CCDs—has recently been completed.

9.1.1 A brief history of imagers

The first imagers were *instantaneous scanners*. They formed a raster on a live scene or film frame by scanning it with a bright spot of light. A phototube picked up the reflected or transmitted light from the raster and generated the electrical signal. The Nipkow disk, the earliest imager, was a mechanical scanner. It was succeeded by the Farnsworth image dissector, an electronic instantaneous scanner that showed a brief promise. A later electronic instantaneous scanner, the *flying spot scanner*, produces a raster on a kinescope and focuses it on a frame of film. All instantaneous scanners suffer from inherently low sensitivity that make them impractical for live pickup.

The first breakthrough in imager technology came with the introduction of *photoemissive storage tubes* in which incident light from an image of the scene causes electrons to be emitted from a photosensitive surface in a pattern that corresponds to the brightness of the image. Because they store the light energy in an electric-charge image of the scene for an entire frame before it is removed by the scanning process, they have sufficient sensitivity for live scenes. Sensitivity is improved a thousand-fold or more compared to instantaneous scanners.

The *iconoscope*, developed by V. K. Zworykin of RCA prior to World War II, was the first practical photoemissive storage tube, and it proved the feasibility of all-electronic TV. But it produced a noisy picture, it was difficult to operate (the term "shading" resulted from one of its operational problems), and its sensitivity was marginal for live pickup.

The iconoscope was succeeded for live pickup by the *image orthicon*, a product of World War II research. It had greater sensitivity, and it replaced iconoscopes except for film pickup. The image orthicon had good sensitivity, but it was also difficult to operate (although easier than the iconoscope) and its SNR was marginal.

The defects of the iconoscope and image orthicon led to the second breakthrough, the replacement of photoemissive tubes with *photoconductive* types. Photoconductive tubes make use of changes in the apparent electrical resistance of a photoconductor when exposed to light. They are smaller, cheaper, easier to operate, and more noise-free than the iconoscope or image orthicon. The first commercial photoconductive tube was the *vidicon*, which was introduced in the United States for film pickup in the early 1950s. Its sensitivity was marginal for live pickup, but its use for film continued for more than thirty years.

Ten years later, a new photoconductive tube, the *Plumbicon* was introduced. It had sufficient sensitivity for live pickup while retaining the other desirable features of photoemissive tubes, and by the early 1970s it had replaced the image orthicon. Another photoconductive tube, the *Saticon*, was introduced in 1974.

As of this writing (1999), the third breakthrough has reached fruition as photoconductive tubes have been replaced entirely by solid-state imagers, specifically *charge-coupled devices* (CCDs). This is another step in the replacement of vacuum tubes by solid-state devices in the electronics industry. Storage tube imagers are still in use, however, and this chapter describes them briefly because of their historical interest and to provide a frame of reference for the performance of CCDs.

Table 9.1 summarizes the introduction dates and technologies of the principal types of storage tubes and CCDs. The introduction dates are not precise but indicate the approximate time period when the imagers became standard commercial items. Widespread market acceptance may have come later; in the case of CCDs, it did not happen until the 1990s.

Table 9.1 Principal video imagers

Name	Introduced	Type	Application
Iconoscope	1939	Photoemissive	Film/live
Image Orthicon	1946	Photoemissive	Live
Vidicon	1952	Photoconductive	Film/live
Plumbicon	1963	Photoconductive	Film/live
Saticon	1974	Photoconductive	Film/live
CCD	1980	Solid state	Film/live

9.1.2 Camera categories

In TV, there are two major camera categories—live and film. Other video camera uses, such as surveillance or videoconferencing, are similar to low-end TV cameras. A very different category is the digital still-picture camera, which is described in Section 9.13.

Live TV cameras generate video signals from the optical images of indoor and outdoor scenes, which may be under the controlled conditions of a studio, a sports venue, or at any location in the field, such as a news event, or in the home. Camera designs range from the "studio" camera, designed to operate on a tripod or pedestal, to "field" cameras carried on the operator's shoulder, or even hand-held. Studio cameras are designed to provide optimum performance in a controlled indoor environment, usually at the expense of size, weight, and portability. These cameras also have all the features necessary to operate in a system with other cameras and recorders. A similar system environment often exists for cameras used at sporting events, but cameras for that use must be packaged so that they are transportable or even portable because such cameras are regularly moved from one venue to another. Cameras designed for electronic newsgathering (ENG), electronic field production (EFP), or satellite newsgathering (SNG), provide portability, even to the extent of being hand held. In most cases, a portable camera is combined with a video recorder—the combination is called a *camcorder* (see Chapters 10 and 11). Most home TV cameras are camcorders.

TV film cameras (called *telecine* cameras) generate TV signals from film or slide images. The utilization of film as a source of video programming is diminishing as an increasing number of programs are produced and recorded on videotape, a result of the improved performance of CCD cameras and recorders. Also the development of portable high-performance TV cameras and video recorders has caused a total shift from film to video tape for on-the-spot newsgathering. But both film and TV have advantages as production and storage media, and 35-mm film is still preferred by some for the production of prime-time programs and commercials (see Section 4.17.5). As a result, vital applications for high-performance telecines will continue.

Typical cameras of each type are shown in Figure 9.1.

9.1.3 Live camera configurations

The apparatus configuration required to produce a video signal from an optical image is known as a *camera chain*. For studio cameras, it consists of the camera head together with the rack or console-mounted unit or units that perform the control, signal processing, composite signal multiplexing, and monitoring functions. Other camera types are generally self-contained or (at the high end) may have a portable control unit that is used for set-up and optionally for operation.

Figure 9.2 is a functional diagram of a studio camera chain. The basic functions of portable and studio cameras are similar, but the portable camera must have the ability to control its own operating parameters such as gain and sensitivity automatically because the camera user cannot be required to make any technical

214 Chapter Nine

Figure 9.1 Typical cameras. (a) Studio camera, courtesy of Sony Corp., (b) professional camcorder, courtesy of Sony Corp., (c) hand-held camcorder, courtesy of Panasonic, (d) telecine camera, courtesy of Cintel.

adjustments. The camera is set up initially with the use of external monitoring equipment, and its stability and automatic features permit it to operate for considerable periods without manual change of the set-up adjustments. Many cameras today are set up by the manufacturer and will never again require setup unless component replacement occurs.

9.1.4 Film camera configurations

Film camera systems must cope with the difference between the standard 24 frames/s film rate and 25 frames/s for most PAL systems or approximately 30 frames/s for NTSC. It was once believed that the 4% difference between 24 and 25 frames/s would be barely noticeable for either video or audio (except to trained

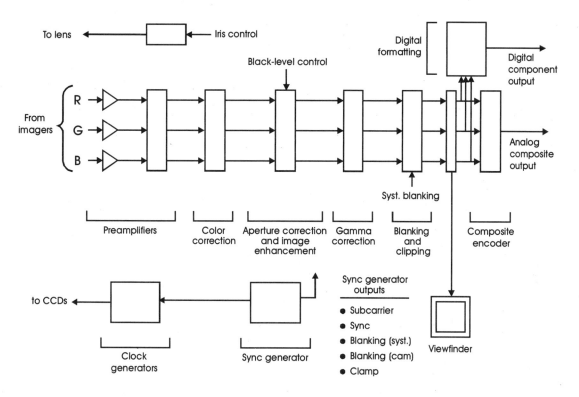

Figure 9.2 Functional block diagram of a 3-imager camera chain.

musicians with absolute pitch) and that film could simply be played at the 25/s frame rate of PAL. More recent European experience has indicated the difference is noticeable to ordinary listeners to dialogue because of the change in voice quality. To avoid this, digital algorithms have been developed to restore the audio to its original frequencies.

Since frame rate conversion was not mandatory for the picture in PAL systems, flying-spot scanners used with continuous-motion film projectors were the preferred film pickup systems. When CCD technology was developed, an alternative to the flying-spot was a single-line CCD device. These approaches provide a high-performance telecine system that is easy to set up and maintain.

For NTSC systems, the projector/camera combination must have the facility of converting from 24 to 30 frames, and a number of complex electronic and mechanical systems were developed to accomplish the conversion. They are now being replaced by digital frame store devices similar to those used for standards conversion (see Section 3.6.2). The signal is generated at 24 frames/s, converted to digital form, stored in a digital memory, and read out at 25 or 30 frames/s, thus accomplishing even the small correction necessary for PAL systems.

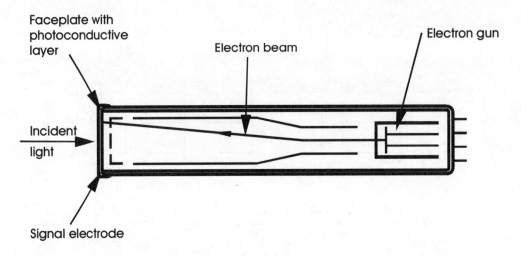

Figure 9.3 Photoconductive imaging tube construction.

9.2 Photoconductive Storage Tubes

Although photoconductive storage tubes have been replaced by CCDs for most applications, they are described briefly here for historical purposes.

The vidicon was the first of a family of commercial photoconductive tubes that eventually included Plumbicons, Saticons, Newvicons, Chalnicons, and silicon diode tubes. The term "vidicon" is sometimes used generically to indicate any photoconductive tube.

The vidicon utilizes the photoconductive properties of antimony trisulfide, and it is an excellent tube for film pickup. However, when its operating conditions are adjusted for high sensitivity it suffers from severe lag or smear, and its use for live pickup is primarily limited to industrial applications.

The Plumbicon is a heterojunction photoconductive tube that uses a lead oxide semiconductor as the photosensitive material. It greatly reduced the vidicon's lag problem and was dominant for live pickups until it began to be replaced by CCDs.

The Saticon is also a photoconductive tube that uses a selenium alloy semiconductor as the photoconductive material. It was introduced later than the Plumbicon and enjoyed certain advantages although it did not replace the Plumbicon. Whereas the response of the Plumbicon to red light is higher after special doping of the semiconductor, the response of Saticons is higher in the blue end of the spectrum.

In addition to these three principal types of photoconductive tubes, there were others that were used for special purposes. They include the Newvicon, which uses zinc selenide as the photoconductor, the Chalnicon, which uses cadmium selenide, and silicon diode tubes. These tubes have greater sensitivity, particularly in the red and infrared portions of the spectrum, and they were used mainly for surveillance cameras.

9.2.1 Photoconductive tube construction and operation

The construction of a photoconductive storage tube is shown in Figure 9.3. The faceplate, mounted on the front of the tube, is an optically flat glass disk with its inner surface coated with a thin layer or layers of photoconductive materials. A ring of conductive material is sealed to its circumference forming a structure known as the target. The target is sealed to the front end of the tube and has a connection to the conductive ring, which is the signal electrode.

An electron gun at the rear of the tube generates a finely focused beam of low-velocity electrons that scans the target in a raster. A lens (not shown) forms an optical image of the scene on the photosensitive surface of the target, which produces a charge pattern in the photoconductive target having density variations corresponding to the brightness variations of the image. As the scanning beam passes over the charged target and locally discharges the charge pattern, an electrical signal appears at the output terminal.

9.2.2 Photoconductive Tubes for HDTV

HDTV puts severe demands on the performance of imagers, and the industry initially responded by developing storage tubes with significantly improved performance. One example was a photoconductive tube with a HARP (high-gain, avalanche rushing amorphous photoconductor) target [1]. This tube offered a major improvement in sensitivity and aperture response as compared with Plumbicons and Saticons. The improvement was not great enough, however, to match the growth of CCD technology, and the HARP tube has not found wide application.

9.3 Charge-Coupled-Device (CCD) Sensors

The development of integrated circuit technology allowed the interconnection of thousands of tiny semi-conductors in a rectangular array to create imagers, known as *charge-coupled devices* (CCDs) [2]. These have important advantages over storage tubes including wider dynamic range, absence of lag, absence of need for operator registration controls, longer life, and superior basic performance characteristics.

The use of solid-state imagers in television began on an experimental basis in the 1960s and was accelerated by the invention of the CCD in 1969. Cameras using CCD imagers are now almost universally used in broadcast, home, and related television applications and also in nontelevision applications such as still imaging systems, electronic photography, and ultrasensitive detectors for astronomy and military surveillance.

A CCD *sensor* is a pixel-sized solid-state photosensitive element that generates and stores an electric charge when it is illuminated. It is a building block for a CCD *imager*, which is a rectangular array of sensors upon which an image of the scene is focused. In most configurations, the sensor includes the circuitry that stores and transfers its charge to a shift register, which converts the spatial array of charges in the CCD imager into a time-varying video output current.

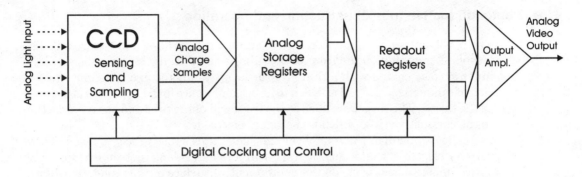

Figure 9.4 Functional diagram of CCD on-chip operations.

9.3.1 Sensor construction

CCDs have achieved a high degree of sophistication as the result of intensive research and development that has covered the gamut of technologies from semiconductor physics to solid-state manufacturing. A complete description of the wide variety of designs that these efforts have produced is beyond the scope of this volume. The description of a representative design follows, and the reader may consult the references for further information.

Four functions—sampling, photosensing, charge storage, and charge transfer—are performed by CCDs (Figure 9.4). The construction of the sensor depends on

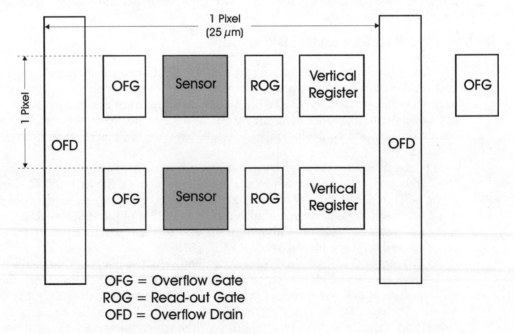

Figure 9.5 Diagram of CCD sensor element layout.

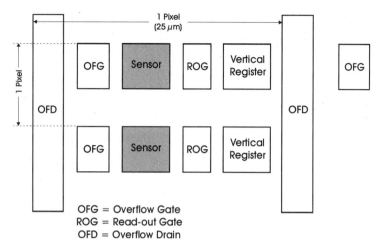

Figure 9.6 Typical CCD sensor element cross-section.

the location of the elements that perform these functions as determined by the architecture of the imager in which it is used (see Section 9.4.2).

Figure 9.5 is a diagram of a CCD element of a type used by an imager with interline transfer CCD architecture (see Section 9.4.2.2). It includes the photosensor itself and the semiconductor circuit elements that control the storage and read-out of the electric charges generated by the photosensor. The functions of these elements are described below.

Figure 9.6 is a cross-sectional view of a CCD sensor showing the details of its construction and its complexity. This design is representative only, and the construction of different models varies widely.

The structures of CCD sensors are complex, and they present challenging manufacturing problems as a result of their small dimensions—each pixel is typically less than 40 μm on a side. The manufacturing difficulties are increased further by the fact that an imager includes 500,000 or more sensors, nearly all of which must work perfectly to avoid unacceptable defects in the picture.

9.3.2 Sensor operation

The operation of a CCD sensor and its associated circuitry can be understood by reference to a "map" of the electrical potentials of its components (Figure 9.7). Electrons freed by the photosensor during a television field are stored in a *potential well* that is collocated with the sensor and positively charged—see Figure 9.7(a). If this portion of the picture is very bright, the well will be filled to overflowing, and the excess electrons will flow into a drain through the *overflow gate* as shown in Figure 9.7(c). At the end of the field, the potential of the read-out gate, which is slightly more negative than the overflow gate when closed, is lowered and the electrons pass into the shift register—see Figure 9.7(b).

The operation of the sensor including the photosensor, gates, shift registers,

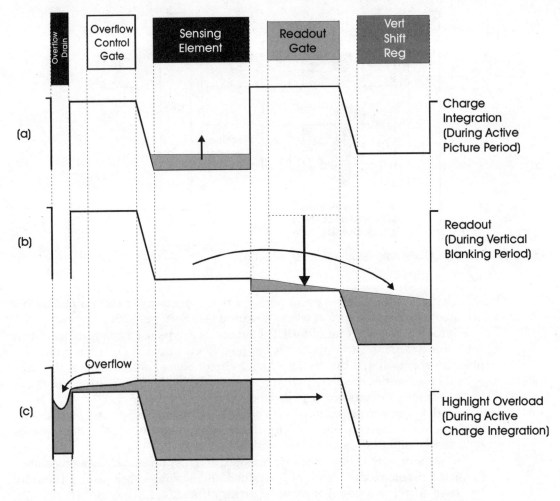

Figure 9.7 CCD potential map diagrams showing (a) charge storage, (b) readout, and (c) overflow gate operation.

and wells is controlled by design, and commercial CCDs offer a wide variety of choices for performance specifications.

9.3.3 On-chip lenses (OCLs)

Because the sensitive area of a CCD pixel is only a fraction of its total area, the sensitivity of CCDs can be increased by mounting a layer of tiny lenses, *on-chip lenses* (OCLs-also called microlenses), in front of its sensors, one for each sensor. They concentrate the light of the optical image into the sensor area of the pixel and thus increase sensitivity.

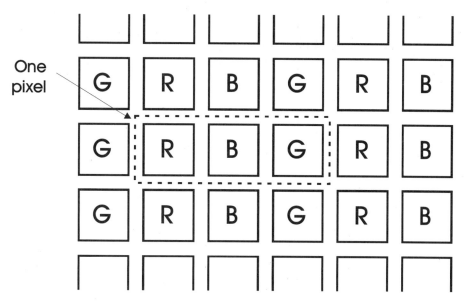

Figure 9.8 Sensor layout of a tricolor CCD.

9.3.4 Tricolor sensors

The standard CCD camera configuration uses a separate CCD for each primary color and an optical beam splitter. For applications that do not require the highest picture quality, however, it is possible to achieve satisfactory results with a single-chip camera that has a single tricolor imager.

The construction of a typical *tricolor imager* is shown in Figure 9.8. Each pixel has three side-by-side photosensors, one for each primary color and each a part of a column of similar photosensors. The signals from these R, G, and B columns are combined at the output to form the luminance and color difference signals as described in Chapter 2.

Since the columns are one-third the pixel width, the aperture response is excellent, although not as good as with a three-CCD camera using spatial offset (see Section 9.4.3). For example, the specified limiting horizontal resolution of a commercial single-CCD camera with 768(h) x 494(v) pixels is 470 TVL. The specified limiting resolution of a three-CCD camera, each having the same number of pixels is 700 lines.

The sensitivity of a single-CCD camera is lower because of the smaller total sensor area and because the filters absorb more than 2/3 of the incident light. For example, the specified sensitivity of the single-CCD camera described above is 2000 lux (lumens) at f/5.6 while the sensitivity of the three-CCD camera is twice as great: 2000 lux at f/8.0. This is because the three-CCD camera uses prism optical separation of the color components that does not absorb as much light as the filter separation used in the single-CCD.

9.4 CCD Imagers

A CCD imager consists of a matrix of individual sensors, one sensor per image pixel (three for a tricolor CCD), mounted in a rectangular array on a silicon substrate. They develop electric charges that are proportional to their illumination by an image of the scene focused on the array. The charges are stored temporarily in the potential "wells" of the sensors and are transferred to shift registers during the field blanking interval, which then transmit them to the imager output in the proper time sequence to generate the video output signal. The purpose of the CCD transfer process is similar to that of a scanning beam in a photoconductive tube—to read out the pixels' stored charges sequentially to a video output.

9.4.1 Photosensor arrays

The number and dimensions of the photosensors in the imager array are basic factors in establishing its performance. The number of rows of sensors must at least equal the number of active scanning lines, and for the lowest cost, it is preferable that these numbers be equal.

The number of vertical columns of sensors establishes the imager's Nyquist limit (see Section 1.5), the television line number where horizontal aliasing begins. Aliasing will occur for horizontal frequency components in the image that exceed one-half the number of vertical columns of photosensors per picture height.

The pixel count, the number of rows (v) and columns (h), is a basic specification of a CCD imager. A variety of counts are available commercially. The dimensions of the total imaging area and the number of active pixels in each dimension for imagers that are representative of designs for NTSC and SMPTE 274M HDTV service and which were the basis of the analysis in Chapter 4 are:

System	CCD Dimensions (mm) H	V	Number of Pixels H	V
NTSC	8.8	6.6	818	513
HDTV 274M	14.0	7.8	1920	1080

The dimensions of the sensing area relative to the pixel area are a tradeoff between sensitivity and aperture response. Maximum sensitivity is achieved when the dimensions of the sensor approach those of a pixel. On the other hand, the aperture response can be increased significantly by reducing the sensor dimensions (see Chapter 6). Spatial offset (see Section 9.3.3), a commonly used technique for improving the aperture response while reducing or eliminating aliasing, requires the use of sensors having a horizontal dimension less than one-half the width of the pixel.

9.4.2 CCD transfer and readout architectures

The essence of CCD imager operation is the conversion of a spatial charge pattern on an array of photosensors into a time-varying video signal as illustrated in

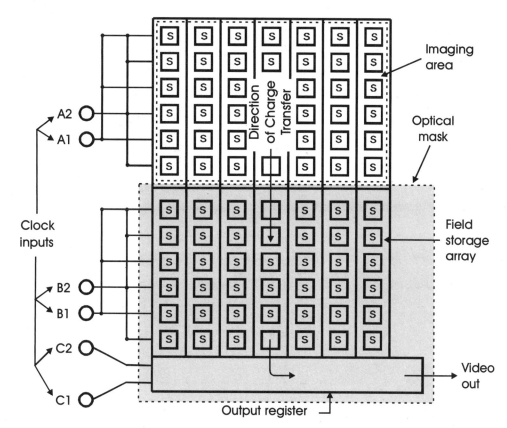

Figure 9.9 Layout of a frame-transfer CCD.

Figure 9.4. The three commonly used architectures for this purpose are *frame-transfer* (FT), *interline-transfer* (IT), and *frame-interline-transfer* (FIT).

9.4.2.1 Frame-transfer architecture

Frame-transfer architecture is illustrated in Figure 9.9. The CCD structure is divided into three sections, an imaging area, a field storage array, and an output register. The field storage array and the output register are covered with an optical mask to prevent stray light from affecting these registers. Charges are allowed to accumulate at each photosensor in the imaging area for a complete frame. During the vertical blanking interval, a command from a clock causes the charges in each column of pixels in the sensing area to be shifted to a corresponding column in the storage area. This frees the sensors to accumulate charges from the illumination of the next field. From the storage area, and again at the command of a clock, the charges are shifted during the next field scan one line at a time to the output register that generates the video signal.

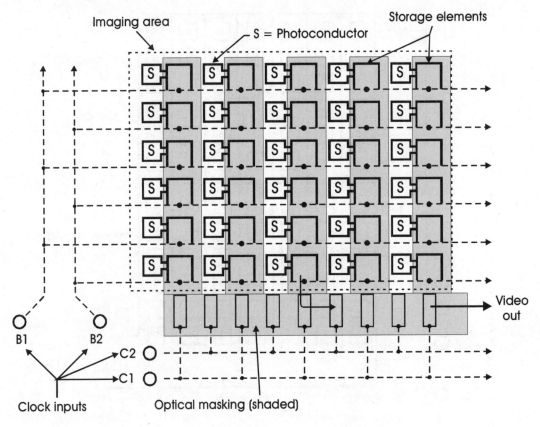

Figure 9.10 Layout of an interline-transfer CCD.

The entire imaging area is available for the sensors, and frame-transfer imagers do not suffer the sensitivity loss of interline-transfer imagers due to masking a portion of the imaging area for the storage areas.

On the other hand, the frame-transfer architecture has the problem of *transfer smear*—vertical streaks above and below bright spots in the image. It is caused when the charges in the image array move across the optical image during the transfer of charges at the end of each frame. This happens very quickly, but the sensor array elements pick up small spurious charges during the transfer that cause streaks in the picture that are most visible on picture highlights.

This type of smear can be prevented by a rotating mechanical shutter that blocks the light from the image during the transfer period. This is a very effective solution, but it has the problem of introducing a mechanical component in an otherwise all-electronic system.

Smear can also be prevented or reduced by the use of interline-transfer architectures.

9.4.2.2 Interline-transfer architecture

Interline-transfer architecture, which is commonly used in consumer cameras, is illustrated in Figure 9.10. The sensors in vertical columns are connected to storage elements (potential wells) in alternate columns. After exposure of a single frame and when commanded by a clock, the photo-generated charges in the pixels in each column are transferred along their rows to the neighboring vertical transfer register. This frees the sensors to accumulate the charges for the next frame as determined by their illumination.

The effective exposure time for each photosensor can be reduced by the use of a clock-controlled electronic shutter that transfers the electrons that are generated during only a portion of each frame. As with photographic cameras, this reduces the blurring of objects in motion but at the expense of sensitivity.

Another clock command causes the charges to be shifted from the vertical to the horizontal register. From this register they are read out as the video signal.

In the basic form of this architecture, the vertical register columns must be covered with opaque masks and only 30 to 50% of the imager surface can be photosensitive. This is an inherent drawback of interline-transfer architecture.

9.4.2.3 Frame-interline architecture

Frame-interline architecture, Figure 9.11, was developed to provide the sensitivity advantage of frame transfer, but with resistance to smearing from strong highlights without the necessity of a mechanical shutter. It is the preferred architecture in most professional cameras. The unique feature of frame-interline-transfer architecture is a row of selection gates between the image and storage areas. The gates are biased so that charges in excess of a predetermined level are drained from the system before being transferred to the storage area.

The sensors for CCD imagers with frame-interline architecture are more complex and hence such imagers are more costly.

Another problem with frame-interline architecture is that the signal from the output of every photosensor must be switched in sequence during each frame, and a very high switching clock rate is required—for the 1036×1920 CCD it is 74.25 MHz. The input impedances of the CCD circuits are capacitive, and driving them at this frequency takes considerable current. This problem can be alleviated by dividing the photosensor columns into two sections, each section including alternate columns. The sections are then clocked alternately at a frequency of 37.125 MHz, and their outputs are combined to produce a signal that is identical to that produced if all the photosensors were in a single group.

9.4.3 Spatial offset

Spatial offset is a camera construction for three-CCD cameras that sets a special relationship between the positions of the columns of R, G, and B sensors relative to the optical image as illustrated in Figure 9.12. Ordinarily, all three imagers would be positioned so that equivalent pixels of each imager would view the same

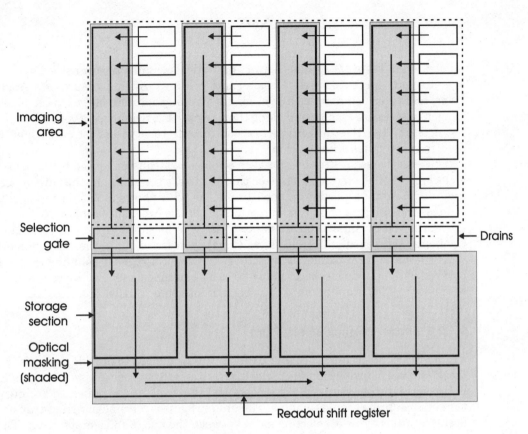

Figure 9.11 Layout of a frame-interline-transfer CCD.

area of the optical image (*registered*). However, in spatial offset, columns of pixels of the green imager are shifted horizontally 50% of a pixel width relative to the equivalent columns of the red and blue imagers. As noted above, this requires that the width of the sensitive areas be less than one-half the width of a pixel, so it is suited to interline-transfer or frame-interline transfer imagers.

The columns of green, red, and blue sensors are on separate imagers, and they are positioned by bonding them with extreme precision to the prism block in the camera (see Section 9.7.2). Such precision in manufacture is a challenging problem, but it has been solved, and no registration adjustments are provided. This greatly simplifies the set-up and operation of the camera—an important additional advantage of this technology.

The operation of spatial offset can be explained by noting that the signals from the R, G, and B imagers are added in the luminance channel, and because of the spatial separation of the R and B from the G sensors the phase of their sampling sideband signals are reversed. If the sum of the amplitudes of the R and B sidebands equaled the G, there would be complete cancellation, and aliasing (which is caused by the overlap of the sidebands) would disappear. In practice, and depending on the color content of the scene, the cancellation is not complete, and some

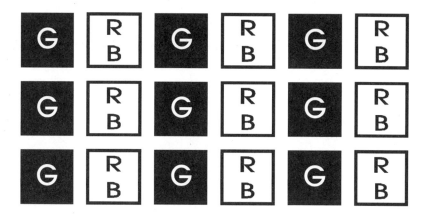

Figure 9.12 Spatial offset CCD.

residual aliasing occurs at image frequency components between the Nyquist limit and twice this number, i.e., the sampling frequency.

9.5 CCD Imager Performance

This section describes the performance of CCD imagers.

9.5.1 Performance criteria and standards of comparison

CCD performance criteria include definition, resolution, SNR, sensitivity, dynamic range and gamma, spectral response, aliasing, smear, lag, artifacts, and other defects. Since the performance of CCDs of different design and under different operating conditions varies considerably, the comparisons here are representative rather than precise.

9.5.2 Image definition and resolution

Criteria for evaluating picture definition or sharpness described in Chapter 6 include the aperture response, as specified by modulation transfer function (MTF), limiting resolution, or the equivalent line number or N_e. The MTF is the most complete specification, but it is presented in graphical form that is not as convenient or as easily interpreted as a single number. Although N_e is a single number and is an excellent basis for analysis, it is not widely used. Limiting resolution is an incomplete specification, and it has the additional disadvantage of ambiguity in systems that involve sampling. These include all television systems in which the scanning process samples the image in the vertical dimension, and CCD imagers in which discrete photoconductors sample the image in both dimensions. For these reasons, the MTF is used in this chapter as the basic criterion for the definition of CCD imagers.

9.5.2.1 CCD MTF

Equation (4.5) for the MTF of a sampling process becomes Eq. (9.1) for a CCD imager.

$$\text{MTF}_{(H,V)} = \frac{\sin x}{x} \tag{9.1}$$

where $\text{MTF}_{(H,V)}$ is the modulation transfer function in the horizontal or vertical dimension, and

$$x = \pi \frac{W_{(H,V)} f_{(H,V)}}{F_{(H,V)}}$$

where $F_{(H,V)}$ is the spatial sampling frequency along the horizontal or vertical dimension (determined by the number of sensors per picture height), $W_{(H,V)}$ is (sensor width or height)/(sensor spacing), and $f_{(H,V)}$ is the spatial frequency component of the image.

Since the number of rows of sensors equals the number of scanning lines, $\text{MTF}_{(V)}$ is equal to MTF for the scanning pattern. The $\text{MTF}_{(H)}$ for CCDs that are representative of NTSC and HDTV systems are shown in Figures 4.11 and 4.13.

9.5.2.2 CCD resolution

CCDs utilize a sampling process in both dimensions, and the ambiguities, which are inherent in specifying and measuring the vertical resolution of all television systems because of scanning line sampling, are present in CCDs for both vertical and horizontal resolution. In fact, there is a question as to the significance of limiting resolution as a criterion of CCD performance.

Nevertheless, the industry is so accustomed to using limiting resolution as a primary specification of imager performance that it is done for CCDs as well.

The horizontal limiting resolution of a CCD imager, which usually is specified as the resolution of the camera, can be approximated by multiplying the number of sensors per picture height by the Kell factor. For the 493 by 768 sensor described in Section 9.3.4, and with a Kell factor of 0.7, the vertical resolution would be 359 lines (513 × 0.7) and the horizontal, 429 lines (818 × 0.7 × 0.75).

The horizontal resolution can be nearly doubled by the use of spatial offset. For example, the specified limiting horizontal resolution of a CCD camera with spatial offset and 818 sensors in each line (Sony Model DXC-M7) is 700 lines as compared with 404 without spatial offset.

9.5.3 Signal-to-noise ratio

Very little noise is generated within photoconductive storage tubes, and their SNR performance is determined almost completely by wideband noise from the preamplifier. High SNRs (as compared with photoemissive tubes) are therefore characteristic of photoconductive tubes.

Unlike photoconductive tube imagers, most of the noise from solid-state imagers is generated within the device rather than in the external circuitry. This is

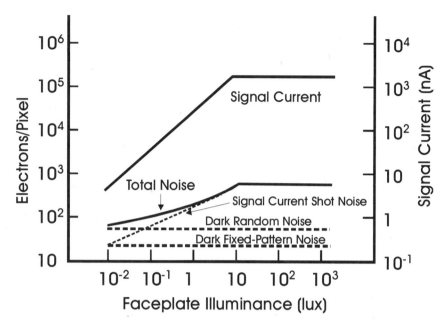

Figure 9.13 Fixed-pattern and random noise (*From [8]*).

indicated by the relative signal currents, which in CCDs are measured in microamperes as contrasted with nanoamperes in photoconductive tubes.

There are two basic noise sources in CCD imagers. The first is coherent, fixed-pattern noise caused by imperfections or irregularities in design or manufacture, e.g., nonuniformity in the dark current (see below) of individual sensors or cross talk from the high-speed clocking signals associated with the imaging process. This type of noise, while troublesome, is not inherent and can be greatly reduced or eliminated by suitable design and manufacturing practices.

The second noise source—thermally generated and random variations in the signal current—is inherent in all electronic communication systems; in CCD imagers it is generated as an inevitable by-product of the collection, storage, and transfer of electrons from and within the CCD. It is produced at three levels-the input circuitry, the transfer circuitry, and the output circuitry. This type of noise predominates except in the darkest areas of the scene as shown in Figure 9.13.

The magnitude of the thermal noise power in a CCD camera can be determined by measuring the *dark current* of the imager, its output when the lens is capped. The SNR is calculated by comparing this power to the power level of the peak-to-peak signal, i.e., black level to reference white.

Noise power must be measured over the entire bandwidth of the camera, and since thermal noise is "white" (see Section 4.15) its magnitude is proportional to the video bandwidth. All else being equal, the noise power in an HDTV system with a 30-MHz bandwidth will be 3.75 times or nearly 6 dB greater than the noise in an NTSC system with an 8 MHz bandwidth.

Although thermal noise is always present in CCD imagers, its magnitude can be reduced by design advances. The SNR of CCD imagers has steadily improved in recent years, and it typically exceeds that of storage tube imagers.

9.5.4 Sensitivity

There are a number of definitions for the sensitivity of television imagers.

The definition of greatest interest to scientists is the *quantum efficiency*, the percent of incident light quanta that create a useful output of electrons from the photosensor.

Two other definitions are more useful for engineering purposes:

1. The speed of the optical system required to produce an output signal of normal amplitude and SNR with a specified scene brightness.
2. The minimum illumination required to produce satisfactory (but not optimum) signal quality with the camera lens set at maximum speed and full gain in the video channel.

An alternative criterion of sensitivity that allows comparison between electronic and photographic imaging is the *exposure index*, E_I, borrowed from photography.

E_I is given by Eq. (9.2), which combines the lens speed, exposure time, and illumination in a single number.

$$E_I = K \frac{f^2}{I X_T} \qquad (9.2)$$

where K is a constant = 280 when I is measured in lux, f is the lens stop, I is the illumination to produce a reference white signal, and X_T = the exposure time.

A high exposure index indicates high sensitivity. Increasing the photosensor size increases the sensitivity of CCD imagers because larger photosensors collect more light from the optical image. The increased sensitivity is achieved at the expense of a lower MTF at higher optical frequencies [see Eq. (9.1)] and the design of CCDs requires a tradeoff between sensitivity and image definition. Fortunately, recent improvements in CCD technology such as on-chip lenses (see Section 9.3.3), an amorphous silicon layer over the sensor array (see Section 9.3.4), and hyper-HAD (hole accumulated layer) sensors [3] have increased sensitivity significantly, and this tradeoff no longer requires major compromises in definition.

9.5.5 Spectral response

The spectral response of a CCD imager of recent design as compared with photoconductive tubes is shown in Figure 9.14. The CCD response exceeds that of phototubes except in the extreme blue end of the spectrum. Note that Figure 9.14 shows the response of the imager. The colorimetry of the camera is adjusted by placing red and infrared filters in its optical system.

The major difference between the spectral response of CCDs and phototubes underscores the importance of including the color temperature of the illumina-

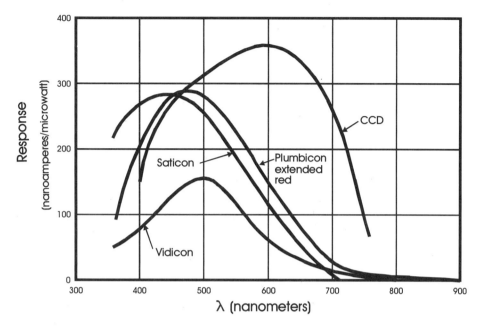

Figure 9.14 Spectral responses of imagers.

tion as part of the sensitivity specification of a camera.

9.5.6 Aliasing

Because CCD imagers use sampling, aliasing is an inherent concern (see Section 4.7). A number of effective methods have been developed to eliminate aliasing with CCD imagers or to reduce it to an acceptable level. These include spatial offset (see Section 9.4.3) and optical filters that remove the high frequency components from the optical image (see Section 9.7.4). An even better approach is to use more pixels in the CCD (both horizontally and vertically, if possible) than that required by the scanning standards and digitally converting to the standards at the output. Of course, this will significantly increase the cost of the CCDs, which may not always be acceptable.

9.5.7 Transfer function and gamma control

The transfer function (see Section 4.14.1) of CCD sensors within their dynamic range is essentially linear. Because of their excellent SNR, which permits operation down to low light levels, the dynamic range of CCD imagers is excellent, often as much as 80 dB (10,000 to 1)—much more than is needed in any displayable video signal. This extra dynamic range can be used very effectively for highlight control (see Section 9.5.10). Further, because of CCDs' excellent SNR, deliberate alterations of the transfer functions for gamma correction and range compression can be made without the introduction of excessive noise.

The SMPTE HDTV production standard 274M specifies a transfer characteristic that is linear at low light levels and has a gamma of 0.45 at higher levels, specifically:

$$V_C = 4\,L_C \text{ for } L_C < 0.0288 \tag{9.3}$$
$$V_C = 1.1115\,L_C^{0.45} - 0.1115 \text{ for } L_C > 0.0288$$

where V_C is the video signal output as a fraction of the reference white signal and L_C is the light input.

Linear operation is specified at low light levels to limit noise amplification. Although it is not specified by this equation, it is normal practice to introduce signal compression at light levels above reference white to avoid system overload (see Section 9.8.4).

9.5.8 Lag

Lag usually originates in the camera imager and is caused by the failure of the scanning process to eliminate the stored charge pattern completely during a single scan. As a result, motion in the image, whether caused by movement in the scene or of the camera, leaves a blurry trail.

Lag was a relatively unimportant problem with image orthicons but was a serious problem with vidicons. Unfortunately the vidicon operating mode that gives it maximum sensitivity (high target voltage) also produces maximum lag. As a result, vidicons had limited use for live pick-up, mainly in closed-circuit applications where lag could be tolerated. For the pick-up of film programs, however, sufficient light is available to reduce the lag of vidicons to an acceptable level, and they were widely used for that purpose.

The Plumbicon and other later tube imagers had a much lower lag than vidicons at low light levels, and they were widely used for studio and field cameras. Even the Plumbicon, however, had a residual signal of about three percent on the third field following the removal of the image from the photosensitive surface. A major advantages of CCD imagers is that they have almost no lag.

9.5.9 Smear and other defects

Vertical smear can be caused by the process of transferring the charges from a CCD storage area to the signal output circuit (see Section 9.4.2.1). Horizontal smear is caused by inadequate frequency or phase response of the video channel. It is an extreme example of low definition, the edges are widened, and because of poor phase response the trailing edges of pulses may differ from their leading edges.

Blooming, streaking, and comet tails are primarily defects of storage tubes. They are related to lag in that they are spurious signals that result from the transducer's response to excessive light levels. They are caused by overloading of the photosensitive surface or storage mechanism with extremely bright highlights.

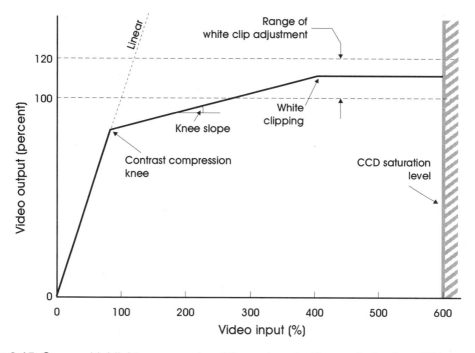

Figure 9.15 Camera highlight compression. (*Reproduced with permission from [6].*)

9.5.10 Highlight control

Highlight is the name generally given to an area of the scene where the light level goes above the level chosen for reference white. This can come from light bulbs, car headlights at night, reflections, or simply a bright sky creeping into a scene that is otherwise in shadows. Highlights can go as much as 100× above white level.

Of course, the overflow drain mechanism of CCDs can prevent highlights from spreading, but there may still be small artifacts surrounding a highlight and there will be no detail seen in a highlight that is using the overflow drains. Since the CCD has more dynamic range that needed for the final picture, the practice in high performance cameras has been to set reference white below the full-well condition by as much as 6:1. The camera processing circuits are also designed to handle levels up to 6 times reference white, at least up to the highlight processing circuits.

Highlight processing can then compress the highlights into the final output range. One example of this is shown in Figure 9.15, which shows the transfer characteristic of a highlight compressor based on setting the CCD overload point (saturation level) at 6 times reference white level. Two break points are provided that can be adjusted to suit operating conditions. The first break point or *knee* is generally set below 100% video level; it reduces the slope of the transfer characteristic so that highlights up to about 400% are still reproduced, but reduced in magnitude. This means that highlights in that range will still show some detail. At 400% a white clipper comes into play to hold any further highlights to 110% of normal level. At 600%, of course, the CCD overload mechanism takes over and prevents increase of the level beyond 600%. This type of highlight handling can do a very good reproduction of an indoor scene containing a window looking out into bright sunlight. Rather than simply clipping the window, the highlight compression allows some details of the outside scene still to be visible.

9.6 Camera Chain Packaging

Early broadcast camera chains were large and heavy, even though most of the processing equipment was located in a control room and connected to the camera head by a heavy camera cable containing often a hundred or more conductors. This awkward and massive packaging was necessary in those days because the components for building cameras were so large. Today, the situation is different—a complete camera chain can be made small enough to hold in your hand and weighs less than 1 kg.

9.6.1 Studio cameras

In studio cameras, however, there are important advantages in making cameras larger and in locating some of the equipment remote from the camera head. The reasons are:

1. A larger, heavier camera is mechanically more stable and, with the proper mounting equipment, it can be moved more smoothly than a small, lightweight camera.
2. A larger camera can support larger lenses, which provide faster optical speed and greater zoom ranges. This is especially important for sporting event pickup and is the reason why large cameras are used there even though there is also a portability requirement.
3. When a studio audience is present, larger cameras look more imposing and professional.
4. Studio cameras are always connected into a system, which requires a large number of connections to support synchronization, video and audio going both from and to the camera, control, intercom, and power. It is awkward and expensive to provide a multiconductor cable for all these functions. Studio cameras generally use a small cable, such as triax (see Section 9.8.8), with all the signals multiplexed on it.
5. In a studio setting, it is desirable for the camera operator to devote all of his or her attention to framing and focusing the picture and moving the camera. All technical matters should be handled elsewhere; thus, studio cameras usually have a remote control position where a technically skilled operator cam oversee

the camera's technical operation and do whatever is required to optimize the picture performance and quality. Many camera chains provide the remote control operator with access to all of the internal functions of the camera.

Most of the above reasons also apply to cameras used for pickup at large sporting events. These events generally have a system environment with a control room containing camera remote controls, recorders, instant-replay equipment, and video switching equipment. Therefore, studio type cameras are often carried into the field for these events. This places a limit on the size and weight of the cameras and most manufacturers have come up with studio-type equipment that is reasonable to take into the field when necessary, yet it still meets all the studio objectives.

9.6.2 Professional camcorders

Most cameras used for newsgathering operate by themselves and are not part of a real-time system. They have their own internal recorder and can operate from batteries and completely free of any cable connection when necessary. However, for professional and broadcast uses, studio-quality performance is still desired. This can be accomplished in a very small package, but the considerations of camera handling and operation generally caues these cameras to also be made larger than the minimum that might otherwise be possible. The reasons for this are:

1. A fast lens with a wide zoom range is desired. The package must be suitable for supporting such a lens. Generally, professionals want interchangeable lenses, so they can choose a different lens for different operating conditions. This also makes the camera larger.
2. Generally, shoulder mounting of the camera is desired because this gives the operator more control over the camera movement and gives a higher shooting angle. Shoulder mounting means that the camera has to be large enough to present the viewfinder to the camera person's eye while the camera body sits on his or her shoulder. This makes the camera longer that it otherwise could be. The camera shown in Figure 9.1(b) is typical of this design.
3. Other facilities, such as lights, microphones, radio intercom, should be mountable on the camera.
4. Camera operations, such as iris (exposure) control, white balancing, and other picture-quality adjustments should be completely automatic to free the camera operator from these details. However, manual control should also be available for the cases where it is needed.
5. The camera must be rugged and able to withstand field operating conditions over a long product life.
6. Considerations such as recording time and battery life are also important.

Thus, the requirements for a professional camcorder are very demanding, both mechanically and electrically. Owners of these cameras will pay for these features.

9.6.3 Home camcorders

The overriding objective for a home camcorder is the price. If that is not right, consumers cannot buy the product. Within a home price objective, however, much can be done with modern technology.

Most home users want a hand-held camcorder. This can be achieved in many different formats with a product including batteries weighing less than 1 kg. Generally, home cameras use a single CCD for color pickup, primarily because it is lower in cost than a three-CCD approach, but it can still deliver performance good enough for home use. The processing electronics is contained in a few ICs.

Home camcorders are described more fully in Chapter 11.

9.7 Camera Optical Components

The optics of a camera includes lenses, beam splitters, and filters. These are discussed in this section.

9.7.1 Lenses

The technology of television camera lenses has advanced almost as rapidly as electronic technology during the past 40 years. Forty years ago, the quality of continuously variable-focal-length zoom lenses was not adequate for color cameras and early color cameras used fixed-focal-length lenses mounted in a turret. The focal length could be varied only in fixed increments and by interrupting the signal. The widespread use of zoom lenses began in the early 1960s, and they have greatly enhanced the versatility of cameras for program production. Their quality has improved steadily, and their performance now approaches that of high-quality fixed-focal-length lenses.

Zoom lenses are available for both studio and ENG cameras, with speeds ranging from f/2.2 to f/1.2 and focal-length ratios ranging from 12:1 to 30:1 or more. Lenses for ENG cameras are much smaller, in part because of lower optical speeds and in part by the use of smaller imagers.

The zoom function is controlled electrically, and the focal length can be adjusted either by the camera operator or remotely.

9.7.2 Optical beam splitters

Early color cameras used image orthicon imagers that required a complex array of lenses and mirrors to split the light into its three primary components. The advent of the smaller Plumbicon tube made it possible to use prisms rather than mirrors for light splitting, and this has continued with CCDs.

Figure 9.16 shows a camera optical system. Prism light splitters simplify the optics, they occupy less space, and they are about one f/stop faster than mirror systems. The advantages of prism systems are so great that they are almost universally used.

The key technology of a prism beam splitter is the *dichroic* surface, which re-

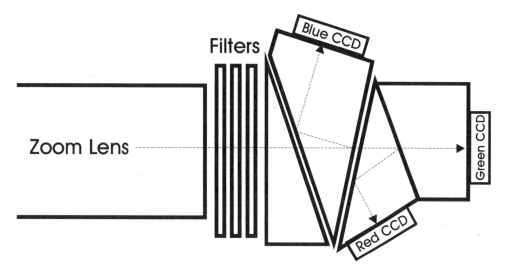

Figure 9.16 Prism light splitter for a 3-CCD camera.

flects light in one region of the spectrum and transmits the rest. The dichroic surface is coated with a monomolecular film that is ¼-wave thick at the center of the reflected spectral band. The phase of reflected light in this portion of the spectrum is shifted 90° in each of its two traversals of the surface and 180° upon reflection, thus emerging in phase.

Notice that there are two air gaps in the prism assembly. This is to cause total reflection of the blue and red light at the glass-air interfaces.

In systems that have spatial offset to reduce or eliminate CCD aliasing (see Section 9.3.4) the imagers are cemented to the output ports of the prism beam splitter. They must be located with extreme precision because the registration of the primary color images is determined by their relative location. The location tolerance is measured in microns, and a sophisticated manufacturing technology has been developed to achieve it.

9.7.3 Color and neutral density filters

A variety of optical filters is used in modern color cameras; they include spectral trim filters, filters for daylight and sunlight, neutral density filters, and optical low-pass filters.

The spectral pass bands of the outputs of the dichroic prism in the beam splitter are broad, and it is necessary to employ trim filters to sharpen their edges. With trim filters, fairly close conformance to the standard primaries can be achieved by purely optical means, although additional electrical correction (masking) by matrix circuits is usually employed.

Filters are also used to adapt the colorimetry of the camera to the color temperature of the illumination. Typically, the color temperature of incandescent light is 3200K, while daylight is 5600K. The human eye with its marvelous power of adaptation automatically compensates for this over a wide range of color temperatures, and the difference between natural and artificial illumination is not noticed while viewing a scene directly. This power is lost when viewing a television image, and pictures of a scene illuminated by incandescent light from a camera adjusted for daylight will appear reddish. Color filters are used to compensate for differences in the color temperature of the scene illumination.

A typical complement of color filters adapt the camera to scene illuminations of 3200K and 5600K, incandescent and daylight. This complement permits the camera to be switched from daylight to artificial illumination without major readjustment of the controls.

Lens assemblies are usually equipped with neutral density filters to reduce the need for large variations in f/stop of the lens and hence depth of focus with widely varying scene brightnesses. They are rated by the amount of light that they transmit, typically 1/4 and 1/16. They are most often used under conditions of intense sunlight.

9.7.4 Optical frequency low-pass filters

The reduction of aliasing by the use of low-pass optical frequency filters that attenuate the high optical-frequency components in the image is described in Section 4.9.1.2. The filter is placed between the camera lens and the optical input of the beam splitter as shown in Figure 9.16.

9.8 Camera Signal Processing Circuits

Some of the camera signal processing circuits are discussed in this section. Both analog and digital applications are covered, although the focus today is on doing these things digitally.

9.8.1 Preamplifier

The output from an imager, tube or CCD, is analog. The preamplifier (preamp) must accept the signal output from the imager and bring it up to a level suitable for subsequent processes. The requirements for tube and CCD preamplifiers are quite different because of the different levels of signal currents coming from the imager. The low output from tube imagers requires special considerations in the design of the preamplifier because it makes the preamplifier a significant source of noise. By contrast, most of the noise in CCD cameras originates in the imaging device, and the output level from the device is high enough so that no special noise precautions are required in the preamplifier.

Camera blanking, the removal of the signal during the blanking intervals, can also be added at the preamplifier. This is augmented by the addition of *system blanking*, which determines the *setup* or system black level, later in the chain.

9.8.2 Analog-to-digital conversion

It is desirable to do all camera signal processing digitally, especially because operations that are often handled separately in analog circuits can be done all at once in a digital circuit. Therefore, ADC should be done as early in the system as possible, usually in the R, G, and B channels right after the preamplifiers. Then, all subsequent processing is digital, including the composite encoder, if used. When the rest of the system is analog, a single DAC after encoding provides the proper analog output. Of course, if the external system is digital, there is no need for any conversion at this point and the digital signals are formatted to the output transmission standard (see Section 3.7.5).

The quantization levels in camera signal digitization should be consistent with the SNR of the CCD sensors used so that the digital channels will handle the full capability of the sensors. For modern broadcast camera sensors, this means that 10 to 12 bits/pixel must be used. This is true even though the dynamic range may later be reduced to suit the capability of displays, where 8 bits/pixel is usually enough.

9.8.3 Aperture correction and image enhancement

The gain-frequency characteristics of aperture correction and image enhancement circuits are described in Sections 4.9 and 4.10. Aperture correction increases the gain of the higher-frequency signal components sufficiently to make the MTF curve of the camera flat to about 250 lines in NTSC systems and 400 lines in HDTV systems. Additional correction, known as image enhancement, can be used in the 200 to 300 line region to increase the sharpness of the picture further, but it must be applied with care to avoid unnatural overshoots.

Aperture correction will enhance noise as well as the picture sharpness. Thus, various techniques are used to minimize the effect of aperture correction on noise. In one refinement, enhancement is applied only in the midrange of the gray scale. This avoids degradation of the SNR in the dark areas of the picture where noise is most visible and also reduces a tendency for bright skin tones to appear rough.

It is important that the aperture correction gain enhancement be achieved without phase shift to avoid enhancing overshoots. One means of achieving this is a transversal filter, which can be achieved in either analog or digital circuits.

Contours-out-of-green is another technique for increasing picture sharpness. Enhancement signals are generated from the green signal at each edge in the image by separating its high frequency components. The green channel is used to generate the enhancement signal because it has the best SNR. The "white" contours—changes from dark to light—are added at the aperture correction circuit while the "black" contours are added after gamma correction.

9.8.4 Color correction (linear matrixing)

Electronic color correction is employed for two purposes—correct residual errors in the colorimetry of the optical system and introduce deliberate distortions of the

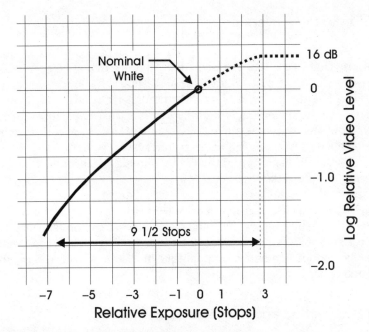

Figure 9.17 SMPTE 240M transfer characteristic.

colorimetry for aesthetic reasons or to more precisely match the colorimetry of different cameras that will be used together. Electronic correction is accomplished by matrixing (intermixing) the color signal components, and is called linear matrixing or *masking*. It can be done in either analog or digital circuits.

9.8.4 Gamma correction and adjustable knee

Gamma correction is usually introduced in each of the three color channels to compensate for the high gamma of the receiver picture tube. This stretches the signal in the black area where noise is very visible, and the uncorrected signal must have a good SNR. The SMPTE Standards 240M and 274M [see Eq. (9.3)] specify the gamma correction for HDTV production systems.

Additional gamma correction can be applied to the luminance (Y) channel of the encoder. This has the advantage that it does not upset the color balance of the signal.

To handle extremely large variations in scene brightness, some cameras provide an adjustable "knee" in the transfer characteristic. The characteristic follows the normal gamma curve up to an adjustable breakpoint, above which increases in illumination produce only a minor increase in signal level. This is a form of range compression (sometimes called contrast compression), and it avoids overloading the system in subsequent stages while retaining some detail in the brightest areas. Such a characteristic is illustrated in Figure 9.17; it follows the SMPTE Standard 240M over its normal operating range, but with a knee above which there is a major reduction in its slope. The knee is set slightly above refer-

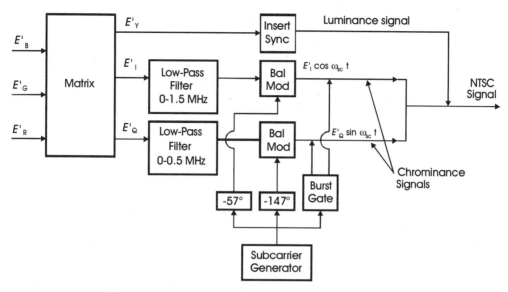

Figure 9.18 Block diagram of an NTSC encoder.

ence white, and the image contrast is greatly compressed but not eliminated above this range. Also see the discussion about highlight control in Section 9.5.10.

9.8.5 Sync generation and blanking

In a studio installation where cameras, recorders, and editing and processing equipments are synchronized by a common source, a central sync generator supplies these signals to all units. When a camera operates by itself, as in a self-contained field camera, the camera must contain its own sync generator. A third possibility is *genlocking*, where the camera derives its timing pulses and subcarrier from an external video signal.

Analog output signals require horizontal and vertical blanking to be inserted to provide clean intervals for the synchronizing pulses. Digital formats, however, often do not have blanking intervals because synchronization can be done from the video data itself. Thus, digital systems do not have to waste the almost 25% of transmission time devoted to blanking in analog signals.

9.8.6 Encoder

Encoding, the conversion of the three primary color signals into composite NTSC, PAL, or SECAM signals, is the final function for camera chains that deliver composite outputs. Encoding can be done with either analog or digital circuits. The characteristics of encoded signals were described in Chapter 2.

Figure 9.18 is a block diagram of an NTSC encoder. The gamma corrected primary color signals, E'_R, E'_G, and E'_B, are passed through matrices to form the E'_Y, E'_I, and E'_Q signals. Sync is added to the E'_Y signal to form the luminance signal.

After filtering and the necessary delays, the E'_I and E'_Q signals modulate subcarriers, which are delayed by 57° for the I component and 147° for the Q, with balanced modulators to generate the color subcarrier. The color subcarrier is combined with the Y' luminance signal to form the composite color signal.

The burst is also added by keying the subcarrier with the burst gate signal. The subcarrier can be generated by an atomic clock, a temperature-controlled crystal oscillator, or derived from an incoming signal with a genlock.

9.8.7 Viewfinder

Electronic viewfinders have been a standard feature of studio cameras from their earliest days, and they are functionally superior to the optical viewfinders on film cameras, since they show the exact signal that is being generated by the camera. Viewfinders on modern cameras provide a number of useful features in addition to their basic function:

- Cursor—A remotely controlled cursor provides a means of communication between the director and camera person.
- Safety zone—An area encompassing approximately 90 percent of the image marking the useful area of the picture.
- External video connection—A connection that makes it possible for the camera operator to see the output of the camera in relation to other signals. Warning and status signals can also be displayed.
- Color—Viewfinders are now available in color.

9.8.8 Triax cable

The first color camera heads were connected to the control unit with large and heavy multiconductor cables. With the smaller size, lower power requirements, and greater stability of solid-state components, and the use of CCD imagers, it has been possible to concentrate more of the circuitry in the camera head and reduce the number of connections between camera head and camera control unit. Thus, all the video, control, and power connections can be transmitted in a single small *triax* (double-shielded coax) cable. The control signals are digitally multiplexed, and the video circuits are transmitted by subcarriers as shown in Figure 9.19 for an analog triax system.

When the signal electronics of cameras is digital, it is possible to perform all triax multiplexing digitally and some newer cameras are doing that. Fiber-optic cable has also been introduced for camera cable applications.

9.9 Digital Camera Electronics

The camera signal processing described in Figure 9.1 and Section 9.8 includes operations that deal with sensor limitations, scene characteristics, and general matching of the signal to the properties of the system that follows. They are fundamental processes that must be done in any camera, whether analog or digital.

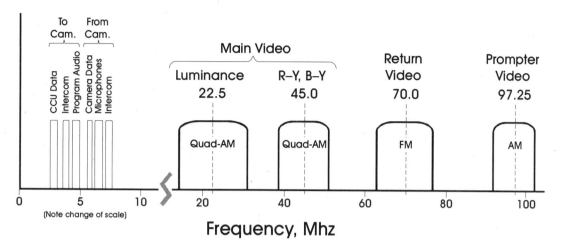

Figure 9.19 Frequency plan for analog triax multiplexing. (*From Sony BVP-700/750.*)

The digital revolution, which is making radical changes in every aspect of video technology, has now come to camera signal processing.

Digital processing offers a number of significant advantages and it is replacing analog processing in most applications. These advantages include:

- Greater stability and repeatability of adjustments. The drifting of analog circuits is a long-standing problem (although there have been great improvements in circuit stability in recent years). With digital processing, circuit parameters are stored in a memory and stay constant for any desired length of time. If it is necessary to make changes, the memory is updated with the new parameters.
- Precision of camera alignment and matching. Cameras can be aligned and matched precisely by equalizing their parameter values.
- Range and flexibility of camera adjustments. Camera parameters can be adjusted over wider ranges and with greater flexibility for optimum control of image quality.
- Highlight handling [4]. The output of CCD imagers from highlights can be 600% or more of the reference white level and it is desirable to reproduce image detail over this entire range. With analog processing, the signal is squeezed into the available gray scale range by the brute force combination of compression and clipping. This results in severe loss of detail in the image highlights. With digital signal processing, detail can be reproduced over the entire range with a combination of analog and digital compression. For example, the CCD signal from a 600% highlight can be compressed in the analog preamplifiers ahead of A-D conversion to about 220% of reference white. This is sometimes called "pre-knee" compression. This signal is then converted to a 10-bit digital signal which can be further compressed by variable amounts over the contrast range- for example, a high degree of compression can be applied to signal levels between 130 and 220% of reference white and a lesser amount of compression can be used at lower signal levels.

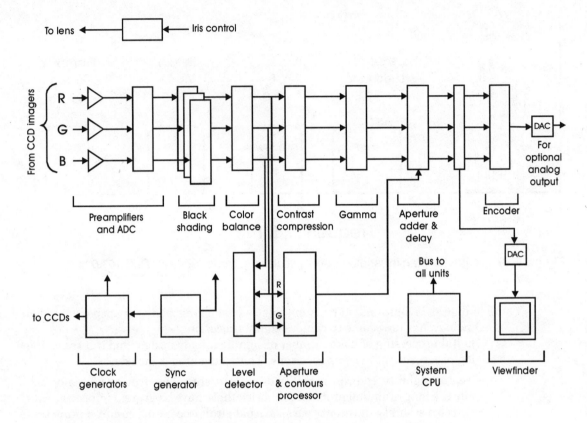

Figure 9.20 Block diagram of an all-digital camera processor.

9.9.2 Basic digital processing

Figure 9.20 summarizes a digital processing system for a fully digital camera [5]. This may be compared to Figure 9.2 for an analog camera.

ADC occurs immediately after the preamplifier in each channel. Then black shading is performed. This is an automatic process that involves capping the camera lens and reading the sensor black signals into a frame memory for each channel. The contents of that memory is then continually read out to correct the black level of each pixel while the camera is in operation.

The next step is to set color balance by adjusting the gain of each channel. This is performed by use of a RAM lookup table that contains the desired pixel value for each of the possible quantizing levels (a 10 bits/pixel system would have 2^{10} (1024) locations in this memory). In a lookup table, the input pixel values are used as the address into the memory and the memory read values are the output. The values in the RAM are calculated and updated by the system CPU during blanking intervals. Adjustments are thus immediately applied to the system.

Contrast compression and gamma are applied in the same way through a RAM lookup table. In this case, the RAM values represent the desired nonlinear transfer function. Again, the system CPU calculates the curves and manages the updating of the RAM.

Aperture correction is applied by extracting signals before the nonlinear processing, creating the appropriate correction signals, and adding them to the channel after the nonlinear processes. This is to avoid enhancement of noise in the nonlinear processes. Aperture is provided both horizontally and vertically, and various algorithms can be implemented. For example, in an NTSC camera, the contour signal can be dynamically derived from either the red or green channel depending on which has the greatest amount of detail present. This avoids the problem of loss of red detail when contours are derived from green only, while still not enhancing registration or lateral chromatic aberration errors.

The presence of the system CPU and its bus allows all control functionality and algorithms to be implemented in software. This provides the same flexibility for offering options or updating as enjoyed by personal computer users.

9.10 Camera Operating Controls

Early color cameras had an enormous array of operating controls, and many stations employed a video operator for each camera to keep it in adjustment. The advent of CCD imagers, digital solid-state circuitry, automatic controls, and general advances in technology have greatly simplified the control problem. Automatic exposure controls that adjust the lens aperture make it possible for the camera to operate over reasonable ranges of scene illumination without adjustment. Digital technology has increased the stability of control circuits so that they do not require frequent adjustment. As a result, hand-held ENG cameras frequently operate without any readjustment of the controls for considerable periods of time and multicamera studios can now be handled by a single camera operator. Digital technology has also made it possible to put a number of preset control positions in a memory to be recalled at will.

On the other hand, the ever-increasing demand for the highest possible picture quality has necessitated more sophisticated control of the cameras' adjustments, including precise camera matching. Further, the adjustments may be dictated as much or more by aesthetic judgment as by rigorous technical considerations. The adjustment of the controls by the appearance of the picture rather than by objective test procedures is called painting. Needless to say, a carefully adjusted color monitor must be used as the reference.

Table 9.2 shows the operating controls (as opposed to setup controls that would be adjusted during routine maintenance procedures) for a typical studio camera. Function controls that have the option of presetting or automatic operation are indicated. The "painting" category includes adjustments that may be influenced by subjective judgments as well as objective measurements.

Table 9.2 Camera operating controls

Item	Options	
	Automatic	Preset
Basic picture quality		
Master gain		X
Master black		X
Exposure		
Iris	X	
Color filter		
Neutral density filter		
Registration		
Not required with CCDs		
Painting		
White balance	X	X
Black balance	X	X
Flare compensation	X	X
Shading		X
Aperture response		X
Image enhancement		X
Gamma		X
Knee	X	X
Color balance		X

9.11 Camera Specifications

Table 9.3 is a typical specification for a current broadcast top-of-the-line camcorder.

9.12 Video Film Camera Systems

Film cameras, called *telecine* cameras, convert motion picture film to video.

9.12.1 A brief history

Before it was possible to record video signals on magnetic tape, film was the only practical medium for the storage of video program material. Video film cameras and their companion projectors were critical components of TV station equipment complements, and in many TV stations most local programming originated on film. The *videotape recorder*, first introduced in 1956, provided an alternative medium with the additional feature that recordings could be made easily and quickly by individual stations. The advantages of the tape format increased steadily

Table 9.3 Representative published camera specification

Item	Specification
General specifications	
Pickup system	3 × 2/3-in FIT CCD
CCD active pixels	980 × 484
Optical system	f/1.4 prism
Filter wheels	4-position color filters
	4-position ND filters
Shutter speeds	1/100 to 1/2,000 s
Camera weight (approx.)	7 kg (with VF, battery, tape, lens)
Dimensions (approx.)	37 × 27 × 13 cm (without lens)
Environmental specifications	
Operating temperature	0°C to 40°C
Storage temperature	−20°C to 60°C
Humidity	Less than 85%
Video Performance (camera)	
Sensitivity	f/8 @ 2,000 lux (3,200°K, 89,9% refl.)
Minimum illumination	Approx. 1.9 lux @ f/1.4 with +30 dB gain
SNR	62 dB (typical)
H. limiting resolution	900 TVL at center
V. limiting resolution	400 TVL
Horizontal MTF	70% at 400 TVL
Smear level	−140 dB
Registration	0.05% (without lens)
Viewfinder	
CRT	1.5-in monochrome
H. resolution	600 TVL
Video performance (VCR)	
Video SNR	51 dB
Audio frequency response	20 to 20,000 Hz (+0.5, −2 dB)
Audio dynamic range	80 dB
Audio distortion	less than 0.5%

as progress was made in recording technology, and today, a transfer from film to tape is often made before a film program is put on the air. This is simply because videotape recorders have become so easy to operate. Video film cameras are now used for film-to-tape transfer as well as to generate signals that are broadcast directly. Film continues to be an important production and storage medium, and film cameras will continue to be an essential part of the TV industry's equipment complement for the foreseeable future.

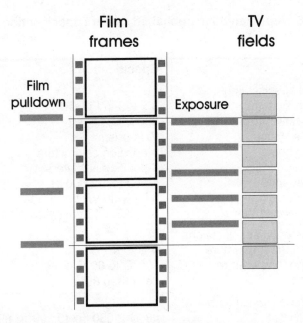

Figure 9.21 3:2 pulldown for conversion of 24 frames/s film to 30 frames/s TV. (*Reproduced with permission from [7].*)

9.12.2 Frame rate conversion

In countries where the standard TV frame rate is 25 per second, film frame rate conversion is sometimes not used. The film is simply speeded up by 4 percent from its normal rate of 24 per second. The difference in the picture rate is barely noticeable, and the increase in sound pitch is noticeable only to the relatively small number of individuals who have absolute pitch. However, with digital technology, frame rate conversion is easy to accomplish so that even this small discrepancy can be corrected (and generally is).

In 30 frame-per-second countries such as the United States, the difference in frame rates is too great to be ignored for both pciture and sound, and frame rate conversion is mandatory. The conversion of the 24 frame-per-second film frame rate to the 30 frame-per-second TV rate is a basic function of film camera systems in these countries.

Standards conversion was originally effected by (1) the use of a flying-spot scanner with a projector that generated a continuous image of film frames that lap-dissolve into each other; or (2) by a storage tube camera with an intermittent projector employing a 3:2 pull down. This is a method where film frames are exposed alternately for 2 TV scanning fields and 3 TV fields, as shown in Figure 9.21. This gives an exact conversion of 24 frames/s to 60 fields/s (30 frames/s).

The most commonly used film systems used 3:2 projectors and storage tubes, usually vidicons or plumbicons. The film frames were pulled down at alternate intervals of 2/60 and 3/60 seconds for a total of 1/12 second or two film frames for each cycle. The storage capability of the camera made it unnecessary for the exposure and readout to be simultaneous, and five television fields were generated from the images of two film frames, thus effecting the rate conversion.

9.12.3 The CCD/continuous-projector system

Two major technical developments brought a third system into use that will probably replace the others. It combines a continuous projector, a CCD camera, and a digital standards converter (see Chapter 4).

The continuously moving film is illuminated by a light source, and its image is projected onto a horizontal row of red-, green-, and blue-sensitive solid-state sensors arranged in triads, one triad for each pixel. Horizontal line scan is accomplished by sampling the sensor voltages sequentially. Vertical scanning is accomplished by the motion of the film. The output line-scan signals, which are generated at the 24 frames-per-second film rate, are converted to a digital format and stored in a memory. Standards conversion is accomplished by reading them out of the memory at the desired television frame rate.

The functions performed by the electronics for a CCD film camera are similar to those for live cameras but with some modifications necessitated by the line-scan format and standards conversion.

9.13 Digital Still-Image Cameras

The technologies of digital video cameras can be adapted to the capture of digital still pictures in much the same way as photographic film cameras. The difference is that there is no film or development—captured photos can be viewed immediately on the camera viewfinder and they can be digitally transferred to a PC for storage, manipulation, or printing. The advantage of instantaneous viewing, no film cost, and PC manipulation and printing are the features driving this market.

The disadvantage of this compared to 35-mm film is that picture quality is not as good as film. However, a substantial market has developed for these cameras and nearly all the traditional photographic camera manufacturers are offering digital cameras.

Except in extremely expensive professional digital cameras, still-image camera resolutions range from 640 × 480 pixels up to about 1280 × 1024 pixels. This is limited by the capability of the CCD sensor in the camera—higher resolutions are available only at great expense. Such cameras are currently priced below $800 and prices are still falling as more are being sold. It is a substantial new market.

A block diagram of a digital still-image camera is shown in Figure 9.22. The signal from the CCD is digitized, processed and compressed for storage. At the same time, it is displayed on the camera's built-in LCD display. Local storage is usually by a flash memory in the camera—often this is a removable card that can be plugged into a special interface on a computer for transfer. Alternatively, transfer occurs over a serial connection. Some cameras also have a composite video encoder so the pictures can be viewed on a TV monitor or receiver.

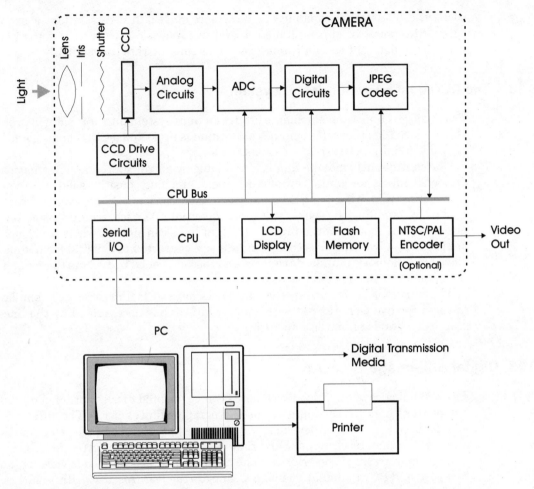

Figure 9.22 Block diagram of a digital still-picture camera and its external environment. (*Reproduced with permission from [6].*)

Since there is such similarity between motion video cameras and still-image cameras, some cameras are being designed to do both. However, this is a compromise, because most applications of still-image cameras really need more resolution than is necessary in a motion-video camera.

9.14 References

1. M. Kurahigi, N. Egami, K. Tanioka, and K. Shidara, "Super-Sensitive HDTV Camera Tube with the Newly Developed HARP Target," *SMPTE J.*, July 1988.

2. Koichi Sadashige, "An Overview of Solid-State Technology," *SMPTE J.*, February 1988.

3. L. Thorpe, "Focus on CCDs," Part I to Part V, *TV Technology*, September-December 1987, February 1988.

4. L. Thorpe, E. Tamura, and T. Iwaski, "New Advances in CCD Imaging," *SMPTE J.*, May 1988.

5. Tom Leacock et al., "An HDTV Digital Camera Processor," *SMPTE J.*, pp. 580-585, September 1994.

6. A. Luther, *Video Camera Technology*, Artech House, Norwood, MA, 1998.

7. A. Luther, *Principles of Digital Audio and Video*, Artech House, Norwood, MA, 1997.

8. Y. Ide et al., "A Three-CCD HDTV Color Camera," *SMPTE J*, July 1990, pp. 532–537.

Chapter 10
Professional Video Recorders

10.1 Introduction

Video program production as it exists today would be impossible without video recorders. This chapter covers some of the fundamental technologies used in nearly all video recorders. It then concentrates on high-end recorders used in broadcasting and program production. Chapter 11 continues the subject of recording and discusses home recorders and related products.

10.1.1 A brief history of video recording

For more than ten years following the introduction of commercial television at the end of World War II, the problem of developing a satisfactory video recorder baffled the industry's major research laboratories. Such a product was desperately needed, but the technical difficulties proved too daunting until 1956. In that year the situation changed dramatically when a small group of young engineers at the Ampex Corporation announced a combination of technologies that provided a satisfactory magnetic tape recording solution called the *quadruplex* recorder.

The key component of the quadruplex recorder was the headwheel, a wheel about 50 mm in diameter with four magnetic recording heads spaced 90° apart around the rim. The direction of tape travel was parallel to the axis of the wheel and the 50.8-mm-wide tape was partially wrapped around the rim of the wheel. The tape was pulled past the wheel at a speed of 381 mm/s, and most of the head-to-tape motion was imparted by the transverse rotation of the wheel at a rate of 240 revolutions/second. The rotation resulted in a head-to-tape speed of 39.6 m/s, far higher than would be practical for a longitudinal recorder. The combination of the rapid transverse motion of the head from the wheel's rotation and the much slower longitudinal motion of the tape produced a series of recording tracks, almost normal to the tape edges and occupying a time interval of approximately 16 scanning lines.

Quadruplex recorders were capable of producing excellent pictures when skillfully maintained and operated, but they were subject to a number of problems that led to a rapid deterioration of picture quality if maintenance was less than perfect. The most serious was "banding"—variations in the appearance of the 16-line bands recorded by adjacent heads due to small differences in their frequency or phase response. In addition, the quadruplex format lacked flexibility and could not be used for slow-scan or stop-frame applications. Aided by year-by-year design improvements, the quadruplex recorder had a long life, and it was in use for more than twenty years until the late 1970s.

However, another type of rotary-head recorder remained in the background—the *helical-scan* format. In the 1970s, these recorders were introduced into consumer and semi-professional markets and became successful.

The helical-scan format solved many quadruplex recorder problems, but it introduced new ones. The most serious was its *time-base instability* resulting from variations in the long recording tracks on the elastic tape medium. This instability could be tolerated in consumer-level recorders but not in complex high-performance broadcast systems. The problem was eventually solved by the digital time-base error corrector, which provided an effective and economical means of stabilizing the time base. (Analog time-base correctors were used in quadruplex recorders, but that technology could not achieve the range of correction required for helical-scan recording.)

After digital time-base correction was introduced, the replacement of quadruplex by helical scan was almost inevitable, even for the most demanding broadcast and teleproduction applications. The transition was not sudden, owing to the huge libraries of tapes recorded in the quadruplex format, but it was nearly complete by the mid-1980s. As of this writing (1999), helical-scan recorders are almost universally used.

The helical-scan format is remarkably versatile, and variations of it now exist for the entire gamut of video recorder uses, from inexpensive mass-produced VCRs to the most complex and sophisticated broadcast and teleproduction systems.

10.1.2 Types of video recorders

The video recording market has grown to include a broad range of products with a broad range of applications. One way to classify the products is by their end use:

Home

Education and training

Broadcasting and production

Multiformat production

These categories are listed in order of signal performance capability, which, not surprisingly, is in inverse order of price. However, there are other measures of system capability besides price-performance and these mean that even the lowest-price units require extremely sophisticated design and production. The following paragraphs discuss each of the categories above.

10.1.2.1 Home video recorders

The home environment is not friendly to sophisticated electronics. Any home product must be completely reliable without the help of any operator skill and it must do that over a period of many years. If it does not, the consumer faces a difficult and expensive service problem, which causes most people to give up and the equipment ends up in a closet. If that happens too soon in the product life, the consumer will feel taken and the product will quickly get a bad reputation and will not sell. Reliability is the most important design objective after price.

The signal performance of a home recorder can have some compromises compared to other categories of video recorder. The pictures must look good and the audio must sound good, but exact performance numbers are secondary. What this means in terms of numbers is shown in Table 11.1.

10.1.2.2 Recorders for education and training

This market is sometimes called "semi-professional" and includes other classes of use that require the same attributes. These users can afford no compromise in reliability or ease of use but they require higher signal performance than the home market. That is because recordings must be good enough to support the production-postproduction style of program creation and tapes will be duplicated to multiple users. At least three generations of recording will often occur in producing and distributing tapes. Other features such as computer control interfaces also become important to these users. At the same time, the market can only afford modest price increases compared to the home market.

10.1.2.3 Broadcast recorders

Signal performance comes first with broadcast video recorders. Standard specifications must be met and elaborate system integration features are required. Skilled technicians will be available, but the need for their time must be minimized. For all this, the broadcaster will pay significantly more to obtain the high-performance reliable equipment that they need. Broadcast equipment prices are often 10 or more times higher than the lesser markets—this is a result of the more exacting requirements and also because the number of products sold is much lower.

10.1.2.4 Production master recorders

Producers of high-quality programs are interested in delivering their programs in as many formats as possible in order to achieve the widest distribution. Conversion to other analog video formats may be required, such as going from NTSC to PAL (or back), video disc, motion picture film, or any other video format. Since the cost of recording is small in the total cost of the production, producers will pay to guarantee that the recording medium will not limit distribution choices. Component digital recorders (see Section 10.11) were first developed with this kind of use in mind.

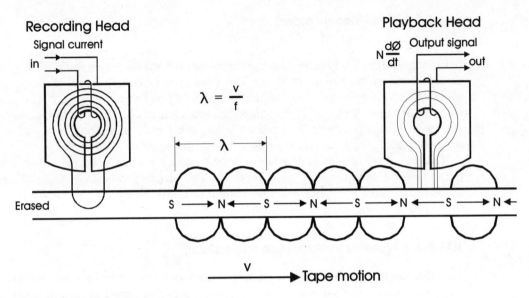

Figure 10.1 Magnetic recording.

Production recorders need the highest possible performance and should be capable of as many generations as required for any type of production. This chapter discusses the technology of broadcast and production recorders; Chapter 11 will cover the home and educational products. However, there is a lot of technology overlap between the classes, so there are many cross references between these two chapters.

10.2 Principles of Magnetic Tape Recording

Magnetic recording is accomplished by providing a means of relative motion between a plastic tape that has a coating of magnetic material and a head structure that is capable of generating a magnetic field for recording and sensing a magnetic field for playback. This is shown in Figure 10.1. The signal current passes through a winding around the core of the record head assembly, which provides a magnetic circuit for the lines of force that are induced by the signal current. The lines of force, whose density depends on the magnitude of the signal current, fringe into the tape at the low permeability gap, leaving a pattern of alternating N–S poles on the magnetic surface of the tape. The pattern is called the *record*.

Figure 10.2 shows the detail of the *magnetizing force*, H, in the vicinity of the pole-tip gap. The solid curve is the locus of points where H equals the coercivity of the magnetic tape coating, H_C. The *coercivity* is the value of H that is required to magnetize a tape or to return its magnetization to zero. The area within the contour $H = H_C$ is known as the record zone, and no magnetism is produced outside this zone. The depth of magnetization and the diameter of the zone is less for high-coercivity tape, and this results in the ability to record shorter wave-

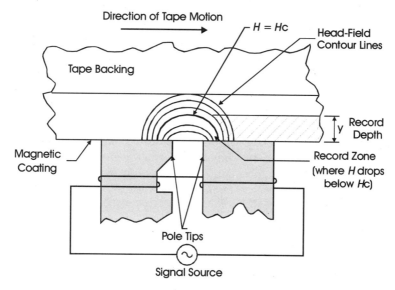

Figure 10.2 Record zone. (*From [6], Fig. 15.39.*)

lengths and higher frequencies.

The reverse process occurs on playback. Magnetic lines of force from the pattern on the tape pass into the core of the head and induce a voltage in the winding that is proportional to the rate of change, the derivative with respect to time, of the magnetic field strength.

$$e = C \frac{dB(t)}{dt} \tag{10.1}$$

where C is a constant and $B(t)$ is the magnetic field strength across the gap.

If $B(t)$ is a sinusoidal function, the voltage is proportional to the frequency and it increases by 6 dB per octave until effects described in Section 10.7.1 reduce the efficiency of the energy transfer and the response declines with increasing frequency.

10.3 Video Head Design

Magnetic video recording imposes stringent requirements on the performance of the video record/playback heads, and improvement of their design has been an ongoing effort that has continued during the past 40 years. In combination with improvements in the tape itself, head development has been one of the most important factors contributing to the steady improvement in recorder performance.

Among the requirements that video recording heads must meet are the following:

1. They must be able to operate effectively at very high frequencies—10 MHz and upward for FM-modulated analog signals and 108 Mbs and upward for PCM-modulated digital signals. Among other constraints, this requires that the gap between pole tips be somewhat less than the *wavelength* λ of the highest recorded frequency f as calculated by Eq. (10.2):

$$\lambda = \frac{\text{head-to-tape writing speed}}{f} \qquad (10.2)$$

   Typical modern recorders operating at an FM carrier frequency f of 8 MHz and a writing speed of 25 m/s have a recorded wavelength of 3 μm. The head gap length must be smaller than this value and must be accurately maintained. The mechanical precision required is awesome.

2. The pole tips must be sufficiently narrow to record and play back on very narrow tracks to provide a high recording density. Typical present-day values range from 25 to 120 μm.

3. The head material must have high *permeability*, the ratio of the magnetization flux density, B, to the magnetizing force, H, so that a satisfactory flux density is induced with a modest magnetizing force.

4. They must have a high ohmic resistance to minimize losses from eddy currents.

5. They must be capable of being machined or otherwise formed with very close dimensional tolerances.

6. They must withstand the mechanical environment of head-to-tape contact and provide a long service life. This means that they must be resistant to wear from abrasion and to mechanical damage from cracking or chipping.

The major research and engineering effort that has been devoted to head materials and designs has resulted in an array of different designs, many of which have proprietary features.

Because the requirements for pole-tip and core-structure materials are different, many heads have a composite construction with one material for the pole tips and another for the core. Figure 10.3 shows the essential features of a typical composite head design. It has a supporting core structure to which pole tips are bonded. As with the resistance of electric conductors, the reluctance of magnetic conductors is inversely proportional to their cross section, and this is large except in the area of the pole tips. The pole tip is narrow to generate a narrow recorded track.

It is also possible to produce heads from a single material that combines suitable mechanical and magnetic properties. The material selection is more difficult, but manufacturing becomes easier and this approach is used in most head designs for high-volume production.

In composite head designs, the core material is usually ferrite, a mixture of iron oxide (Fe_2O_3) and manganese, nickel, or zinc, which has high permeability and resistivity. Ferrites come in two varieties, hot-pressed and single crystal. Single crystal ferrites can be machined or otherwise shaped more accurately, but the crystallographic orientation must be carefully controlled.

The pole-tip material has the additional requirement that it must be resistant to wear by abrasion and mechanical damage. This was a problem for ferrites that

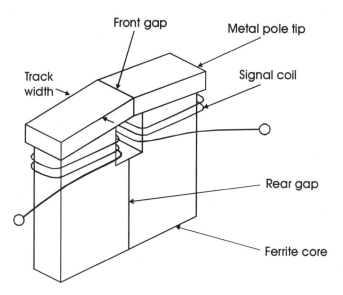

Figure 10.3 Magnetic recording head. (*From [6], Fig. 15.40.*)

was reduced by a glass bonding technique in which a layer of glass is bonded to the surfaces in the vicinity of the gap. It also led to the use of iron/silicon/aluminum alloys known by such names as alfesil, alfecon, and sendust. These alloys had the problem of low resistivity, but this was overcome by fabricating them from powders in which the particles were held together by an insulating cement. This reduces the permeability and a balance must be struck. The electrical and magnetic properties of representative head materials are tabulated in Table 10.1.

10.4 The Helical-Scan Principle

The process of presenting the tape to the heads is basically a mechanical one and mechanisms for this purpose are called tape transports. They handle the unreeling and rereeling of tape, and provide a tape path where the tape passes over video, audio, and auxiliary recording heads.

A *capstan drive* is used for accurate control of tape speed. A capstan is a rotating metal shaft that contacts the tape under pressure from a rubber pinch roller. The capstan is driven by a servo motor whose speed is electronically controlled to achieve the exact tape speed and synchronization.

The preferred means for achieving the relative motion between tape and heads at speeds suitable for video recording is the helical-scan method, illustrated in Figure 10.4. The tape is wrapped in a helix around a drum on which record/playback heads are mounted, and the drum rotates in a direction opposite to the longitudinal tape motion. The result is a series of diagonal tracks recorded across the tape as shown in Figure 10.5, which is for the SMPTE Type C format. In addition to the diagonal tracks used for video and sync (video and sync are recorded sepa-

Table 10.1 Magnetic properties of head materials

Material	Permeability*	Resistivity (μW/cm)†
Alfesil‡	100–1,000	90
Ferrite		
Hot-pressed, Mn, Zn	300–500	10^5
Single-crystal	300–500	10^5

* Permeability = $\dfrac{\text{flux density } B}{\text{magnetizing force } H}$

For a complete definition, see Fink and Christiansen, p. 6-65. Permeability is an important specification for head materials because it indicates the flux density and signal voltage generated by the magnetic pattern on playback.
† High resistivity is desirable to minimize eddy current losses.
‡ The resistivity of alfesil is increased and its permeability reduced by fabricating from powders.
SOURCES: K. Blair Benson, *Television Engineering Handbook*, McGraw-Hill, New York, 1986; Donald G. Fink and Donald Christiansen, *Electronic Engineers' Handbook*, McGraw-Hill, New York, 1989.

rately in this format), longitudinal recording is used for the audio tracks and for the control track on which signals are recorded for synchronizing the tape deck.

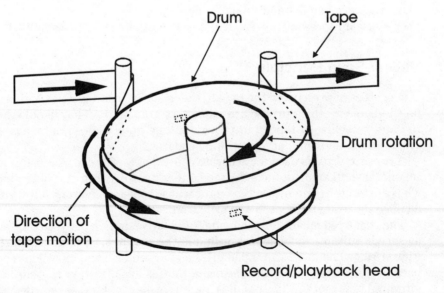

Figure 10.4 The helical-scan principle.

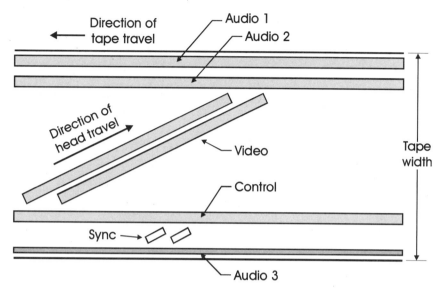

Figure 10.5 SMPTE Type C record format. (*From [1]*).

10.5 Tape Transport Configurations

A variety of helical-scan tape transport configurations have been utilized in commercial recorders to meet the varying requirements of different applications.

The tape can be stored on reels for reel-to-reel recording or in cassettes (see Section 10.5.1). Cassettes are easier to handle and load but their tape capacity is not as great as open reels. Capacity limitation has become less of a problem with improved and thinner tapes, and the trend is toward cassettes. All consumer recorders use cassettes because of their ease of loading. The SMPTE Type C recorder, the most widely used professional format, is reel-to-reel, but cassettes are becoming increasingly common as their performance is improved.

A basic classification of helical-scan systems is the way the tape wraps around the drum. *Omega wrap* (as in Figure 10.4) and *alpha wrap* are named for their resemblance to the Greek letters and the classification is based on whether the tape path loops all the way around the drum (α) or whether the tape approaches and leaves the drum along a radius (Ω). Most recorders use omega wrap and one or two video heads per channel.

With two heads, the tape only wraps a little more than 180° around the drum but wth single-head recorders the tape must wrap entirely around the drum except for a very small gap. On recording, the rotation of the drum is synchronized with the video signal so that the head crosses the gap (when recording is interrupted) during the vertical blanking interval.

Another classification of helical-scan systems is based on the fraction of a TV field recorded in one pass of a head, field-per-scan, or segmented scan. Field-per-scan is advantageous because no head switching occurs during the picture and

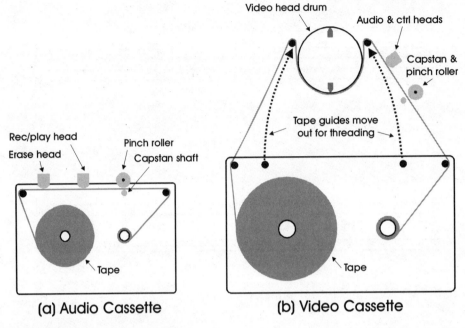

Figure 10.6 Operation of tape cassettes.

slight differences between heads are more easily tolerated. Segmented-scan formats require two or more heads per channel and very precise matching of heads or the use of digital recording.

10.5.1 Tape cassettes

A *cassette* is an enclosure that holds two tape reels and a tape that can be reeled from one reel to the other by an external drive that engages the cassette reels. This concept was originally developed for audio recorders, which have stationary heads. In audio recorders, the tape is presented to the heads by having an opening in the side of the cassette that exposes the tape and simply presses that part of the cassette against the heads. The only other step is to provide a means to engage the capstan drive for the tape, which is usually done by dropping the cassette over the capstan shaft and using a rubber pinch roller outside the cassette to press the tape against the capstan. This is illustrated in Figure 10.6(a).

The cassette principle is more difficult to implement for video recorders because the tape has to wrap around the scanning drum as well as press against stationary heads and capstan. The usual approach is to use a two-head-per-channel configuration and provide a mechanism that draws the tape out of the cassette to the scanner. Figure 10.6(b) shows this. In spite of the difficulties, all new video recorders, both professional and home, in the last five years use cassettes. The widely used acronym VCR means *videocassette recorder*.

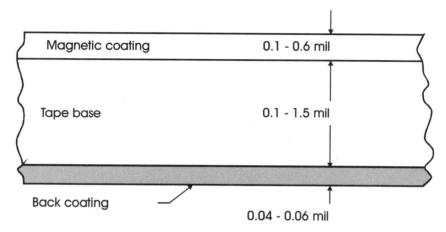

Figure 10.7 Magnetic recording tape. (*From [6], Fig. 15-47.*)

10.6 Magnetic Recording Tapes

Magnetic recorders originally used a fine iron wire as the recording medium, but this was superseded by tape soon after the tape format was developed by German engineers during World War II. As with magnetic recording heads, magnetic tape has been the subject of an enormous amount of research since World War II, not only for its uses in audio and video recording but also for data processing and storage.

A cross-section of a magnetic recording tape is shown in Figure 10.7. It consists of a plastic tape base film coated on the front with a magnetic material and on the back with a conductive backcoating to prevent the build-up of static charges.

10.6.1 Tape base material

The tape base is a polyester chosen for its strength and dimensional stability with respect to temperature, humidity, and time. The thickness of the base film is chosen to be as thin as possible to minimize the reel diameter but thick enough for mechanical strength and to prevent print-through, which is the tendency for recordings to interfere between adjacent layers of the magnetic coating when the tape is rolled up on the reel. The smoothness of the magnetic coating surface must be controlled carefully also—as smooth as possible to provide close contact with the recording head but still rough enough to provide proper friction for tape handling.

10.6.2 Tape magnetic material

The layer of magnetic material is a powder suspended in a binder that bonds it to the base film. The preparation and application of this layer involves processes,

Table 10.2 Magnetic properties of tape materials

Material	Permeability	Coercivity
Ferric oxide	5,000	300–350
Chromium dioxide	6,000	300–700
Metal particles	10,000	1,000

SOURCE: Donald Christiansen, *Electronic Engineers' Handbook*, McGraw-Hill, New York, 1989.

often proprietary, that differ from manufacturer to manufacturer.

The desired electromagnetic properties are high permeability and coercivity (see Figure 10.2). These may seem to be in conflict since permeability is a measure of the ease with which a material can be magnetized while coercivity is a measure of the threshold magnetizing force. An ideal material is one in which the threshold is high so that the depth of magnetized layer and the minimum wavelength is low but in which small increases in magnetizing force above the threshold produces large increases in magnetization.

Three magnetic materials, ferric oxide, chromium dioxide or CrO_2, and metal particles have been commonly used for the magnetic layer. Each has advantages that make it suitable for particular applications. A fourth material, iron oxide doped with cobalt, has desirable magnetic properties but suffers from strong temperature dependence. The permeability and coercivity of these materials are tabulated in Table 10.2.

Ferric oxide was originally used for the magnetic tape and is still the most common. It is relatively inexpensive and is satisfactory for consumer applications that do not require the ultimate in performance.

Chromium dioxide has another desirable property, a high saturation magnetization, the maximum possible value of B. With the increased magnetization, a SNR 5 to 7 dB higher can be achieved than with ferric oxide. A disadvantage is that it is more abrasive.

Metallic particles suspended in lacquer to reduce rusting have high coercivity, high output signal level, high SNR, and a wider frequency response. It is the common choice for applications with demanding performance requirements. The recording characteristics of these materials are different, and they require different head designs for optimum results.

10.7 Tape and Record/Playback Head Performance

This section describes the combined performance of magnetic tape and record/playback heads with respect to frequency response and linearity for analog signals and bit rate for digital—the signal criteria that are directly determined by the tape and head characteristics.

10.7.1 Frequency response

Achieving an adequate frequency response for the FM-modulated video signal was one of the most difficult problems faced by the pioneers in video tape recording technology. The bandwidth of the original quadruplex recorder RF channels was limited to 6.5 MHz, an enormous technical achievement at the time. Highband recorders, introduced in 1964, had an RF bandwidth of 10.0 MHz, another breakthrough. Present professional analog recorders have RF bandwidths to 15.0 MHz or higher. (Consumer-type analog recorders typically have RF bandwidths of 7 MHz; see Chapter 11.)

The equivalent frequency response (half the bit rate for PCM digital channels) of digital video recorders is even greater. The SMPTE D-2 standard for composite NTSC signals specifies a $4f_{SC}$ sampling rate with 8-bit words (see Table 3.2). This requires a bit rate of 114.6 Mb/s, approximately equivalent to an analog bandwidth of 57.3 MHz. The D-2 recorder uses two parallel tracks to achieve this bit rate.

The standard sampling rates for 4:2:2 component signals are 13.5 Mb/s for the luminance signal and 6.75 Mb/s for each of the two color-difference signals. The corresponding bit rates are 108 Mbs and 54 Mbs. Combining these signals serially on single track would require an equivalent bandwidth of 108 MHz, an impractical rate even with the most modern technology. As a consequence, in the D-1 recorder, these signals are recorded with four parallel tracks.

At lower frequencies the frequency response of the tape-head combination increases at the rate of 6 dB per octave as shown by Eq. (10.1). However, as the frequency is increased further and the wavelength [Eq. (10.3)] approaches the gap length in the direction of tape travel, the other effects described below reduce the response so that it falls off with frequency.

The fall-off in frequency response at higher frequencies has four sources:

1. The finite gap width. The response from a gap of width d is:

$$e = k \frac{\sin\left(\frac{2\pi d}{\lambda}\right)}{\left(\frac{2\pi d}{\lambda}\right)} \qquad (10.3)$$

where k is a normalizing constant, d is the length of the pole-tip gap, and λ is the recording wavelength.

This equation has the $\sin(x)/x$ format that describes many aperture processes. Note that there is a null where $d = \lambda$. There are practical limits to how small the gap length can be made. Gap lengths smaller than about 1.0 μm are difficult to maintain in mass production. Considering maximum realizable tape speeds, there is an upper limit to the bandwidth of a single recording channel—around 20 MHz. If a system requires more channel bandwidth than this, some means of multiplexing two or more channels must be used. This is necessary in most digital video recorders (see Section 10.11).

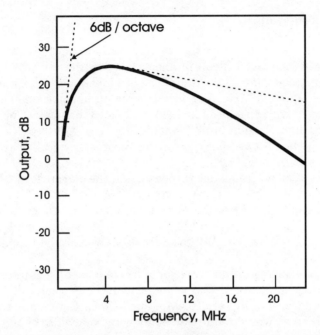

Figure 10.8 Representative frequency response of a magnetic playback head.

2. The depth of recording. The deeper the recording, the wider the record zone and the poorer the response at high frequencies. The response begins to fall off at the point where its wavelength is about one-third of the record depth. Since the recording depth is less with high coercivity tape, its high-frequency response is better. Record depth can also be controlled by limiting the thickness of the magnetic coating.
3. The spacing between the head and the recording layer. High-frequency response falls off with separation between tape and head according to Eq. (10.4):

$$\text{Separation loss (dB)} = 54.6 \, \frac{d}{\lambda} \tag{10.4}$$

For example, at a wavelength of 2.5 μm, there would be approximately 5 dB of separation loss for a head-to-tape average spacing of 0.25 μm. Such a small number, which is comparable to the tape surface roughness, is achievable only by forcing the head into actual contact with the tape. All videotape recorders use in-contact recording, which means that some amount of head and tape wear is always present. (Note that computer hard disks operate with a deliberate amount of separation between head and disk—see Section 10.17.)
4. Azimuth loss. This occurs when the gap is not aligned at the same angle with respect to the recording track on playback as when recording. It depends on the wavelength, the track width, and the angle of misalignment. Heads must be precisely aligned to control this effect.

In each of these cases, the high-frequency response is a function of the relationship between the relevant critical dimension and the wavelength. Figure 10.8

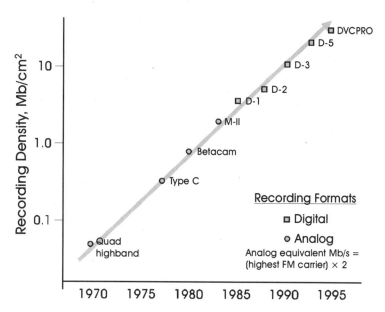

Figure 10.9 Progress in video recording density.

shows the frequency response of a typical head-tape combination. The increase in high-frequency response that has been achieved in recent years is the cumulative result of technology improvements that affect all of these parameters. Gap lengths have been reduced by precision fabrication techniques. Tape writing speed has been increased by moving the head past the tape rather than vice versa. Improved tape materials have increased in coercivity and permeability. Improved tape manufacturing techniques have reduced surface roughness and permitted closer head-to-tape spacing.

10.7.2 Linearity

The linearity of the magnetic recording process is determined by the shape of the highly nonlinear B/H magnetization curve. The nonlinearity is so severe that direct recording of a baseband analog audio or video signal is not satisfactory. The problem is solved for analog audio signals by adding a high-frequency bias waveform to the audio waveform. The problem is solved for analog video signals by the use of frequency modulation. In digital recorders, the linearity issue is dealt with by special modulation methods (see Section 10.12.3).

10.7.3 Recording density

As the fundamental performance of video recorders has improved over the years, the amount (area) of tape has become less for a given recording time at a given quality level. This is expressed by the parameter of recording density. In digital recording, it is expressed in terms of the number of bits recorded in a square inch of tape. The same parameter can be used for analog recorders by applying the approximation that an analog bandwidth of W is equivalent to a bit rate of 2W bits/second. Figure 10.9 shows how recording density has changed with time. This graph clearly displays how the recording industry has achieved steady improvement in performance over many years. It is similar to the graph of Figure 8.1, which shows steady improvement in the solid-state integrated circuits industry.

Improvement in recording density obviously reduces the volume of tape needed for a given playing time. Over the years, this has made possible cassette tape handling (early tape reels were so large that a two-reel cassette would have been impractical). It also has reduced tape costs and storage space for tape libraries.

10.8 Analog Recording

Video signals are analog by nature, and until the late 1970s they were universally recorded in analog form. The advantages of digital recording of analog video signals are numerous, however, and advances in basic recorder technology made digital video recording practical as well. Digital recording has important advantages for many applications and it is replacing analog in all markets.

10.8.1 Analog recording standards

There are three standards for analog helical-scan recorders with performance satisfactory for professional use:

SMPTE Type C: This was the original standard for professional helical-scan recorders, which was approved in 1983 after reconciling the proposals of competing manufacturers. It specifies a reel-to-reel format with 1-inch (25.4 mm) tape.

M-II: This is a standard that specifies ½-inch (12.65 mm) metal particle tape contained in a cassette and is used in the same applications as Type C. M-II is an upgrade of the VHS home format (see Section 10.4). Two video tracks are used to record luminance and chrominance separately.

Betacam: This standard also uses ½-inch tape in a cassette. It is an upgrade of the Betamax home VCR format and is widely used for ENG, particularly in camcorders, combined cameras and recorders. Like the M-II, it also has two video tracks and records analog components.

10.8.2 Analog signal modulation

Direct magnetic recording of analog video signals is not satisfactory because of inherent nonlinearity. The linearity problem is solved by the use of a high-frequency carrier frequency modulated (FM) by the video signal. FM for video recording was introduced by Ampex in the original quadruplex product, and it has continued to be the predominant modulation format for analog signals.

FM video signals differ from FM communication signals in three important respects:

1. The carrier frequency is only a little higher than double the highest frequency component of the signal.
2. The frequency deviation relative to the width of the baseband is much lower.
3. The DC component (see Figure 1.7) in the signal is retained by setting the frequency representing the sync tip at a fixed value regardless of the content of the signal.

The SMPTE Type C recording standards are an example of FM recording. The specified instantaneous frequencies in megahertz are:

	NTSC	PAL
Peak white	10.0	9.3
Blanking level	7.9	7.8
Sync tip	7.06	7.06

The useful picture information is transmitted by a carrier deviation of only 2.1 MHz (7.9 to 10.0 MHz) for NTSC and 1.5 MHz (7.8 to 9.3) MHz for PAL. The deviation for the complete composite video signal, including sync and blanking, is 2.94 MHz for NTSC and 2.24 MHz for PAL. With a maximum video signal frequency component of 4.2 MHz, the modulation index for the composite NTSC signal is only 0.7. This compares with a typical modulation index of 2.5 for microwave and satellite systems. The small modulation index is required to minimize the lower second order sidebands ($f_C - 2f_M$) that would overlap the baseband or even extend into the negative frequency region (phase reversed, sometimes described as "folded back").

The frequency spectrum of an FM carrier modulated by a composite color signal in accordance with SMPTE Type-C standards is shown in Figure 10.10. Both sidebands of the FM signal are recorded, but most recorders suppress part of the upper sideband on replay to improve SNR. This must be done with a precise response curve to avoid differential gain distortion. The removal of part of one FM sideband introduces an amplitude-modulated component that is removed with a limiter.

10.8.3 Signal-to-noise ratio

The characteristics of FM video signals make the calculation of SNR complex and imprecise. The problem is complicated further by the fact that significant noise components can be generated by three sources, the tape, the head, and the electronics. A mathematical analysis of SNR in FM video recording systems is beyond the scope of this volume, but it is important to consider the factors that affect it:

1. The low modulation index adversely affects the SNR in two respects: S, the signal power, is proportional to the square of the modulation index; and the FM improvement factor (see Section 17.8.2), the ratio of the SNR of the demodulated baseband signal to the SNR of the carrier, is much smaller than in communication systems.
2. The SNR of the demodulated signal decreases linearly with the carrier ampli-

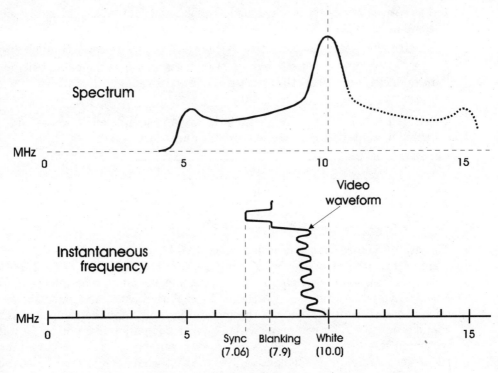

Figure 10.10 Frequency spectrum of SMPTE Type C NTSC video records.

tude down to a critical threshold at which point the SNR drops very rapidly and the signal becomes unusable.

3. As with FM communication systems, the SNR can be improved by preemphasis of the high-frequency components of the video signal before recording with complementary deemphasis on playback. The preemphasis standard is CCIR 567 (based on Report 637).
4. Also in common with all FM systems, the amplitude of the frequency components of noise generated by the recording and playback process (not noise in the original baseband signal) rises linearly with frequency. This leads to a greater difference between weighted and unweighted noise than with flat or white noise (see Section 4.15.3) because high-frequency components are not as visible. A typical weighting factor is 12 dB.
5. The quality of the tape has a major effect on the SNR. Metal particle tape with its high coercivity improves SNR by several decibels compared with conventional iron oxide tape.

The unweighted SNR currently being achieved on professional helical-scan recorders is approximately 49 dB (Table 10.3).

Table 10.3 SMPTE Recording Standards and typical performance, Type C, M-II, and Betacam analog recorders

	Type C	M-II	BetaCam
	Standards		
Tape handling	Reel-to-reel	Cassette	Cassette
Track geometry	Full field	Full field	Full field
Tracks/TV field	1	2	2
Tape width, in.	1	0.5	0.5
Tape type	Oxide	Metal particle	Metal particle
Tape speed, cm/s	24.4	6.6	11.7
Tape usage, cm^2/s	62	8.4	14.8
Number of record heads	2*	4	4
Video track width, μm	20	9	9
Azimuth recording	no	yes	yes
Drum diameter, cm	13.7	6.2	7.44
Drum rotation speed, rev/s	59.94	29.97	29.97
Video writing speed, m/s	25.4	5.8	7.0
	Typical Performance		
Playing time	126 min., 30-cm reel	90 min., cassette	90 min., cassette
S/N ratio, unweighted			
Composite, dB	49	47	47
Luminance, dB		49	49
Bandwidth, MHz			
Composite	4.1	4.2	4.1
Components			
Luminance		4.5	4.1
Chrominance		1.5	1.5
Differential gain	<4%	<3%	<3%
Differential phase	<4°	<3°	<3°

* Entire active picture recorded on one head, second head records only the vertical blanking interval.

10.9 SMPTE Type C Recorders

Although Type C recorders are no longer being manufactured, they are in widespread use and are described here as a state-of-the-art example of analog recorder design.

10.9.1 Tape transport geometry

The SMPTE standard layout[1] of the Type C record drum assembly is shown in Figure 10.11(a) (plan view) and 10.11(b) (elevation). The assembly includes a ro-

272 Chapter Ten

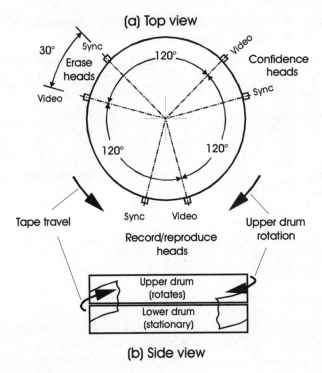

Figure 10.11 Type C helical-scanning drum.

tating upper drum on which the head pole tips are mounted and a slightly smaller stationary lower drum that maintains contact with the reference (lower) edge of the tape to provide accurate alignment as the tape follows its helical path around the drum.

The upper drum makes a complete revolution each TV field or approximately 3,600 rpm for the NTSC system, and each track records the complete active field. This gives the Type C recorder two important advantages over segmented systems: The problem of matching heads is eliminated and full-quality variable-speed playback is possible (variable-speed operation of a segmented system results in a bar across the picture).

There are three pairs of heads on the upper drum, each pair having one head for video and one for sync. The head pairs have different purposes. The first pair performs the basic record/playback functions. The second pair provides simultaneous playback of the recorded material for monitoring the recording. They are sometimes called the "confidence heads." The third pair provides the facility for selective erasure of portions of the record for editing purposes.

A servo system synchronizes the rotation of the drum with the field rate so that vertical blanking occurs during the short time gap at the end of a video track as the head completes one track and begins the next. During the gap in output from the video head, the output is switched to the signal from the sync head, which is interrupted at a different time because of its 30° offset location on the drum. Some early recorders used a single head and reconstructed the sync and blanking signals during this gap, but recording the vertical sync signals on a separate track provides more positive synchronization.

10.9.2 Video, audio, and control track records

Figure 10.5 shows the record pattern for the Type C recording standard. The following dimensions should be noted:

The width of the video and sync tracks is only 20 μm. To maintain this tolerance on playback would be extremely difficult without automatic tracking (see Section 10.9.4.2).

The video tracks are more than 40 cm long, and the timing errors introduced by an elastic medium of this length necessitates the use of time-base error correction. The video track is nearly parallel to the reference edge of the tape, the slant angle is only 2.5°. (This is not shown to scale in Figure 10.5.)

The tape speed (24.4 cm/second) is a little faster than would be indicated by the video track width and the track angle; this causes an unrecorded space between tracks on the tape. This is known as a *guard band* and it is necessary to avoid crosstalk between adjacent tape tracks, especially when there is some mistracking. (More recently developed record standards have solved this problem a different way—see Section 10.14.1.)

The sync tracks are about 4 cm long, which gives sufficient time for the transition of the video pole tip from one track to the next during the vertical blanking interval. Recording of the sync track is switched off during the remainder of the field.

10.9.3 Record electronics

Figure 10.12 is a functional diagram of the record electronics for a Type C recorder. In this design, the video input is AC-coupled, and DC restoration is provided by the *automatic frequency control* (AFC) circuit that maintains the instantaneous frequency of the blanking pulse to a reference value at the FM modulator. The gain of the video amplifier is adjusted to set the frequency deviation of the modulated carrier at the values specified in Section 10.8.2. The video signal is preemphasized and signal excursions that exceed the established white level are clipped.

The gain of the record amplifier is adjusted to bring the magnetizing field, H, to the level that produces maximum carrier output from the confidence head. This is called *optimizing* the record current. The record amplifier is carefully designed to have a frequency response consistent with the magnetic characteristics of the head and it must have good amplitude linearity to avoid distorting the FM signal and generating spurious frequency components.

The record electronics also generates the control signals that go to the drum servo to synchronize the rotational speed and phase of the drum to the recorded signal, and to the capstan servo to control the tape speed. The synchronizing

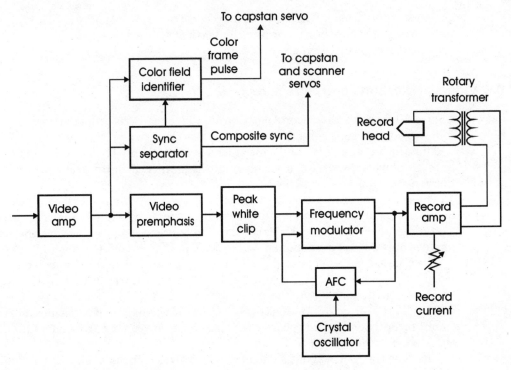

Figure 10.12 Type C record electronics. (*From [6], Fig. 15-60.*)

signals include the sync pulses and the identification of the first field in the sequence of four for NTSC or eight for PAL (see Sections 5.7.3 and 5.8.3).

10.9.4 Playback electronics

The electronics for playback systems are more complex than those for recording since they perform a large number of functions that support and enhance the basic playback function. These include the detection and correction of *dropouts* (momentary interruptions of the FM carrier caused by tape defects that interfere with head-to-tape contact), time-base and velocity error correction, and slow-scan and stop-motion modes. Figure 10.13 is a simplified functional diagram of the playback electronics of a helical-scan playback system.

10.9.4.1 Playback signal channel

Because of the rotation of the head drum, the record/playback head is connected to the preamplifier by means of a rotary transformer. The head/transformer/preamplifier combination has a major effect on playback SNR, and it must be designed carefully for compatibility of the components. The input stage of the preamplifier may be an FET or a bipolar differential amplifier. Its input impedance and the

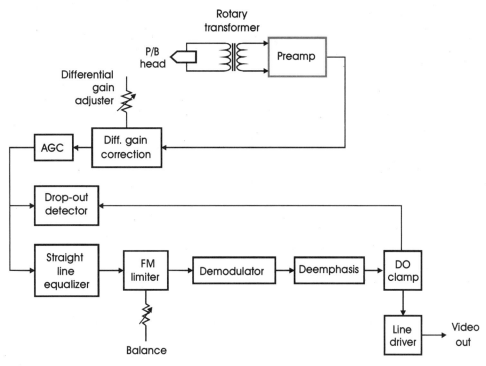

Figure 10.13 Type C playback electronics. (*From [6], Fig. 15-61.*)

turns ratio (step-up) of the transformer are among the parameters that must be matched.

The preamplifier is followed by an equalizer with rising response at higher frequencies like an aperture corrector. Typically its response is a constant plus a cosine function beginning at 180° (–1.0) at 0 MHz and continuing to 360° (+1.0) at the upper end of the frequency band. This compensates for the losses in the record/playback process shown in Figure 10.7. The equalizer is followed by an automatic gain control (AGC) amplifier and a dropout (DO) detector.

The AGC amplifier is connected to the limiter and the demodulator in the main signal channel through a *straight-line equalizer* circuit. This circuit converts the FM signal to a vestigial-sideband format and it improves SNR by attenuating the upper sidebands that have more noise coming from the tape. The video signal from the demodulator passes successively through a low-pass filter to remove spurious signal components, a deemphasis circuit to complement the preemphasis in the recorder, and the DO corrector.

The sync and burst waveforms are separated from the output signals and are used to synchronize the output of the recorder with the studio sync systems by means of the drum and capstan servos and the time-base corrector.

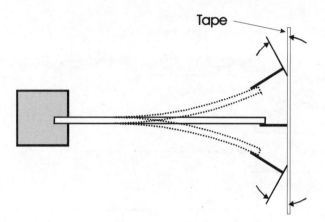

Figure 10.14 Bimorph mounting of the playback head. (*From [6], Fig. 15-107.*)

10.9.4.2 Automatic scan tracking

A means must be provided to cause the playback head to follow the long and narrow recorded tracks that may not be exactly straight. This is accomplished by a technique known as *automatic scan tracking* (AST).

The video playback head is mounted on a *bimorph*, which is made up of two thin flexible strips of a piezoceramic material bonded together (see Figure 10.14). The strips bend when a voltage is applied across them, and they are mounted so that this causes the head to move at right angles to the recorded track. During the fabrication of the bimorph, the strips are polarized so that a bias voltage is required to maintain them in an undeflected position. Deflection in either direction is achieved by adding the deflection voltage to the bias.

Mistracking of the head is detected by a reduction in its output, but this does not indicate the direction of the error. One method of identifying the direction of the error is to apply a dither voltage, typically 500 Hz, to the bimorph strips. The resulting waveforms are shown in Figure 10.15. When the head is on-track, the detected envelope has double the dither frequency. When it is off-track, it has a component at the dither frequency but of opposite phase for locations of the head above and below the track. The phase difference is detected by a synchronous detector, and a correction voltage is generated to return the head to the track. (A similar principle is employed for the tracking of laser disks.)

Although it is not necessary to employ scan tracking for playback at normal speeds, it is an additional assurance.

10.9.4.3 Synchronization

It is usually necessary to synchronize the scanning pattern and burst phase of the recorder playback with the studio sync generator so that recorder signals can be

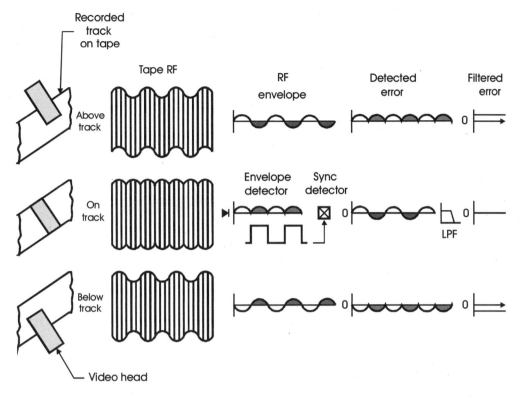

Figure 10.15 Mistracking detection using dither. (*From [6], Fig. 15-109.*)

combined with other studio signals for mixing and editing. Three events must be timed and synchronized: the beginning of the four-field subcarrier sequence, the beginning of each line, and the burst subcarrier phase.

The synchronization of fields and lines is performed by the servos that govern the rotation of the drum and capstan (the shaft that imparts the linear motion of the tape). The servos are controlled by comparison of the timing of signals on the control track with the sync generator output.

10.9.4.4 Time-base correction

The precision required for synchronization of burst frequency and phase is measured in nanoseconds and is orders of magnitude better than can be achieved by electromechanical servos alone. Providing this precision is one of the functions of the time-base corrector. As described earlier, the digital time-base correction was a key technology in the development of helical-scan recorders for professional use. The operation of time-base correctors is described in Section 3.6.1. The time-base correction circuitry may be incorporated in the recorder, or it may be a separate unit.

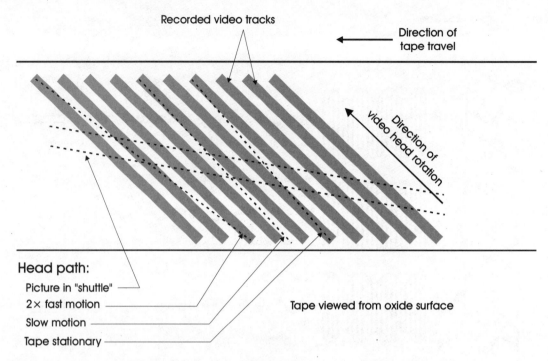

Figure 10.16 Slow-motion, still-frame, and picture-in-shuttle record tracks.

10.9.4.5 Drop-out compensation

The first step in compensating for DOs is detecting them. The amplitude of the envelope of the FM carrier is continuously monitored, and a drop in level of the order of 16 dB or more compared to the average carrier level is considered to be a DO. The DO detector produces a pulse lasting for the duration of the DO. This pulse is then transmitted to the DO compensator, DOC, which fills the gap in the signal produced by the DO by interpolating the signal from a corresponding position on an adjacent line. Since the immediately adjacent line is in an alternate field, the phase of its color subcarrier is reversed, and this would cause an objectionable color error. This problem can be solved by interpolating a signal from an adjacent line on the same frame, two lines removed, or by processing the subcarrier from the adjacent line to reverse its phase.

10.9.4.6 Slow-scan and still-frame operation

The helical-scan format is particularly adapted to slow-motion and still-frame operation. These effects are achieved by slowing or stopping the tape capstan drive while continuing the drum rotation at its normal speed. Since the angle of the track with respect to the reference edge of the tape is determined by the relationship of the tape and rotational speeds, changing the tape speed automatically causes mistracking.

Figure 10.16 shows the relationship between the record tracks and the playback path under three conditions:

1. Tape motion stopped; still-frame operation
2. Tape motion slowed moderately; slow-motion operation
3. Tape motion slowed or accelerated greatly; "picture-in-shuttle"

When the tape motion is stopped on playback, the head does not follow the record track perfectly because the track angle has no tape-motion component. In the absence of tracking correction, the end of the playback paths will coincide with the ends of the preceding record tracks. This mistracking is corrected by the automatic scan tracking system described in Section 10.9.4.2. The mistracking in the slow-scan mode is even smaller, and this, too, can be corrected by automatic scan tracking systems.

The picture-in-shuttle mode can be used either to greatly accelerate or to reverse the tape motion. The mistracking is so severe that tracking correction is not possible or practical, but the resulting picture contains enough information to be recognizable for cueing or editing, even when the tape is run backward or moved at 20 or 30 times its normal forward speed. Picture-in-shuttle, therefore, is a useful mode for rapid search of picture content for the location of a particular frame as required for these purposes.

10.10 ½-inch Cassette Recorders

Major improvements in tape performance, particularly the introduction of metallized tapes, together with improved heads and circuitry, have made it possible to produce recordings on ½-inch (12.65-mm) tape with a quality level that approaches that of 1-inch (25.4-mm) tape systems.

When the narrower tape became practical, the logical next step was to adapt the cassette configuration for many applications. ½-inch cassette recorders of professional quality are now widely used, both in portable configurations for electronic news gathering (ENG) and in studio configurations.

Two standards, M-II [2] and Betacam, exist for professional-quality, ½-in cassette analog recorders (see Section 10.8.1). Some of the significant recording parameters and typical performance specifications of the Betacam and M-II formats were shown in Table 10.3.

The Betacam and M-II records have parallel dual tracks recorded by closely spaced heads using the azimuth recording method (see Section 10.14.1). The luminance signal, Y, is recorded on one track, and the color-difference chroma signals, R–Y and B–Y, on the other. The components, R–Y and B–Y, are combined for recording by time-division multiplex (see Section 10.10.1).

All signal components use frequency modulation. However, a carrier frequency lower than that used for composite signals can be tolerated for component recording because the absence of a high-energy color subcarrier greatly reduces the cross talk between the baseband and modulated carrier sidebands. A lower deviation can be tolerated because of technical advances that have improved the SNR, particularly the use of metal tape.

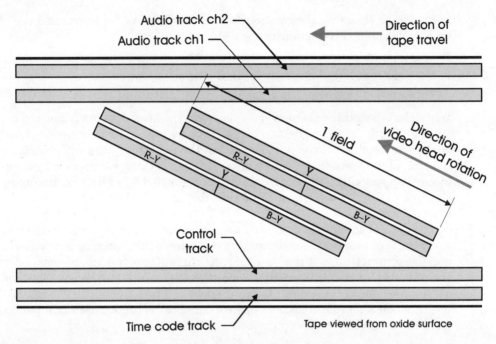

Figure 10.17 M-II record format. (*From Panasonic Corporation,* Instruction Manuals, 17-11 recorders.*)*

Figure 10.17 is a drawing of the M-II record format. Note the parallel pairs of Y and C tracks. Also note that, unlike Type C, the head rotation and tape travel are in the same direction. To facilitate the cassette configuration, which is difficult to implement with single-head recorders, both of these recording systems use two heads per channel. The tape only wraps 180° around the drum, so automatic tape threading is much easier.

10.11 Digital Recording

The nearly transparent characteristics of digital video recording are particularly important in complex editing systems and other situations where many rerecordings or "generations" are necessary. Its relative freedom from SNR and linearity degradations as compared with the analog format make it almost a necessity for these applications.

Digital recording was hindered initially by the high-bit-rate requirements of high-quality video, but this problem has been overcome by steady improvements in tape and head technology. Another problem was the establishment of standards. This has been particularly difficult not only because of the many choices that are available for digital recording, but also because of the growing number of broadcasters, manufacturers, and engineering laboratories whose views had to be considered. The problem was addressed seriously in the late 1970s when representatives of all industry segments began standardization meetings under the aegis of the SMPTE in the United States and the EBU in Europe.

10.12 Digital Recording Standardization Policy

Early in the standardization process, it was realized that there were serious conflicts of interest between competing manufacturers with respect to equipment standards. Fortunately it was possible to resolve many of these conflicts by standardizing the tape and its record rather than the apparatus for achieving it—a major policy change. This assured interchangeability of tape records (the principal purpose of standardization) while retaining the benefits of competition.

In the United States, the SMPTE has led the effort to establish standards for digital recording and created the Digital Television Tape Recording group, DTTR, for the purpose of establishing standards for 525-line systems. At the same time, the EBU appointed a group called MAGNUM to establish standards for 625-line systems.

10.12.1 Standard formats

When digital recording became technically practical, the industry reaction was to go all the way in picture quality improvement and use a component format that would eliminate the artifacts, cross color, cross luminance, and other defects of the composite NTSC, PAL, and SECAM systems.

The first digital recording format to be standardized was for component signals and was designated D-1 [3] (see Table 3.3), using ¾-inch tape. SDTV sampling is according to ITU-R Rec. BT.601-5, 4:2:2—13.5 MHz for the luminance component and half that or 6.75 MHz for each of the two-color difference components.

The D-1 format is an excellent choice for production use and has made possible all-digital production facilities, but it is overkill for general broadcast use. The component format also makes D-1 machines difficult to integrate into broadcast facilities that are mostly analog composite format. Therefore, a need was recognized for a digital machine that would more easily fit into analog facilities and still provide the advantages of digital recording. This could be met by digital recording of composite video, and the D-2 [4] composite format was the answer.

The D-3 [5] and D-5 formats came along later—they are composite and component formats, respectively, and they use ½-inch tape, which allows smaller, lower-cost equipment with good performance. The D-6 format is a component recorder for 1125/60 and other HDTV signals. Each of these is discussed further below. Technical parameters for all the standards are given in Table 10.4 and tape record diagrams are shown in Figure 10.18.

The digital recording market is becoming so competitive that other formats are being offered, exploiting the trend to smaller tape sizes and the use of video compression. Three of these that are discussed here are Betacam SX by Sony, and DVCPro and HD-D5 by Panasonic.

10.12.2 Scope of standards

Standardizing the tape and the recording formats rather than the apparatus requires detailed specifications to assure interchangeability. For example, the D-1 format is defined by five SMPTE standards:

SMPTE 224M - Tape Record

SMPTE 225M - Principal Properties of Magnetic Tape

SMPTE 226M - Dimensions of Tape Cassettes (D-2 also)

SMPTE 227M - Signal Content of Helical Data Records of Associated Control Records

SMPTE 228M - Signal Content of Cue and Time Code Longitudinal Records

10.12.3 Data rates

The data rates given in Table 10.4 are the raw rates that result from ADC—sampling frequency is multiplied by bits per sample for each video channel. However, the data rate actually recorded in the tape channel is higher than these numbers because of the addition of audio data and the overhead bits needed for error management.

10.12.4 Channel coding for digital recorders

As with digital transmission, recorders require some form of modulation to optimize the use of the record-replay channel. This is often called channel coding. Because of the extreme nonlinearity of magnetic recording, the multilevel amplitude-modulated codes described in Section 3.7.2 for transmission channels are not applicable. Instead, recording codes are based on frequency or phase modulation. Because the record-replay process differentiates the signal, it is convenient to think in terms of transitions between recording states (reversal of record current) that produce pulses on replay.

Other considerations of code selection include minimization of the DC component, self-clocking on replay, and control of the minimum and maximum separation of transitions. This latter consideration is called *run-length limiting* (RLL) and it determines the bounds of the frequency spectrum for the signal. The details of channel coding are beyond the scope of this discussion, but the code used in each recording standard will be discussed

10.12.5 Error protection in digital recorders

Error correction and concealment are of critical importance in digital recording since the ability to make multigenerational copies of high quality is one of digital's most important advantages.

The error-detection and -correction techniques described in Section 3.8 for digital transmission channels are applicable to digital recording, including Reed-Solomon coding and interleaving. R-S coding is applied twice, by operating orthogonally on a large block of data such as a video field. For example, the *outer code* may operate on the rows of data and the *inner code* then operates on columns of the data array. In addition, digital recorders use a form of random interleaving, called *shuffling*, which is applied between the outer and inner R-S coding steps.

The purpose of shuffling is error concealment. If a large defect causes oveload of the error management system, the errors will all appear together in the picture or (with interleaving) they will form a pattern determined by the R-S blocking algorithm. Shuffling spreads the residual errors randomly, making them much less visible.

10.13 The D-1 Component Format

The D-1 tape width is ¾-inch, and it is cassette-mounted. The video format is 4:2:2 component according to ITU-R Rec. BT601-5, which means that the sampling rate is 13.5 MHz for the Y or luminance signal. (See Section 4.7.1 for considerations involved in the choice of this rate.) The bandwidth and SNR exceed the performance of analog formats. The absence of cross color, cross luminance, and color artifacts that is characteristic of the component format is an important feature.

Because of the high total video data rate of 216 Mb/s resulting from the 4:2:2 sampling, four parallel video channels are used on the tape. Each video track consists of two video sectors, each having a length of 77.71 mm, separated by a group of four audio sectors, one for each audio channel and each having a length of 2.55 mm. The total length of each track is 170 mm.

The audio segments are located in the center of the tracks because the center of the tape is less prone to dimensional changes and physical damage. The performance of the audio system is excellent. The modulation is digital with a 48-kHz sampling frequency. Quantization is 16 bits per sample, and the result is frequency response to 20 kHz, less than 0.05% distortion, and a 90-dB dynamic range. Noise is virtually unmeasurable.

The D-1 system uses a modified NRZI (nonreturn-to-zero, inverted) channel code. A normal NRZI code has a wide spectrum and poor clock content; both of these are corrected by modifying the code to add transitions to the channel signal.

10.14 The D-2 Composite Format

The sampling rates that have been used for composite signal transmission are $3f_{SC}$ and $4f_{SC}$, where f_{SC} is the color subcarrier frequency. $4f_{SC}$ or 14.4/17.7 MHz (NTSC/PAL) gives superior performance and is used by the D-2 standard. D-2 is also a ¾-inch tape format using the same physical cassette as D-1 but with different tape in it.

From the information in Table 10.4 and Figure 10.18, it can be seen that the D-2 format uses a larger head drum than D-1, but there are fewer heads on it. This is because D-2 is a two-head-per-channel system, but there are only two channels instead of the four used in D-1. The drum rotates slower and less tape is used because composite recording (even at $4f_{SC}$) has a lower data rate than 4:2:2 component recording. The result is a considerably more economical system to purchase and operate. For this reason, the D-2 format is widely used.

D-2 channel coding is Miller-squared (Miller^2), which is a modified FM code.

10.14.1 Azimuth recording

The D-2 format (and many other formats developed since 1985 or so) uses a technique that eliminates the guard bands between video tracks (see Section 10.9.2) without causing crosstalk between tracks. This technique works when there are two or more heads per track, and it depends on the heads being deliberately offset in azimuth so that their gaps are not perpendicular to the track. As explained in Section 10.7.1, misalignment of the head gap with the track causes a serious loss of signal when the track is replayed by a properly aligned head. Therefore, if adjacent tracks on the tape are recorded with substantially different (but standardized) head alignments, on replay (with the same heads), one head will experience azimuth loss whenever it plays any of the adjacent track due to mistracking. This technique is called *azimuth recording* and it allows guard bands to be eliminated, which increases the recording density. Also, the sensitivity to tracking errors is reduced, in some cases the tracking servo can be removed, which is a further simplification because the control track is no longer necessary.

10.15 The D-3 Composite and D-5 Component Formats

As professional-level analog recorders came into use with the ½-inch tapes that were originally developed for the consumer market, it was natural that digital formats would also be developed for ½-inch tape. The D-3 and D-5 formats from Panasonic resulted. As with other ½-inch tape recorders, the equipment is smaller and somewhat less expensive than the larger tape systems.

Both D-3 and D-5 are based on the same tape cassette (D-3) and the same scanner dimensions, which provide for two heads per channel and nominally 180° tape wrap. The D-3 (composite) is a two-channel system, so with record, play, and two-track erase heads it has 10 total heads. The D-5 (component) system has four channels to achieve the necessary greater bit rate of component recording and it has a total of 18 heads on its scanner.

D-3 and D-5 channel coding is a *group code*, called 8,14. The numbers mean that 8-bit data bytes have been remapped by the coding to 14 bits. This would appear to be a large amount of overhead but, in fact, it actually reduces the data rate to be recorded. This happens because there are 64 times as many possible codes in a 14-bit word than in an 8-bit word. Therefore, only 256 codes can be selected from the 16,384 available to achieve special features, for example, always have at least two like bits together. That by itself cuts the channel data rate in half because only bit transitions are recorded. The code choice is actually much more complex than that in order to control DC content and maximum and minimum run lengths.

10.15.1 HD-D5 Format

As HDTV comes into broader use, there is a need for a less expensive HDTV recorder than the D-6 system. This can be done by using some compression and recording on one of the other digital recorders.

The D-5 system has a data rate capability of 216 Mb/s, and a modest degree of compression can bring HDTV within the capability of D-5. That is what the HD-D5 system does. Using intraframe compression at about 4:1, high-quality HDTV signals can be recorded at their full resolution of 1920 × 1080.

The HD-D5 tape deck and track layout is identical to the D-5 SDTV recorder. HDTV capability is achieved by adding an HDTV processor within the cabinet to receive SMPTE 292M serial HDTV bit streams and perform the necessary compression functions for recording. On replay, the HDTV processor performs decompression and delivers a standard HDTV component bit stream to its output.

A down-converted board is also available to convert HDTV recordings to a standard 525-line SDTV output. This is useful when both HDTV and SDTV will be broadcast from the same recording.

10.16 The D-6 HDTV Component Format

The D-6 format was developed for uncompressed component recording of 1125/60 or 1250/50 HDTV production signals. This calls for a total bit rate of 1200 Mb/s (1.2 Gb/s), which was achieved by using the D-2 scanner dimensions in an 8-channel configuration. The scanner is also speeded up by about 70 percent. The same tape cassette used for D-1 and D-2 is used, but special tape must be loaded for this use. D-6 channel coding uses an 8,12 group code.

The D-6 is an expensive solution, but it is the only system available for uncompressed HDTV recording.

10.17 Betacam SX

Sony has made two digital versions of their popular 12.65 mm (½-in) Betacam format analog recorders—Digital Betacam and Betacan SX. The Digital Betacam uses minimal compression (1.6:1) of 4:2:2 component SDTV signals to achieve the highest quality performance. Compression is intraframe, so it causes no limitation on editing locations. The recorded data rate is 125 Mb/s.

The Betacam SX format addresses lower cost markets. That is achieved by using greater compression (10:1) to record a data rate of 18 Mb/s. For that degree of compression, interfield compression must be used, which places some restriction on edit locations with this format.

The Betacam SX tape deck is mechanically compatible with the analog Betacam formats, and those formats can also be played on some of the Betacam SX products. This is a great convenience for customers making the transition from analog to digital recording.

Table 10.4 Digital recording formats and typical performance

	D-1	D-2	D-3	D-5
Video format	SDTV 4:2:2	NTSC/PAL	NTSC/PAL	SDTV 4:2:2
Tape width (mm)	19	19	12.65	12.65
Cassette	D-1 standard	D-2 standard	D-3 standard	D-3 standard
Tape speed (mm/s)	287	132	83.8	167.6
Tape usage (cm^2/s)	54.5	25.1	10.6	21.2
Number of channels for video	4	2	2	4
Number of record heads/channel	1	2	2	2
Total heads on video drum	12	10	10	18
Video track width (μm)	45	39	20	20
Azimuth recording	No	Yes	Yes	Yes
Drum diameter (mm)	74.9	96.5	76.2	76.2
Drum rotation speed (rps)	150	89.9	89.9	89.9
Video writing speed (m/s)	35.6	27.4	21.5	21.5
Video sampling rate (MHz)	13.5	$4f_{SC}$	$4f_{SC}$	13.5/18.0
Video bits/sample	8	8	8	10/8
Total video data rate (Mb/s)	216	115/142	115/142	270/270
Video channel coding	R-NRZI	Miller^2	8,14	8,14
Number of audio channels	4	4	4	4
Typical Performance				
Maximum record time (min)	94		124	124
Video bandwidth (MHz)	5.75	5.5	5.5	5.5
Video SNR (dB)	56	54	54	54
Audio sampling (kHz, bps)	48, 20	48, 20	48, 20	48, 20
Audio dynamic range (dB)	105	105	105	105
Audio bandwidth (Hz)	20–20,000	20–20,000	20–20,000	20–20,000

	D-6	HD-D5	Betacam SX	DVCPro
Video format	HDTV 4:2:2	HDTV 4:2:2 compr	SDTV 4:2:2 compr	SDTV 4:1:1 compr
Tape width (mm)	19	12.65	12.65	6.35
Cassette	D-2 standard	D-3 standard	Betacam SX	DVCPro
Tape speed (mm/s)	497.4	167.6	59.6	33.8
Tape usage (cm^2/s)	94.5	21.2	7.5	2.15
Number of channels for video	8	4	1	1
Number of record heads/channel	2	2	2	2
Total heads on video drum	34	18	8	4
Video track width (μm)	20	20	32	18
Azimuth recording	Yes	Yes	Yes	Yes
Drum diameter (mm)	96.5	76.2	74.4	21.7
Drum rotation speed	150 rps	89.9	74.9	150
Video writing speed (m/s)	45	21.5	17.5	10.2
Video sampling rate (MHz)	74.25	74.25	13.5	13.5
Video bits/sample	8	10	8	8
Total video data rate (Mb/s)	1.2 Gb/s	270 (compressed)	18 (compressed)	25 (compressed)
Video channel coding	8,12	8,14		
Number of audio channels	10	4	4	2
Typical Performance				
Maximum record time (min)	64	124	184	123
Video bandwidth (MHz)	30	30	5.75	5.5
Video SNR (dB)	54	60	54	60
Audio sampling (kHz, bps)	48, 20	48, 20	48, 16	48, 16
Audio dynamic range (dB)	105	>100	90	80
Audio bandwidth (Hz)	20–20,000	20–20,000	20–20,000	20–20,000

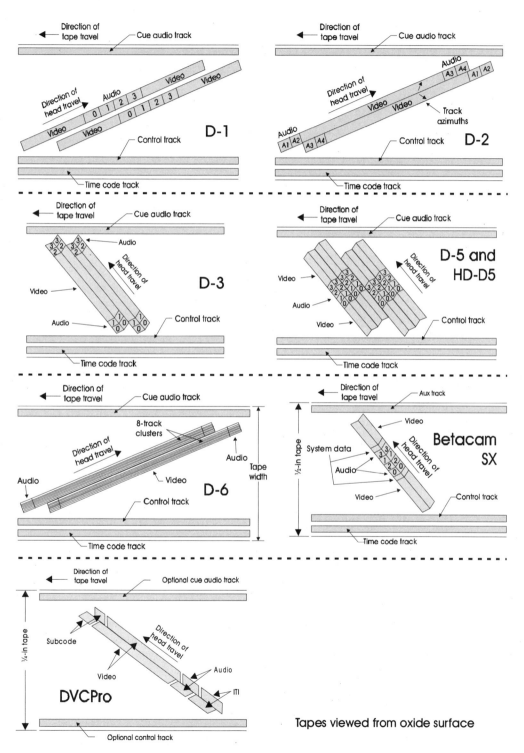

Figure 10.18 Record patterns for digital recorder formats.

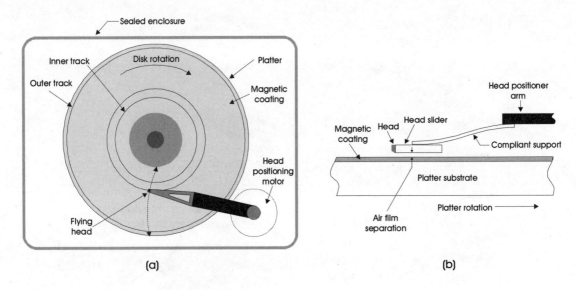

Figure 10.19 Computer hard disk drive: (a) drive, (b) detail of flying head.

10.18 DVCPro

A 6.35-mm (¼-in) digital video tape format called *DV* was developed for the consumer market (see Section 11.9). Both Sony and Panasonic have created professional versions of this format: DVCam and DVCPro, respectively. As an example of these formats, DVCPro by Panasonic is described here.

DVCPro is based on a data rate of 25 Mb/s, although there are two other formats, DVCPro50 and DVCPro100, that run at 50 or 100 Mb/s by trading degree of compression for picture quality (or even HDTV, in DVCPro100). Recording of the greater data rates is achieved by increasing the helical track width, which of course, reduces playing time accordingly.

10.19 Computer Hard Disk Video Recording

Computer hard disks are used for video postproduction editing (see Section 11.5). This technique is rapidly taking over the editing business, but it is only one of the possible applications for video on hard disks. To understand this potential, the following section describes hard disk technology and compares it to digital tape technology.

10.19.1 Hard disk technology

A hard disk uses the same magnetic recording principles described in Section 10.2, but the magnetic medium is a coating on a rotating disk rather than a tape.

The disk substrate can be either flexible (floppy disk) or rigid (hard disk). A single head operates on each surface of the disk; it is mounted on a moving arm that allows the head to take specific positions along a radius as shown in Figure 10.19. To create the necessary relative motion between medium and head, the disk rotates at a speed in the range from about 60 to 150 rps.

During record or replay, the head remains stationary when processing a single track (one revolution of the disk), but it jumps rapidly from track to track if more than one track's worth of data is required. All these actions occur under computer control. Because the movement of the head from track to track takes approximately 1 ms or less, data from any track on the disk surface can be rapidly accessed.

Because the head can be stationary on a single track for long periods of time, it is not practical for the head to be in contact with the magnetic surface as is usual for tape recording. A deliberate and controlled separation is produced by designing the head to "fly" on the film of moving air created by the disk rotation. The design of the head mounting that will do this is a real art and it has taken many years for the technology to progress to its situation today where head-disk separations of less than 0.2 μm are routinely maintained. This requires precise control of the disk surface, the head materials and mounting, and the environment in which the disk runs. Even a single dust particle getting between disk and head will cause a disk "crash" that may destroy both head and disk surface. Modern hard disk drives are sealed in their own atmosphere and can operate continuously for many years without crashing.

A disadvantage of a hard disk is that it is limited to the recording capacity of a single disk surface per head. Even though many hard disk drives have multiple disks (called platters) with a head for each surface, the data capacity is still far less than a similar-sized tape cassette. Present capacities are in the range of 4 to 30 GB, which supports reasonable amounts of video only with fairly high compression ratios. However, the technology is continuing to improve and the production volumes are increasing rapidly, leading to higher disk capacities and lower costs in the future.

10.19.2 Hard disk applications

Most hard disk video applications use video compression to get reasonable playing time from practical disk capacities. For example, recording the ATSC DTV compressed video at 19 Mb/s gives a playing time of 7 minutes per gigabyte. That's not too bad, considering that it is a medium that has almost instant random access to any frame of the video, it is easily editable, and it is going to get better in the future.

Because more than 20 GB of storage is now available in a small removable hard disk package, and video compression can be boiled down into one or two custom ICs, it is practical to consider a camcorder that uses a hard disk for its storage. Editing can be done in the field or the disk can be taken back to the studio for editing and replay. This is just the beginning of the use of general-purpose digital storage devices for video (see Section 10.21.3).

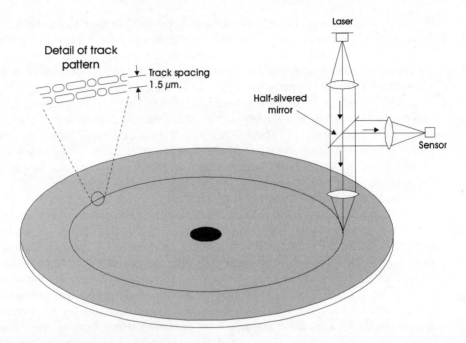

Figure 10.20 CD-ROM replay hardware.

10.19.3 Video Servers

The random-access capability of hard drives allows a system to be built to make any selection from a large library of video material to be almost instantly available to one or more users. This is known as *video on demand* (VOD) and the system is called a *video server*. By assembling an array of hard drives, a server containing 100 GB or more of video can be built. With MPEG-2 compression at a data rate of 4 Mb/s, a 100-GB server can hold up to 55 hours of video, any frame of which can be accessed in a few milliseconds. By rapidly cycling the hard drives between different video segments, multiple users can view different video from the same server. The limit is the data rate of the server—that must be at least equal to the number of simultaneous users times the data rate of one user. Video servers are becoming widely used in postproduction editing systems and broadcast stations. This is described further is Chapter 12.

10.19.4 Hybrid recorders

Because tape recorders are very suited for initial capture of video, but hard disks are more suitable for nonlinear editing, a product that combines the two methods has some advantages. Such a unit has been developed by Sony, their DNW-A100 Digital Video Hybrid Recorder. This is a Betacam SX tape recorder combined with computer hard drive storage. It offers high-speed (up to 4×) transfer between tape and hard disk recordings, so the process of storing video on the hard drive for editing is speeded up. The recorder also includes an edit controller, although it can also be connected to an external controller.

10.20 Optical Video Recording

Analog optical video recording has been on the market for more than 20 years in the laser videodisk. Although it was originally introduced as a consumer product, the laser disk never caught on in that market but it has seen wide use as a random-access prerecorded video store in education and industrial training. When teamed with a computer, a laser video disk player was, for many years, the only viable way to have interactive video.

10.20.1 The Compact Disc

The Compact Disc (CD)—a trade name, note the different spelling of "disc"—was introduced to the audio market in the early 1980s as a replacement for records and cassettes of prerecorded music. Because it is digital, it offers nearly perfect sound reproduction and the small disc makes a very convenient package. As the market grew, the price of CD players dropped to less than $50 and it is now the preferred medium for prerecorded audio. The signal processing of an audio CD player is described in Section 5.5.3.

It was a natural extension to apply the CD for distribution of computer data. Because most CDs are permanent recordings, the computer version is called the *CD-ROM* (CD-Read-Only Memory). The CD audio players were adapted for computer interfacing and are now available built-in on most new personal computers.

Figure 10.20 shows how the CD-ROM technology operates in replay. A recorded disc has a pattern of tiny pits in its surface that represent bits in a FM format. The optical system senses the difference between the pits and the lands (between pits) to produce the playback signal. As indicated by the track-to-track spacing of only 1.5 μm, the density is extremely high. Because of that, the mechanical head positioning used by hard disks is not good enough and the optical playback system must have servos to maintain both focus and tracking (not shown in Figure 10.20).

Focus servoing is accomplished by sensing the symmetry of the focused spot and controlling a magnetic actuator that moves the objective lens. A typical tracking servo optically splits the laser beam into three spots using a diffraction grating. The spots are positioned so that the center spot reads the track while the other two spots read along the left and right edges of the track. The system is on-track when the two side spots deliver equal output. This information controls a moving mirror or other means for adjusting the radial tracking.

With the high area density of optical recording, the data capacity of a single CD-ROM surface is 680 MB. The CD-ROM standard was introduced in 1983—a new standard called the *Digital Versatile Disc* (DVD) with up to 20 times that capacity is now on the market. It has been developed specifically for prerecorded video applications but there surely will be many other uses in both the computer and consumer markets (see Section 10.20.2).

Table 10.5 DVD characteristics and performance

Sides/layers	Capacity (GB)	Video playing time (min)*
Single/single (SS/SL)	4.38	133
Single/double (SS/DL)	7.95	241
Double/single (DS/SL)	8.75	266
Double/double (DS/DL)	15.90	483
CD-ROM	0.65	19

* Based on 3.5 Mbps MPEG-2 SDTV video plus 1.15 Mbps for three channels of Dolby Digital Audio.

When originally introduced, CD-ROMs could only be produced by special mastering facilities that were far too expensive for most users to own. Mastering facilities replicate discs by a pressing process similar to that used for vinyl records and they can produce thousands of copies at costs less than $1 per disc. Now, however, CDs can be recorded one at a time on standard PCs using a drive that costs as little as $300, with blank discs costing $2 or less. Recordable CDs are made one at a time and each one takes from 10 to 30 min, depending on the recording drive's speed. Thus, for mass replication of CD-ROMs, it is still appropriate to use a mastering facility.

The CD-ROM originally had a fixed data rate of 1.2 Mb/s, which limited its use in video to fairly low-quality compressed video. At that data rate, the maximum playing time is 74 min. A lot of development of compression techniques has been directed to that data rate and the results are sometimes remarkable, but really good video requires higher data rates. Recent CD-ROM drives have been speeded up by as much as 40×. However, since the data capacity is fixed, the playing time drops in direct proportion to the speedup. This explains why the industry has developed larger-capacity optical discs.

10.20.2 Digital Versatile Disc (DVD)

With the requirements of high-quality SDTV video in mind, the companies building CD equipment undertook to develop a new standard suited for distribution of movies to consumers. The result was the Digital Versatile Disc (DVD), which is now being marketed to consumers for its original purpose and to PC users for a larger-capacity replacement for CD-ROM, CD-R, and CD-RW. At this writing, all of these markets appear to be developing nicely and equipment prices are falling as production volumes grow.

DVD provides a storage capacity of 4.38 GB per side per layer; it has the capability of using both sides of the disc and can record two layers of information at different depths in the coating on each side. The result is a total of more than 17 GB per disc when all four recording surfaces are used. This is achieved with the same basic technology as the original CD, but the density has been increased by using shorter wavelength light, better optics, better servo systems, and more efficient error protection.

The performance of DVD is shown in Table 10.5. Note that the capacity of the second layer on a side is less than the first layer—this is because the second layer has to be read through the first layer, causing some loss of optical efficiency.

10.21 Trends in Professional Video Recorders

The objective of professional video recording has generally been to make systems that are as transparent as possible to the video and audio signals. With truly transparent systems, signals can pass through the equipment repeatedly without accumulating distortions, which is especially important for recorders used in production and postproduction.

Component digital video systems are the ultimate realization of this objective. However, these systems are expensive to own and operate and there is always pressure for recorders that are less expensive, smaller, and easier to integrate into existing broadcast or production systems. Some of the new possibilities are discussed below.

10.21.1 Use of video compression

The discussion of DTV compression in Chapter 7 made it clear that substantial reduction of data rate is possible by use of video compression. High-quality HDTV pictures can be transmitted at 19 Mb/s. However, this system including the compression and decompression is definitely not transparent, and is intended only for broadcasting.

Distortions caused by compression will accumulate when the compression-decompression process is cascaded. That would not be a problem if compression occurred only once at the camera and all intervening processes were done in the compressed format with a single decompression at the final display. However, it is difficult to edit video in compressed form. Various products are on the market at different points on the curve of degree of compression vs. picture quality. Of course, these products also represent a range of price—as compression increases, the price comes down. The higher-priced products are intended for production-postproduction use, where less (or preferably, no) compression is used.

10.21.2 6.35-mm Tape Formats

New recording systems are now available based on the use of 6.35-mm (¼-in) tape for consumer and professional markets. All applications of this tape size are digital, even for the consumer market. Equipment and cassettes are significantly smaller than larger-tape systems, which is advantageous in portable systems.

The design objectives for 6.35-mm tape decks and cassettes assume that a modest amount of video compression will be used, so the considerations of the preceding section apply to these systems. Professional digital formats have been developed to use less compression by trading off recording time. The 6.35-mm home recorder format (DV) is discussed further in Section 11.9.

10.21.3 The general-purpose digital recorder

The digital tape recorders discussed above are specifically *video* recorders in that they are tied to one or more video signal formats. They would not necessarily handle other kinds of digital data that did not contain the specific structures of scanned video formats. These limitations are necessary because of synchronization requirements of video systems.

As indicated by the use of computer hard disks, CD-ROMs, and DVDs discussed above, it is possible to use general-purpose digital storage devices such as hard disks to handle video. A general-purpose digital tape device would also be usable if it had a high enough data rate. The synchronization requirements could be handled by appropriate software. This probably will lead to digital video tape recorders becoming more video format-independent and more use of general-purpose digital recorders and other storage devices for video.

10.22 References

1. SMPTE Standards 18M-1991, 19M-1991, and 20M-1991, SMPTE, White Plains, N.Y.
2. SMPTE Standards 249M-1991, 250M-1991, 251M-1991, and 252M-1991, SMPTE, White Plains, N.Y.
3. SMPTE Standards 224M, 225M, 226M, 227M, and 228M, SMPTE, White Plains, N.Y.
4. SMPTE Standards 249M-1991, 250M-1991, 251M-1991, and 252M-1991, SMPTE, White Plains, N.Y.
5. SMPTE Standards 263M-1993, 264M-1993, and 265M-1993, SMPTE, White Plains, N.Y.
6. K. Benson and J. Whitaker, *Television Engineering Handbook*, McGraw-Hill, New York, 1992.

Chapter 11

Home and Semiprofessional Video Recorders

11.1 Introduction

The home video recorder is one of the great technical achievements of the last two decades. The underlying technology of home VCRs—helical scan (see Section 9.4)—was originally developed for training and education users who did not require full broadcast-quality signal performance but who did need the lower cost and easier maintenance of helical-scan systems.

Product development based on the helical-scan principle then proceeded in two directions. One path led to recorders of professional quality that surpassed and then replaced the original quadruplex format (see Chapter 10). The other led to VCRs for the home—low-cost, mass-produced recorders that deliver pictures of remarkably good quality, although short of professional standards. The combination of performance and cost that is available from today's home VCRs would have been unthinkable twenty years ago. The seemingly simple VCR, operating in a home environment, uses technology that embodies greater electromechanical sophistication than any other consumer product.

Although most home recorders are analog, digital recorders have been introduced and show signs of capturing this market as prices fall with increased volume. The performance advantages of digital recording are not lost on the home market if they can be achieved at a suitable price.

This chapter discusses home video recorders and their close relatives in the semiprofessional market.

11.2 Comparison with Professional Recorders

Although professional recorders and VCRs use the same basic technology, there are large differences in their design priorities, which result in wide differences in their cost, performance, and operating features (see Section 10.1.2).

Table 11.1 Camcorder Characteristics

	Home VCR	Professional
Price	$200	$30,000
Size	100×350×300 mm	250×420×520 mm
Weight	<6 kg	25 kg
Tape width	6.35 mm (½ in)	6.35 mm (½ in)
Operator skill	Same as TV receiver	Technician
Recording time	Up to 6 hours	Up to 2 hours
Bandwidth		
Luminance	2.5 MHz	4.3 MHz
Chrominance	0.4 MHz	1.5 MHz
S/N (composite)	45 dB	49 dB
Differential phase	10°	3°
Time-base instability	>2 μsec.	<2 nsec.
Number of generations	2*	4
System integration	Poor	Good
Meets broadcast standards	No	Yes

*With marginal performance

Features that are given the highest priority for home VCRs include (in approximate order of priority):

1. Low price
2. Reliability and ease of operation
3. Recording time and tape usage
4. Picture and sound quality
5. Size and weight

Picture quality, while obviously important for home VCRs, is not as critical as for professional recorders. Features that are given the highest priority for professional recorders include:

1. Basic picture quality—bandwidth, S/N, linearity
2. Time-base stability
3. Multigeneration performance
4. System integration features
5. Compliance with broadcast standards
6. Price

The effect of these priorities on product design is shown in Table 11.1, which compares the specifications of representative home VCRs and professional recorders. These specifications are chosen to illustrate the significant differences between the two product types, and they are not based on specific designs. The following sections further discuss these considerations.

11.2.1 Price, size, and weight

Price is the most critical specification for recorders intended for the consumer market. If the product is not affordable by consumers, it will not even be considered. However, other attributes cannot be sacrificed in favor of price and that is where the consumer VCR was initially such a challenge. At the time that home VCRs were introduced to the market at prices under $1,000, professional video recorders were selling for upwards of $100,000, yet the home VCR contained nearly equivalent technology.

The size and weight of rack or cabinet-mounted professional recorders are inappropriate for a home environment. Home recorders are typically of a size that can be placed on top of a TV set, on a bookcase, or on a small table and they weigh less than 6 kg. Remarkably, the achievement of those goals in home recorders has led to professional recorders that are just as small. This is an example of how the technological advances in one part of the recording market quickly appear in all markets.

11.2.2 Reliability and operational simplicity

Although operational simplicity is an important feature of both professional recorders and home VCRs, it can reasonably be assumed that professional products will be operated by trained technicians. Home VCRs, on the other hand, must perform satisfactorily when operated by completely unskilled individuals. This establishes a requirement for a minimum of adjustments and fully automatic operation. Certain operations, however, are inherently complicated—programming a VCR for unattended recording, for example. In spite of many design efforts, this feature still seems to be beyond many home users.

11.2.3 Playing time and tape storage

The *Video Home System* (VHS) home VCR format won a clear victory over the Beta format (see Section 11.4) in the marketplace primarily because of its longer playing time. This is important to home users because of the need to record long programs unattended. That happens when the consumer sets up a VCR to automatically record a program for later viewing.

The VHS system offers three choices of picture quality vs. playing time so the user can choose the one that best suits the situation. The playing time of most home VCRs is at least 4 hours; the playing time of professional recorders can be less because programs seldom run longer than 1 hour without interruption—typically 2 hours for reel-to-reel models and 1 to 1.5 hours for cassettes.

Cassette loading is mandatory in the home market because handling and threading of tape would be too difficult. Cassettes also protect the tape for easy storage. This feature has also proved important to the professional markets and all new professional recorders now use cassettes.

11.2.4 Bandwidth, signal-to-noise, and linearity

For direct recording of composite NTSC or PAL signals, the bandwidth of the record-playback process must be sufficient to transmit the color subcarrier and its sidebands (see Section 5.7.4). The nominal bandwidth requirement is 4.25 MHz for NTSC or 5.5 MHz for PAL, but to record that as-is requires too high a head-to-tape speed for a home VCR. Therefore, the signals recorded on the tape by analog home VCRs are not NTSC but are another format, known as *color-under* (see Section 11.7). With color-under, the chroma subcarrier frequency is converted to about 700 kHz and it is recorded below the lower sidebands of the FM luminance signal.

The color-under signal is reconverted to the NTSC format during playback, but its chroma bandwidth becomes limited by the low frequency of the color-under subcarrier. The narrow chroma bandwidth is adequate for most picture content, but it can cause color smearing of images with high color saturation and sharp hue discontinuities. Improvement of this parameter is an objective of some of the newer home VCR formats.

The bandwidth of the luminance channel is a design choice based on a cost/performance tradeoffone of the many that must be made in home VCR design. The lower limit of the bandwidth range shown on Table 11.1 (2.5 MHz—a common choice) results in a horizontal limiting resolution of slightly more than 200 lines. This is not a high-quality picture but one that is satisfactory to most home viewers.

The SNR and the linearity (as indicated by differential phase in Table 11.1) of VCRs also involve cost/performance tradeoffs. The color-under format may result in coarse, low-frequency noise in areas of high saturation. This becomes especially noticeable in second-generation color-under recordings.

The advent of digital recording in the home eliminates the limitations of color-under, trading them for the artifacts of digital video compression. Viewers generally see that as an improvement.

11.2.5 Time-base stability

As noted in Section 10.1.1, time-base instability is an inherent problem with helical scanning because of the long recording tracks and the elastic recording medium. Without time-base correction (TBC), the playback instability may exceed 2 μs— three orders of magnitude greater than the 2-ns stability demanded for professional recorders. Stability is restored in professional recorders with a digital TBC (see Section 9.9.4.4) but these are (so far) too costly for analog consumer VCRs, although they are inherent in digital VCRs for any market.

Most TV receivers can synchronize successfully to the luminance channel of an analog home VCR signal in spite of the instability of its time base and display a stable monochrome picture. However, the instability of a directly recorded chroma channel is too great for the color circuits of a receiver to handle and no color can be displayed. This problem is solved by the color-under approach because the color-under subcarrier on playback is up-converted back to a stable frequency (see Section 11.7.2).

The result of a color-under playback is that luminance and color are displayed, but they are not synchronized with each other (the receiver is synchronizing separately to each of them). That means the subcarrier interlacing features of the NTSC system cannot work properly, and there are spurious effects that show, particularly on edges of highly-colored areas in the picture. This is tolerated by consumers, but it is another reason for future digital systems to get away from color-under.

Because of the lack of luminance-chrominance synchronization in home VCRs, it is impractical to integrate them in systems that require the mixing of multiple signal sources, unless external TBCs are used.

11.2.6 Compliance with FCC broadcast standards

The signals from home VCRs generally do not comply with FCC broadcast standards. Their greatest defect is timing instability, but they may fail to meet FCC standards in other respects such as sync and blanking level format, and signal and blanking levels. They can usually be upgraded, however, to meet FCC standards by the use of digital time-base correctors and other signal-processing equipment (which can easily cost more than the VCR itself).

11.3 Recording Drum Configurations

Two recording drum configurations are commonly used in home VCR tape transport systems as illustrated in Figure 11.1, the essential difference being their diameters. The diameter of the larger drum is 50% greater than the smaller, and it has two heads compared with four. With these diameters, the length of a 270° track on the smaller drum is equal to the length of a 180° track on the larger. The rotational speeds are controlled so that the smaller drum rotates 270° and the larger 180° during one television field, thus creating record tracks of equal length for one field.

The scanning sequence of the four-head format is shown in Figure 11.2. The two-head and four-head configurations are compatible, and tapes recorded on one can be played back on the other.

11.4 Home VCR Record Formats

Helical scanning is a very versatile technology, and it has been adapted to a variety of formats to meet the widely different requirements of the consumer, audiovisual, and professional markets.

A description of all the home and semiprofessional VCR formats and their variations (some of which are in a continuous process of change) is beyond the scope of this volume. The formats described here include the most commonly used analog ones: U-Matic, VHS, and 8 mm, and the DV digital format.

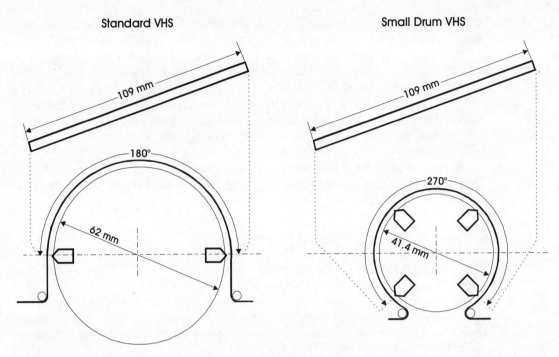

Figure 11.1 Recording drum configurations (VHS).

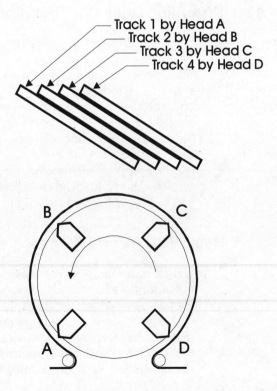

Figure 11.2 Recording track sequence.

Table 11.2 Record parameters—analog recorders

Parameter	(reference)	U-Matic	VHS	8 mm
SMPTE type	C	E	H	
Tape width, mm	25.4	19	12.65	8
Drum diameter, mm	137	110	62/41.4	40
Number of record heads	1	2	2/4	2
Track width, μm	200	132	91/45/30	32/16
Tape speed, mm/s	244	95	33/16.5/11	14/7
Recording time, hours	2	1	2/4/6	2/4
Head-tape speed, m/s	25.4	7	5.8	3.75
FM carrier, MHz				
Sync tip	7.06	3.8	3.4	4.2
Peak white	10.0	5.4	4.4	5.4
Minimum wavelength (μm)	2.1	1.4	1.1	0.6
Color-under subcarrier, kHz	N.A.	688	629	743

11.4.1 Analog record parameters

The record parameters of the most commonly used consumer and audiovisual analog formats are tabulated in Table 11.2.

The parameters of the SMPTE Type C professional reel-to-reel format are also shown for comparison. As indicated by the acronym, VCR, all consumer formats use cassettes, and professional recorders now use them as well (see Chapter 10).

The *U-Matic* was the first helical-scan format to achieve commercial success. It was an audiovisual and industrial product with the characteristics described in Section 11.1, and it is still used in these markets.

U-Matic recorders were followed by a series of lower-priced models designed specifically for the consumer market:

  Beta   1976
  VHS   1978
  8 mm   1983

In spite of its earlier introduction and excellent performance, the Beta format lost the competitive battle with VHS in the consumer market, largely because of the latter's longer playing time. However, it's heritage continues in the professional market as the Betacam series by Sony.

VHS (including its special-purpose variants) is now the most widely used home VCR format, but it, in turn, is receiving competition from the 8-mm format, particularly for camcorders where the smaller size of an 8-mm tape deck is an important advantage.

The VHS and 8-mm formats offer a choice of tape speeds, usually described as standard, LP (long play) and SLP (super long play). Note especially that the drum

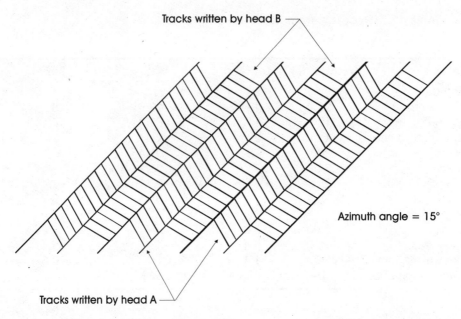

Figure 11.3 Azimuth recording.

rotational speed remains constant, meaning that tape-to-head speed remains constant for different tape speeds. Therefore, changes of longitudinal tape speed result in changes of track width.

Upgraded (and more costly) models of the 8-mm and VHS formats have been developed—*Hi-8* and *Super VHS* (SVHS)—with performance specifications that are adequate for semiprofessional applications. These are described in Section 11.8. The digital formats are described in Section 11.9.

11.4.2 Azimuth recording

Azimuth recording is incorporated in almost all home VCR record formats because of its effectiveness in minimizing the effects of mistracking, reducing cross talk between adjacent tracks, and improving SNR. It is not used with the SMPTE Type C professional format, but it is used in the professional digital formats.

The record pattern for azimuth recording is shown in Figure 11.3. The gap in the record head is tilted from its normal 90° orientation with respect to the head tracks, and the magnetic lines of force in the record are not parallel with it. The gap azimuth angle, the direction of tilt, is reversed on alternate scans, and on playback the signal from adjacent scans is attenuated by the difference between the head gap and magnetic record azimuths (see Section 10.7.1).

The azimuth angle is normally set at about 10° or 15° from a 90° angle with the head path. Once chosen, azimuth must be held to tight tolerances between

Home and Semiprofessional Video Recorders 303

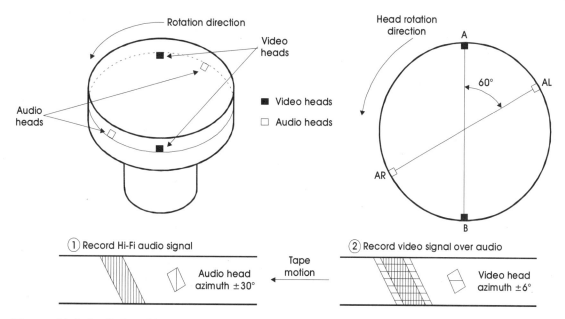

Figure 11.4 Audio head locations.

machines to guarantee interchangeability.

Since signal pickup from adjacent tracks is attenuated by the azimuth mismatch, it is unnecessary to provide guard bands on either side of the recorded tracks. This permits the use of wider tracks and heads that produce a higher SNR and are less sensitive to tracking errors. It also allows the tape speed and track widths to be changed while using the same physical head width.

The head width is wide enough for the widest track width needed at the highest tape speed. When the tape speed is lower, each head pass over-writes some of the track from the previous head pass, automatically resulting in narrower recorded tracks.

11.4.3 Audio recording

Audio signals can be recorded conventionally on longitudinal tracks or by a 1.5 MHz (1.4 and 1.6 MHz for stereo) FM carrier recorded with rotating heads mounted on the same drum as the video (see Figure 11.4). The performance of longitudinal-track audio is limited on home video recorders because of the very slow tape speeds, there is not room on the narrow tape to provide wide enough audio tracks, and because of magnetic crosstalk from the video heads that pass over the audio tracks. FM audio gives higher quality and is usually referred to as hi-fi audio. However, FM audio shares the same tape space as the video, which precludes editing the audio once the video has been recorded.

In the rotating head configuration, the audio head gaps are located 60° around the periphery of the drum from the video, and their azimuths are tilted 30° from normal.

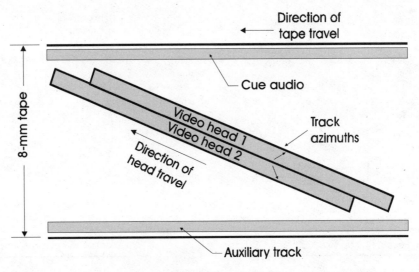

Figure 11.5 The 8-mm record format. Reproduced with permission from [3].

The video and audio records occupy approximately the same area on the tape. The signals are kept separate by:

1. The difference in carrier frequencies
2. The difference in the head azimuths
3. The difference in the strength of the record magnetizing force. The audio heads have a longer gap length and a longer recorded wavelength, which causes the audio recording field to penetrate more deeply into the tape. The audio signal is recorded first and the video record is overlaid on it. The video does not penetrate so deeply into the tape because of its shorter wavelength, so very little of the audio recording is erased when video is recorded over it.

11.4.4 Format patterns

The format patterns for home VCRs are similar in principle to those shown in Figure 10.5 for professional recorders. The diagonal video and audio tracks occupy the center portion of the tape, one field per track, and the fixed-head audio and control tracks are recorded longitudinally along the edges. The 8-mm format (Figure 11.5) shows the essentials features of home VCR record patterns although they have many variations.

11.4.5 Record frequency spectrum

The frequency spectrum of an 8-mm video record is shown in Figure 11.6. The frequency-modulated luminance channel occupies the region from 4.28 to 6.8 MHz. The audio signal occupies a band around 1.5 MHz, and the color-under chroma

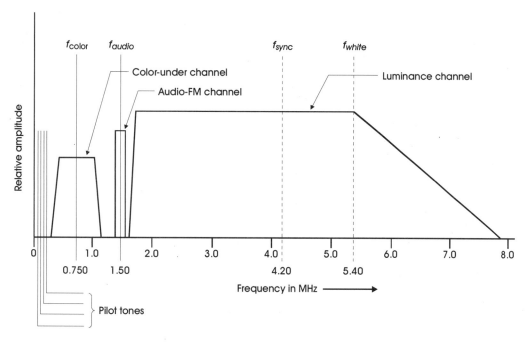

Figure 11.6 Record frequency spectrum, 8-mm system using color-under.

carrier occupies a band around 750 MHz. There may be, in addition, four low-frequency pilot tones that are used for automatic scan tracking.

11.5 Automatic Scan Tracking

Automatic scan tracking (AST) systems that employ bimorph mounting of the heads as used in professional recorders were described in Section 10.9.4.2. Home VCRs may use different systems, one of which is based on the use of low-frequency pilot tones and incremental adjustments of the drum rotational speed.

The pilot tones are recorded on either side of the video tracks, the tone on one side having a higher frequency than the tone on the other. On playback, the head is wide enough to pick up both tones, and if it is properly centered the amplitude of both will be equal. If the tracking is not perfect, the amplitude of one of the tones will be greater, and the capstan servo that determines the tape speed is adjusted by a feedback circuit to restore the tracking.

11.6 Record Information Density

The home VCR marketplace has imposed two conflicting requirements on VCR designs—small size and long playing time. These needs can be satisfied simultaneously only by increasing the recording density (see Section 10.7.3), i.e., the amount of information recorded per unit area of tape.

Table 11.3 Tape area per frame for different analog formats

Format	Area/frame (cm^2)
SMPTE Type C	2.1
U-Matic	0.4
Beta	0.17
VHS	0.14
8 mm	0.04

11.6.1 Record density requirements

Record density can be calculated in terms of the amount of tape area needed to record one frame of the TV signal:

$$\text{tape area/frame} = \frac{\text{tape speed} \times \text{tape width}}{\text{frame rate}}$$

The increase in this parameter in the more recently introduced formats shown in Table 11.2 has been dramatic. The area required to record one frame of an NTSC signal with these formats as compared with the SMPTE Type C format is shown in Table 11.3.

Some of the area reduction is possible because of differences in the bandwidths of the recorded signals. For example, the highest-frequency component of the 8-mm RF signal is about 7 MHz as compared with about 12 MHz for the Type C signal. But most of the reduction of the area per frame is simply caused by recording information closer together (higher density).

The new digital recorders yield even better record density performance (see Section 10.9).

11.6.2 High-density technologies

The information density of the recorded signal is determined by the wavelength of its highest frequency component and the pitch or spacing of the recording tracks (equal to the track width divided by the sine of the track angle for azimuth recording). The track width and minimum wavelength of the common VCR formats were given in Table 11.2.

Reducing the track pitch and minimum wavelength from the values used in the SMPTE Type C format to the values used in the Beta, VHS, and 8-mm standards was a major engineering challenge, and it required the addition or improvement of a number of technologies. The problems included:

1. Reduction in bandwidth and a lower SNR because of the necessity of narrower recording heads.
2. Tracking difficulties because of the narrower tracks.
3. Crosstalk between adjacent tracks.

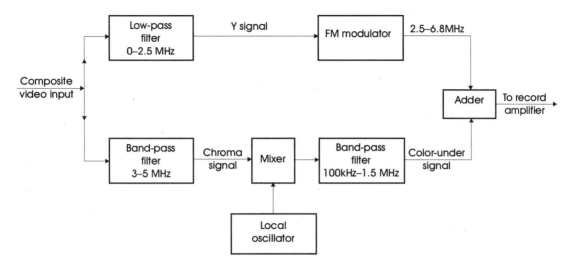

Figure 11.7 Record electronics of the color-under system.

The solutions included:

1. Improvements in tape/head performance (see Section 10.7). The track pitch and the wavelengths in the 8-mm format are so small that metal-particle or evaporated-metal tape is almost a necessity.
2. Azimuth recording (see Section 11.4.2). This reduces the effects of mistracking, permits the use of wider recording heads for improved SNR, and reduces crosstalk between tracks.
3. Automatic scan tracking (AST) (see Section 11.5).

11.7 Color-Under Signals

All the analog home recorders use some variation of the color-under system.

11.7.1 Record frequency spectrum and electronics

The record frequency spectrum of the color-under signal format was shown in Figure 11.6. The essential elements of the record electronics are shown in Figure 11.7.

The chroma component (the color subcarrier) of the incoming signal is separated by a band-pass filter, and the NTSC chroma and luminance components are processed separately. The frequency of the chroma component is downconverted to the 650 to 750-kHz region by heterodyning and its bandwidth reduced (see Table 11.2) to form the color-under component of the recorded signal. This places the color subcarrier and its sidebands below the band occupied by the FM video signal rather than being interleaved with it as with NTSC and PAL; hence the term *color-under*. Since the bandwidth available for the color-under signal is lim-

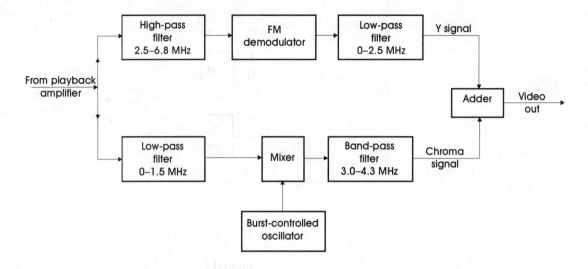

Figure 11.8 Playback signal processing of the color-under system.

ited, it is necessary to pass it through a band-limiting filter.

The luminance component passes through a low-pass filter with a bandwidth of about 2.5 MHz (the luminance component has a wider bandwidth in some high-band models—see Section 11.8) and it frequency modulates the video carrier between the limits shown on Table 11.2.

The audio signal is recorded on a track along the edge of the tape, or alternatively it is multiplexed with the video record (see Section 11.4.3).

If pilot tones are used for automatic scan tracking, they are recorded at the bottom of the spectrum as shown in Figure 11.6.

11.7.2 Playback electronics

Unlike professional recorders, which require a time-base stability of about 2 ns on playback so that the signal can be mixed and edited with other signals (see Chapter 12), home VCRs are not used in complex editing systems and can tolerate a considerably higher degree of instability. The use of time-base correctors is an option that can be included, however; see below.

Figure 11.8 is a functional diagram of the playback electronics of a typical playback system. The chroma and FM luminance signals are separated by high-pass and low-pass filters, and the luminance signal is demodulated. As noted in Section 11.2.5, the time-base instability of the chroma signal is too great to operate satisfactorily with a fixed-reference subcarrier, and the reference subcarrier is generated by a burst-controlled oscillator whose phase varies with the timing errors on the tape. This retains the phase relationship between the color and reference subcarriers to a reasonably satisfactory degree.

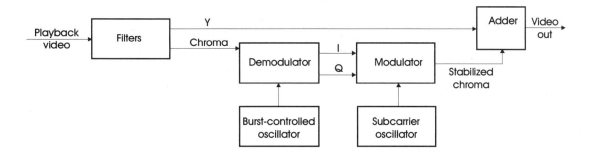

Figure 11.9 Chroma time-base corrector.

Improved performance can be obtained with a relatively inexpensive analog time-base corrector for the chroma signal as shown in Figure 11.9. The playback video is separated by filters into its chroma and luminance components. The luminance component, its bandwidth limited to about 2.5 MHz, passes directly to the output. A burst-controlled subcarrier is generated, as in the playback electronics in Figure 11.8. This subcarrier is then used to demodulate the I and Q or color-difference signals. These, in turn, remodulate a stable subcarrier, which is combined with the uncorrected luminance signal. This gives a stable chroma signal combined with an unstable luminance, a result which is satisfactory for many applications.

11.8 S-VHS and Hi-8

Tape and head performance has improved substantially since the VHS and 8-mm home VCR systems were introduced. Exploiting these advances, higher-performing standards have been developed for both of these systems. The new standards are called Super VHS (S-VHS) and Hi-8; they are based on the same basic tape transports, scanners, and cassettes as the original VHS and 8-mm systems. Specific parameters are shown in Table 11.4, which can be compared to Table 11.2.

Because of their higher performance, these standards are very suitable for semiprofessional applications and product lines of recorders and camcorders have been developed for this market. Prices are slightly higher than home VCRs but far less than broadcast-level products.

11.8.1 Control interfaces

A significant feature of higher-priced home and semiprofessional recorders is a control interface to allow the units to be used with an edit controller. This requires that tape transport controls for play, shuttle, and recording be brought out to a connector for remote control. Two standards for that purpose are called Control-L and Control-M.

Table 11.4 Record parameters of S-VHS and Hi-8

Parameter	S-VHS	Hi-8
Tape width (mm)	12.65	8
Drum diameter (mm)	62/41	40
Number of record heads	2/4	2
Track width (µm)	58/29/19	20/10
Tape speed (mm/s)	33/16.5/11	14.3/7.15
Recording time (hours)	2/4/6	2/4
Head-tape speed (m/s)	5.8	3.8
FM carrier (MHz)		
Sync tip	5.4	5.7
Peak white	7.0	7.7
Minimum wavelength (µm)	0.8	0.5
Color-under subcarrier (kHz)	629	743
Density (cm^2/frame)	0.14/0.07/0.04	0.04/0.02

11.9 Digital Home Recorders

In spite of the demanding price goals of the home market, digital home recorders have been developed and are now on the market. Prices are a little high right now, but the products are being accepted because of their exciting small size, performance, and reliability compared to existing analog recorders. Production volumes are increasing and the normal price reduction scenario that occurs for products entering this market is underway. Prices should eventually be competitive with analog recorders, at which point analog systems will be relpaced by digital.

The digital home recording standard was originally called *Digital Video Cassette* (DVC), however, it soon became clear that this was not just a tape recording system, but an entire new video system. As a result, the acronym was shortentd to simply *DV*, for Digital Video. Although the DV system was developed for the home market, it quickly was adapted to the professional and broadcast markets in the forms of DVCPro (see Section 10.18) and DVCAM.

The DV system handles a data rate of 12 Mb/s, which allows high quality SDTV operation using compression. The video sampling is 13.5 MHz 4:1:1 for 525-line or 13.5-MHz 4:2:0 for 625-line systems. Thus, it is a true component video system, quite remarkable for a home system! The data rate from sampling is about 120 Mb/s, so approximately 10:1 further compression is required, which is done using DCT. Picture quality is excellent—horizontal resolution is over 400 TVL and good SNR is achieved.

The parameters of DV are listed in Table 11.5. Because the format is digital, there is no need to operate the helical scanner at one track per field to avoid head-switching disturbances. A small-diameter drum is run at a high rotation speed (150 rps), which gives a smaller tape deck with a high head-to-tape speed (10 m/s).

The track patterns for DV are shown in Figure 11.10. Although space is avail-

Table 11.5 Parameters of DV

	525-lines	625-lines
Tape width (mm)	6.35	6.35
Tape speed (mm/s)	19.05	19.05
Max. recording time (h)	4	4
Drum diameter (mm)	21.7	21.7
Head-tape speed (m/s)	10.2	10.2
Number of record heads	2	2
Track width (µm)	10	10
Video sampling rate (MHz, bps)	13.5, 8	13.5, 8
Video decimation	4:1:1	4:2:0
Video compression factor	10:1	10:1
Helical data rate (Mb/s)	12	12
Audio recording method	PCM, stereo	PCM, stereo
Audio bits/sample	16	16
No. of audio channels	2	2
Density (cm²/frame)	0.04	0.048

able for two longitudinal tracks, the system can operate without either of them, which can be a further cost reduction. The helical tracks provide for recording of video, audio, subcode (time code or other information), and insert-and-track-information (ITI) codes. The subcode and ITI blocks can be read in high-speed shuttle mode.

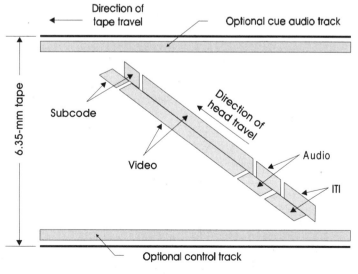

Figure 11.10 Track patterns for DV.

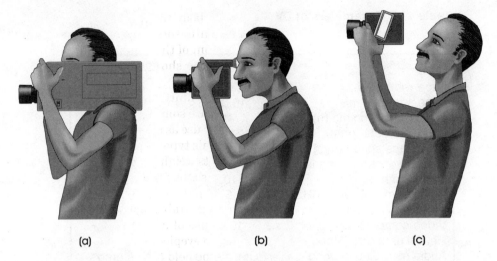

Figure 11.11 Camcorder configurations: (a) shoulder-held, (b) hand-held with an eyepiece viewfinder, and (c) hand-held with an LCD viewfinder.

A digital version of VHS has also been introduced, called D-VHS. This used 12.65-mm (½-in) tape, in cassettes of the same sizes as VHS. This does not appear to be particularly competitive with DV, because of inherently larger size. However, it is too early to say how the market will respond.

11.10 Camcorders

Shooting video requires at least two types of equipment—a camera and a recorder. As equipment became smaller, it was natural to make a combination of these two. Thus, the *camcorder* was born. Camcorders were first developed in the broadcast market, where they were used for a new class of operation called *electronic news gathering* (ENG.) The first ENG camcorders were large and heavy, but the single unit was an advantage in that it eliminated cables and in-field hook-up. An ENG operator could begin shooting as soon as he/she took the camcorder out of its carrying case. Professional camcorders are now available at all levels of performance, even HDTV.

Later, as the home VCR market matured, home camcorders were introduced. Since home VCRs were inherently smaller to begin with, home camcorders were smaller, lighter, and more convenient. Home cameras are now available only as camcorders.

11.10.1 Camcorder configurations

Because of their weight, early portable cameras and camcorders were designed to be carried and used on the operator's shoulder. This allowed the operator to use the stability and inertia of his/her upper body to achieve smooth camera movement. Shoulder mounting inherently requires some large dimensions because the camera viewfinder eyepiece must be in front of the eye while the center of gravity of the unit as a whole must be back at the shoulder. This is shown in Figure 11.11(a).

As it became possible to make a smaller unit, home camcorders that could be held in one hand were introduced. These are sometimes called *palm* camcorders, and they are held up in front of the eye for use as shown in Figure 11.11(b). Since there is no limiting large dimension for this type of support, the unit could be as small and light as possible. Most such units weigh less than 1 kg with battery and cassette. They are easy to carry and to use, but the lighter weight and the out-front holding make it more difficult to keep them steady for stable pictures. This has further led to the development of image stabilization (see Section 11.10.4).

A third configuration depends on the use of an LCD viewfinder that is large enough for the operator to view without an eyepiece. This configuration, shown in Figure 11.11(c), allows the camcorder to be held anywhere within the operator's reach. The viewfinder swivels so that it can be positioned for viewing from any angle.

11.10.2 Camcorder system

The system diagram for a camcorder is not very different from that of individual cameras and recorders except that some common functions are combined and the signal interface between camera and recorder does not have to be standard NTSC or PAL.

In an analog camcorder, the camera can generate the color-under format directly, which eliminates the composite encoder and any degradation that may cause. However, most camcorders have a separate composite output signal for driving a monitor, computer, or another recorder, so most units still contain the encoder.

Digital camcorders using the DV format are full component and have no composite encoder between the camera and recorder. They do still provide composite encoding at their output, however, to feed signals to analog displays or other components. Digital camcorders are beginning to have digital outputs as well, usually IEE 1394. This looks toward a full digital video system for the home.

Sync and power-conditioning circuits for the camera and recorder are combined in a camcorder. This is one sort of duplication that is eliminated by the single-unit configuration.

The camera in a home camcorder is always a CCD, usually only one, although a few three-CCD camcorders are on the market for semiprofessional use (see Section 11.11). Tape transports for camcorders are a special design, optimized for small size and light weight. In the VHS format, a special small cassette (VHS-C) is available for camcorders. VHS-C cassettes are playable in any standard VHS players by using an adaptor cassette.

One might think that a camcorder would not need a playback capability. However, a major advantage of tape (compared to motion picture film) is that the result of shooting is viewable immediately, while still in the field. Therefore, playback to the viewfinder is an essential feature. Playback is also needed because most camcorders have some form of in-camera editing, such as assemble editing (see Section 12.2.2.1). Playback is essential to determine edit points.

Another purpose of playback in the camera for home use is to allow the camcorder to have a different tape format from the bulk of home VCRs, which are VHS. This has allowed 8-mm camcorders to proliferate even though not many homes have 8-mm players. A video output from the camcorder connects to a TV receiver for playback.

11.10.3 Viewfinders

Another advantage of video shooting is that an electronic viewfinder shows exactly the picture being recorded. For this reason, all camcorders have electronic viewfinders, using either CRT or LCD panel displays. The trend is to color LCD displays, and as mentioned above, a large LCD panel display gives added flexibility compared to a viewfinder that requires an eyepiece. Viewfinders, even those with eyepieces, usually have very flexible mounting so that they can be adjusted to the operators preference.

11.10.4 Image stabilization

Holding a small, lightweight camcorder steady enough to deliver a stable picture, especially with a telephoto zoom, is not easy. Many camcorders have *image stabilization* to help with this. There are electronic, optical, and mechanical systems available, but the trend seems to be toward electronic methods. Figure 11.12 shows how this works.

The operation is similar to a digital TBC in that the picture from the CCD is digitized and stored in a field memory. The read-out from the memory is controlled by a means for sensing motion. The output signal is taken from an area that is slightly smaller than the full CCD area and motion correction is obtained by moving the output area relative to the full CCD area. The location of the output area moves according to the motion sensing to essentially cancel the movement of the picture. The output picture must be zoomed up slightly to achieve a full size picturethis can be done in the memory read-out process. This approach loses a little of the CCD's resolution because the output picture does not use all the CCD pixels. Some cameras use an optical means for moving the picture on the CCD—no resolution is lost with this approach.

The trick of this is in the motion sensing. One approach [1] divides the image into four areas and performs frame-to-frame comparison separately for each area to develop four motion vectors. The four vectors are processed by a fuzzy logic system to identify the component of motion that is common to all areas this is assumed to be caused by camera movement. This motion component is used to

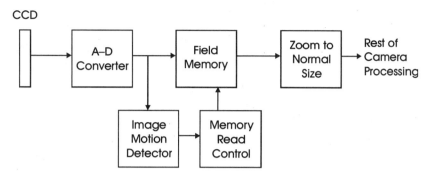

Figure 11.12 Block diagram of one form of image stabilization.

control the location in the field memory that is read out. There is more to it, because the system must deal with motion in the scene itself and it also should not try to correct if the operator is smoothly panning or tilting the camera.

Another approach [2] (not shown in the figure) uses the CCD itself instead of a field memory and applies vertical motion control signals to the CCD scanning. That can correct for vertical instability; a line memory is used to provide horizontal correction.

11.11 Semiprofessional and Professional Camcorders

The camcorder is not just a home product. Upgraded home-market camcorders are available for the semiprofessional market, and full broadcast camcorders are available for the professional markets.

11.11.1 Professional features

Semiprofessional and professional camcorders have better cameras, with more elements in a single CCD or three CCDs. Recorders are compatible with one of the semiprofessional or professional formats, such as Hi-8, S-VHS, Betacam, or M-II. Professional camcorders may have a feature known as docking, where the camera and recorder are separate units so the cam and corder can be taken apart. This is useful in that either part could be upgraded in the future without making obsolete the investment in the other part. It also may be convenient at times to use the camera by itself or with a separate recorder. For this purpose, an adapter module plugs in to the docking interface of the camera. The adapter module provides all the usual system interface features that a standalone studio camera would have. Thus, the camcorders camera could be used in the studio when it was not needed in the field.

11.12 References

1. Y. Egusa, H. Akahori, A. Morimura, and N. Wakami, "An Application of Fuzzy Set Theory for an Electronic Video Camera Image Stabilizer," *IEEE Transactions on Fuzzy Systems*, Vol. 3, No. 3, August 1995.

2. Y. Kubota, Y. Nishi, K. Shintani, T. Urabe, K. Shimada, and T. Katsumoto, "Latest Advances in Camcorder Technology," *Proceedings of the IEEE*, Vol. 82, No. 4, April 1994.

3. A. C. Luther, *Video Recording Technology*, Artech House, Norwood, MA, 1999.

Chapter 12
Video Postproduction Systems

12.1 Introduction

Before the development of electronic video recording, video programs were produced live with everything occurring in real time, or they were created on motion picture film. Live production causes serious limitations in that every element of the program has to occur on schedule when it is called for and there is no opportunity to correct mistakes or to integrate material shot at different times or locations.

Film production inherently separates the tasks of shooting and program assembly because of the need to develop the film. Thus, it avoids the limitations of live television production but it has its own difficulties. The most serious of these is that the actual success of shooting is not known until the film is developed—hours or days later. Film also requires an *editing* process to assemble the individual shots into a program. Although this is a difficult process that calls for a lot of skill, it actually is a major advantage because it allows the separation of the shooting and the program assembly steps. It gave rise to the part of the film industry known as postproduction.

The film-making process logically divides into two parts:

1. *production*—get it all onto film. This is the familiar activity where each piece of each scene in the film is set up in a studio or on location and shot with movie cameras. The shots are separate and no effort is made to assemble them together at this time. Shooting is planned at the convenience of the set-builders and the talent. For example, all the shots on a particular set would be done in the same session without regard to where they will be used in the final movie.

2. *postproduction*—add effects, edit, and assemble. This has grown into a very complex process involving many different kinds of equipment and skills. Many specialty companies exist to support the different tasks of film postproduction. As the film industry matured, different people became masters of the art of

each of the parts of production and postproduction. A successful film required all of them. This style of operation greatly enhanced the artistic portfolio of filmmaking (called the *production values*) and it depended entirely on the ability to capture the elemental program material onto an editable medium. Of course, that was not available for television in the early days except for *kinescope recording*, which was recording from a television screen to film. However, the performance of kinescope recording was so poor that it was never seriously used as an editing medium.

When video recording was developed in the late 1950s, the opportunity to have all-electronic film-style production and postproduction was immediately recognized. However, videotape technology was not ready for that because of the difficulty of editing (it required literally cutting and pasting the tape). But when there's a need, video engineers will find a solution, and videotape electronic editing was quickly developed.

12.2 Electronic Editing of Videotape

In electronic editing, a tape of the complete program is created by recording the individual scenes onto a new tape in the precise sequence required by the program. If the scenes are already on tape, then the process involves copying the scene from its original tape to the new tape. This creates a *second-generation* recording, which has the problem that any distortions of the signals occurring in either generation will accumulate in the final recording. However, it works, and an entire industry for videotape editing has grown up from this idea and its enhancements. At the same time, analog videotape performance has improved to the point that electronic editing processes involving up to three or four generations are quite satisfactory.

By the use of multiple tape recorders and video switching systems, any of the transition effects (dissolves, wipes, etc.) that are possible in live television can also be used in postproduction. In fact, they can be done better because the source materials and timings can be adjusted much more precisely than is ever possible in live production.

12.2.1 Synchronization for editing

For electronic editing to work at all, there must be some degree of synchronization between the signals being edited. Synchronization of composite video signals occurs at three levels:

1. frame/field
2. line
3. color subcarrier

If the scanning of two signals does not match so that field sync intervals occur at approximately the same time within a few lines, all receivers will have to resynchronize their vertical scanning when the video feed is switched from one signal to the other. This causes a "roll" effect, where the vertical synchronization interval will briefly appear in the picture and the picture will slowly move to either the top or the bottom of the screen while it restabilizes. It is very disturbing and is not acceptable for almost any use.

A corollary effect occurs if such a nonsynchronized switch is done at the input of a video recorder. It causes the recorder to lose lock, which usually causes severe interference to the picture until lock is reestablished. Because of these annoying effects, frame/field synchronization was the first editing feature to be developed. It allowed direct switching between signals without disturbing receivers or tape recorders.

However, with frame/field sync alone, it was not possible to combine or mix signals to achieve effects such as a lap dissolve or a wipe. When the signals were combined, one could see that the sync stability was not good enough at the line level—the two signals jittered horizontally with respect to each other, often by a large percentage of the width of the picture. Line sync techniques were developed to correct this problem, which then allowed mixing of monochrome signals. But it didn't work for color signals—the color subcarrier phase still jittered between the signals, making the color of the mixed signal totally unstable. Color synchronization techniques were thus added to correct this problem.

The development of synchronization techniques for tape editing took a number of years, but it eventually became possible to perform fancy effects and editing without concern for synchronization. Today, most semiprofessional video recorders and all professional recorders are capable of full color synchronization. However, most home recorders have only frame/field synchronization, which limits their use in editing systems.

Another aspect of synchronization is maintaining exact frame locations so that corresponding points in different video or audio clips can be identified. This is best accomplished through the use of *time code*, which is a system that gives each frame a number. Time code is covered in Section 12.4.

12.2.2 Videotape editing techniques

There are three basic methods of videotape editing:

1. Assemble editing
2. Insert editing
3. A-B roll editing

The first two methods require one or two tape machines and the third process takes at least three machines. In each case, special control features are required in the system to allow precise positioning of the tape and accurate synchronization between machines when they are run. The degree of sophistication and performance of these features varies widely between professional, semiprofessional, and home recorders, but they exist to some extent in all video recorders today.

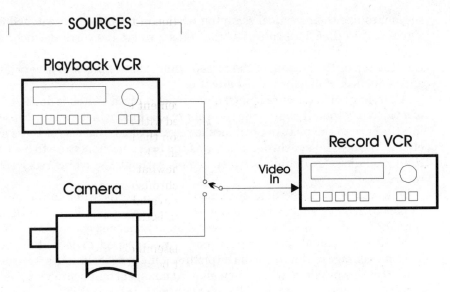

Figure 12.1 Equipment setup for assemble or insert editing.

12.2.2.1 Assemble editing

Assemble editing is often called *add-on editing* because it is a process of building the edited tape by adding one scene after another. Figure 12.1 shows an assemble editing setup. A video source is set up to deliver the signal to be added on, and the recorder making the new tape (called the edited master) is positioned at a point ahead of where the actual edit is going to occur. This is called setting up for *preroll*, and it is necessary to allow time for the recorders to synchronize before reaching the actual edit point. When the edit is made, both the source and the recorder are started up, and at the predetermined edit point, the recorder is switched from playback to recording to record the input. The recording then is made up to a point slightly beyond where the current scene is expected to end. At that point, everything is stopped.

In assemble editing, the source can either be a recorder or a live camera. If it is a recorder, that unit must also be set up for preroll by appropriately positioning its tape ahead of the edit point. Assuming that the source tape is an original (first-generation) recording, the resulting edited master will be a second-generation recording. Assemble editing requires that the shots be added in the order they will appear and the transitions between scenes are usually limited to cuts only.

For simple scenes, assemble editing may be done live by using a camera and one recorder. The process is the same: set up the edit point and the preroll, run the tape, and cue the source talent to start doing their thing at the edit point. A first-generation tape is produced. This works well for low-budget productions, but it offers no opportunity for correcting mistakes once editing has gone beyond a particular shot (unless all shots after the mistake are redone).

For this kind of one-camera live-edited production, it is actually better to use the insert editing technique described below, because that will allow any shot to be redone at any time as long as the duration of the shot does not change in the redo.

12.2.2.2 Insert editing

The insert editing technique is an enhancement of the assemble technique that depends on having the new tape prerecorded with at least black pictures and sync. That allows edits to be done at any point on the tape at any time, so there is no need to edit in the order of the final scene. This is a better technique than assemble for any editing, but it requires somewhat more expensive features in the recorder. It works best with full color synchronization, although it is still useful with frame/field sync. It is not necessarily available in the lowest-priced recorders. The equipment setup for insert editing is the same as shown in Figure 12.1 for assemble editing.

Insert editing is operated the same as assemble editing except that the insert must be stopped at exactly the right point because it will not be overwritten by another edit.

A useful technique with insert editing is to record the audio track for the segment before inserting the video. This allows the audio to be used for the timing of the video inserts. When the insert editing is done, audio-video synchronization has also been completed

12.2.2.3 A-B roll editing

By using more equipment as shown in Figure 12.2 and going to more trouble, many other possibilities are available by using *A-B roll editing*. In this technique, two source VCRs (A and B) are fed through a computer-controlled video switcher to the recording machine. Alternate shots are recorded on each of the source VCRs; these tapes are produced in a preediting session by either assemble or insert editing. When the edit is run, the A and B machines start up together and the video feed to the record machine is switched in real time between them. If the A and B recordings overlap appropriately in time, any transition type that the video switcher has can be used in the edits. This type of editing requires full color synchronization of the source machines. For more complex assembly, additional source machines can be added so that more than two signals can be combined at once.

12.2.3 Off-line editing

A professional-level editing suite with all the capabilities discussed above is expensive, it requires a highly skilled operator, and it costs hundreds of dollars per hour to operate. Since much of the time spent in editing is simply reviewing the production material and locating preliminary edit points, it is valuable to have a low-cost way to do all the preliminary work and then go to the full edit suite only for the final editing. This thinking has led to the concept of *off-line editing*.

Off-line means that the bulk of the edit decision making is done in a different format (and presumably a lower-cost one) than will be used for the final editing.

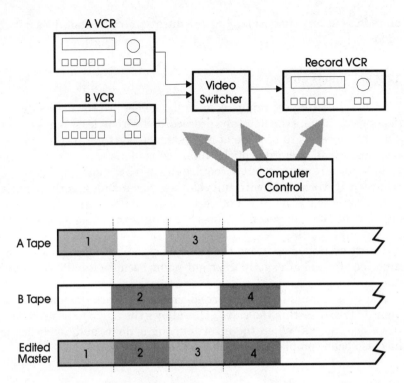

Figure 12.2 A-B roll editing setup and operation.

In film editing, there is the *Moviola machine*, which is relatively simple equipment that allows raw footage to be viewed with precise control so that someone can locate the desired edit points for each shot. A lot of development has gone into developing an electronic equivalent of the Moviola.

One easy way of doing this is to copy all the production shot tapes to a low-cost format such as VHS, and let the producer, director, and anyone else involved in the edit decision-making process take that and use home VHS players (or another low-cost format) to locate edit points (Figure 12.3). That is not as far-fetched as it sounds, because the time codes from the production tapes can be copied into the video frames themselves on the VHS tapes (this is called *burned-in time code*). Then, when a proposed edit point has been located by still-framing the home recorder, the time code value can be read from the TV screen and written into a list, called the *edit decision list* (EDL). When all the preliminary decisions have been made this way, the EDL is taken into a full editing suite, entered into the computer, and preliminary edits can be immediately viewed. Of course, the result may need fine tuning, but a lot of expensive edit suite time has been saved.

The use of home-style VHS equipment does not allow the off-line decision makers to actually view the edits they have chosen. They have to come into the expensive edit suite for that. Other systems have been proposed and even developed to provide more capable off-line editing, but until the growth of digital video, they were all too expensive for the purpose. Now, with personal computers and digital video, practical systems for off-line editing are available (see Section 12.5).

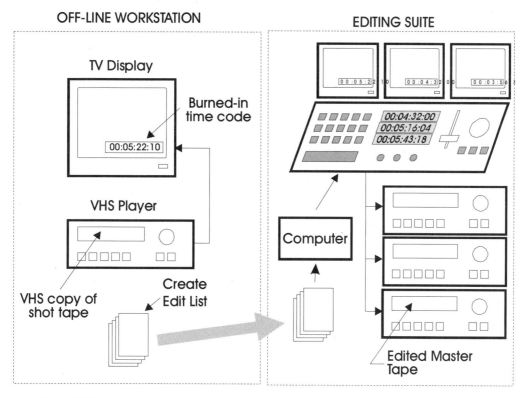

Figure 12.3 Off-line editing.

12.3 Tasks of Postproduction

There is much more than editing in postproduction. Some people refer to the entire postproduction process as editing, but in this book, editing is just the process of assembling separately produced shots into a scene or program. Postproduction can include many other tasks in the process of delivering the highest-quality, most artistic program from the material that has been produced:

- video color correction or enhancement
- transition effects
- video effects
- video titles
- audio sweetening
- audio mixing
- audio-video integration
- video editing
- final format conversion

Each of these tasks could take a chapter itself; here there is only room for a summary (see the bibliography for more information).

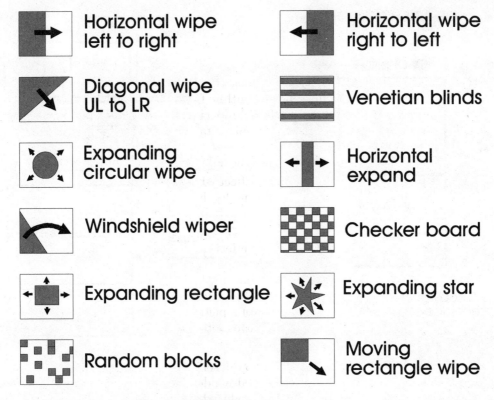

Figure 12.4 Some typical video transition effects.

12.3.1 Video color correction or enhancement

Many things can happen in video production to cause the resulting video to not be up to full quality. Sometimes, those problems can be corrected in postproduction. For example, if the colors do not match correctly in all of the shots to be used, a *color corrector* can usually fix it. Similarly, if a particular shot is noisy, the SNR can be improved by the use of noise reduction. Fixing substandard video or video errors is an important part of postproduction.

12.3.2 Video transitions

The use of video transition effects such as dissolves or wipes has become commonplace in most video production. If a program has video transition effects, it is a clue that the program was probably built in postproduction. Figure 12.4 shows some typical transition patterns. These transitions do not change the picture geometry—they simply replace one picture with another according to the motion shown by the arrows.

12.3.3 Video effects

There are many video effects that are more than transitions. These are sometimes called *special effects*. Examples are:

Chroma keying—this is an analog effect where two pictures are combined by using one picture to create an outline for keying in another picture. A simple example is showing a window behind a newscaster that displays another video. The window exists as a bright blue rectangle in the background of the news set. The chroma keyer is set up to insert the other picture into any area that is bright blue. This effect can also be done digitally (see below).

Split screen—this is an analog effect where different pictures appear in different areas of the screen. For example, the screen may be split between left and right or between top and bottom. It is similar to starting a wipe effect and stopping it in the center.

Digital effects—a whole class of effects are referred to as "digital" because they became available in television when digital video processing was first developed. They include all effects that involve any kind of geometric manipulation of pictures, which is something that cannot be done with analog hardware. Examples are: changing the size of a picture (zooming), rotating pictures, page-turning effects, or placing pictures into odd-shaped areas, such as having pictures on the sides of a 3-D cube.

The chroma key newscaster example described above can be combined with a digital zoom so that the chroma-keyed window image zooms out to full screen for presentation of the clip that is introduced in the window.

Because digital video effects are so powerful, it makes sense in analog systems to use ADC and DAC just to get the effects. Of course, in a system that is already digital, no conversion is necessary.

12.3.4 Video titles

Many programs use titles or other computer-generated overlays. Special equipment in postproduction generates titles and overlays and keys them into the video on demand. When keying computer-generated art into composite analog video signals (NTSC or PAL), special features must be included to avoid aliasing on the keyed edges.

12.3.5 Audio sweetening

The equivalent to video enhancement and color correction is *sweetening* for audio. This includes processes to make the audio sound as good as possible. Examples are noise reduction, equalization, effects (such as reverberation), stereo balancing, and level matching.

12.3.6 Audio mixing

Mixing is the audio process for combining signals. Most audio channels for video are produced as several separate tracks that get combined in postproduction. For example, it is common to produce separate tracks for music, dialog, effects, and ambience. Each track can be optimized by itself and then the audio postproduction editor will mix these to achieve the desired final audio track.

12.3.7 Audio-video integration

It is normal to separate audio and video and operate on them differently during postproduction. This is necessary because the audio and video usually have different continuity. For example, in a dialog scene between two people, the video may be cut or dissolved between shots of the two participants or cut away to supporting material, but the dialog audio runs continuously throughout. By separating the audio, the video editor does not have to worry about the audio when doing the cuts and dissolves. Similarly, the audio mixer does not have to be concerned with the video.

Once both audio and video are completed separately, they can be combined in the final format. The only requirement is to maintain synchronization. This is facilitated by having the entire postproduction process based on time code (see Section 12.4).

12.3.8 Video editing

Techniques for video editing were described in Section 12.2.2 for tape-based editing and are described in Section 12.5 for computer-based editing.

12.3.9 Final format conversion

Usually, the edited master is produced in the format required for the end user. However, if additional copies are to be made from the master, it is better to perform all postproduction operations in the highest-performance format available and drop down to the distribution format only for the copies made from the edited master. This is especially important if the end product is to be a low-performance format such as VHS tape or microcomputer compressed video such as MPEG-1. In any case, the final format conversion is considered to be the last step of postproduction.

12.4 Time Code

Postproduction requires exact positioning of video and audio edit points. Points can be found by playing video and audio on machines that have shuttle (variable-speed playback) and jog (single-frame stepping) features, but once the desired frame is located, there must be a way to save it and come back to it exactly when doing the edit. That is the purpose of time code. All frames are identified with sequential numbers based on hours, minutes, seconds, and frames within a second.

Time codes on video or audio tape are usually placed on a separate longitudinal track that will be readable when the tape moves at any speed. This is called *linear time code* (LTC). Variable-speed readability is important because it is desirable to run the tape as fast as possible when searching for a location and slowing down only as the time code indicates that the tape is getting close to the target point. Time code can be recorded during the shooting or it can be inserted after shooting, at the start of postproduction.

Another possibility for time code in video is to insert digital codes in the vertical blanking interval of the video. There is plenty of room for that—it is called *vertical interval time code* (VITC). However, this has limitations compared to the separate tracks because it is more difficult to make VITC readable at all tape speeds.

12.4.1 SMPTE time code

Standards for time codes have been set by the SMPTE. Both LTC and VITC versions have been defined. They are based on a bit rate of 2400 bits/s, which provides 80 bits per frame in a 30-frames-per-second system and can easily be recorded on an audio track and played back at any speed from one-fifth of normal up to 20 times normal. The 80 bits per frame are allocated as shown in Figure 12.5.

Time is specified in hours, minutes, seconds, and frames; 26 bits are allocated for that. In the case of NTSC, where the frame rate is not exactly 30/second (it is 29.97), it is necessary to periodically drop a frame from the time code to keep the code in sync with clock time. One frame is dropped every 1000 frames, which works out to 108 frames dropped per hour. A flag bit indicates when the dropped-frame mode is in use. Another flag can be used to identify one color frame of the color-frame sequence.

12.5 Digital Postproduction

As digital video technology has developed, postproduction was one of the first places it was used. Digital effects, described above, have had a particularly important impact on television postproduction. But now, with video compression technology, complete digital editing has become a reality. But the real reason for this is not simply the usual advantages of digital systems, but because digital editing allows a completely different mode of operation—nonlinear editing.

12.5.1 Linear and nonlinear

Videotape is a *linear* medium, meaning that it is one-dimensional in time. To get from point A to point B on a tape, one must play through all the intervening material. There is no means of fast random access to points in a tape recording.

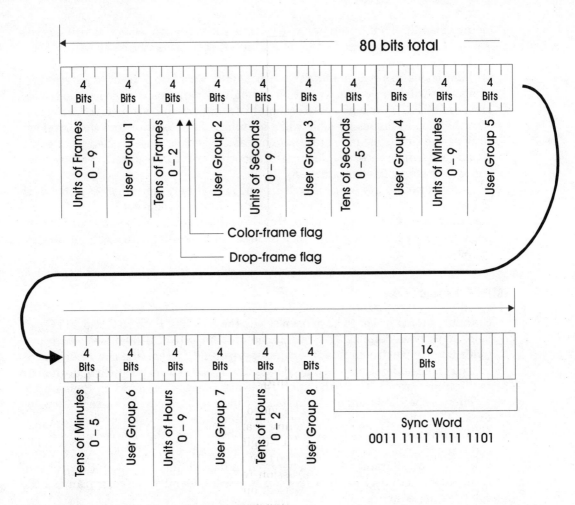

Figure 12.5 SMPTE time code 80-bit format.

However, a computer hard disk is a true random-access device which provides rapid access to any material recorded on the disk by moving the read heads from track to track. Any point on a typical hard disk can be accessed in a few hundred milliseconds or less. That is fast, but not fast enough to provide seamless editing of video, which would require the access time to be less than the width of vertical blanking, about 1 ms for NTSC or PAL.

In a digital system this problem is easily handled by providing *buffer memory* that stores enough video data to cover the seek time of the hard disk. Then, an editing system can effectively switch to any frame of video on the disk "instantaneously." It is an ideal environment for editing. Because there is no restriction of playing through material on a linear medium to find edit points, hard disk editing is called *nonlinear*.

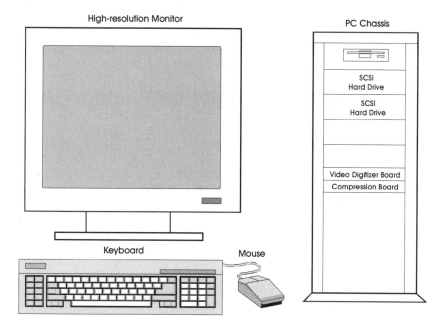

Figure 12.6 Hardware for a nonlinear editing system.

12.5.2 Other requirements of digital editing

The idea of true random access for video editing using computer hard disks and buffering has been around for many years, but there has been a large problem that has prevented much use of it—digital video was just too much data. This severely limited the amount of video that could be stored on any reasonable-sized hard disk or disks. Video compression technology that had been developed for computer video display could be used, but it did not provide full quality performance, so the use of nonlinear editing was only for low-quality services, or for an off-line medium (see Section 12.2.3). In that case, full editing can be done with the nonlinear system, but to obtain a full quality output, the EDL from the nonlinear system has to be taken to a full quality (analog or digital) system to perform the final editing.

However, video compression is a moving target, and performance has improved in the latest systems to the point that affordable nonlinear systems can in some cases deliver full broadcast-quality output. At the same time, performance of hard drive systems has improved and the cost of storing uncompressed (or lightly compressed) video continues to fall.

A full quality nonlinear editor represents more than an order of magnitude reduction of cost for an editing system. As shown in Figure 12.6, all it takes is a powerful PC, a lot of hard disk storage, and add-in hardware for video capture, compression, and decompression. Because of the extreme competition in the markets for these systems, costs are dropping rapidly and performance is improving at the same time. It is the editing system of the future, here today.

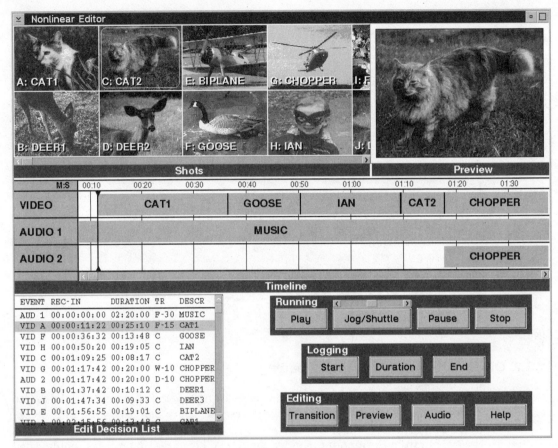

Figure 12.7 Control screen for a hypothetical nonlinear editor.

12.5.3 Typical nonlinear editor

There are many products on the market for nonlinear editing at different quality levels. Because of that diversity, this discussion is based on a fictional system that embodies the principal features of the best real products. The main control screen is shown in Figure 12.7.

At the top of the screen, an area of "thumbnail" images shows the start frame of each shot selected for the edit. At the top right there is a preview window where actual edits (or the whole scene) can be viewed. Below that is a "timeline" view. It shows the edits on a time scale using the thumbnail names to identify them and indicates the timing and duration of each. The time scale can be expanded or compressed and this part of the display can scroll horizontally to show different parts of the scene in as much or as little detail as desired. The timelines also include audio tracks. Below the timelines there is a window that displays the EDL in actual timecode (this is for the purist who wants to fine tune the numbers) and a set of control buttons for the system.

Operation of the system is as follows:

First, all source video is copied from original tape to the hard drive. The copying includes the time code (or adds it if it was not on the tape) and video compression.

Once all the video has been copied, the operator uses the menu to select a source video for logging of shots. (Logging is the process of specifying exact frame numbers for the start and end of each shot.)

The first frame of the video appears in the first thumbnail window. By using the control buttons, the operator can shuttle and jog the video to find an exact frame for starting the shot. By clicking the "Start" button, time code for this frame is logged in the computer, the frame is shown in the first thumbnail, and the source tape control automatically jumps to the next thumbnail space.

If the operator selects the first thumbnail again and clicks the "Duration" button, he or she can shuttle or jog the picture in the thumbnail to log the edit end point. The "End" button is clicked to confirm this.

All shots for the session are chosen by repeating the above procedure. Any points can be revised by selecting the thumbnail, shifting the frame with shuttle or jog, and clicking either the "Start" or "End" buttons. A dialog box is available for each thumbnail by double-clicking on it. Using this, the operator can type in a description of the shot, which then appears on its thumbnail.

Assembly of the scene on the timeline is accomplished by "dragging" from the thumbnails to the timeline view. The shot names are reproduced on the timeline at their locations and their durations are shown. The EDL is automatically created by this operation.

Transition effects are chosen by placing the timeline cursor (dotted vertical line) over the desired location and clicking the "Transitions" button, which brings up a dialog box containing transition choices and their parameters.

At any time, the editing can be viewed by clicking the "Preview" button. It can also be examined by using the jog and shuttle controls.

Nonlinear editing greatly simplifies and speeds the editing process.

12.6 Summary

The use of postproduction allows the shooting of video and audio to be unencumbered by the considerations of program assembly while the assembly process gains far more flexibility than could ever be handled during shooting. Because of these advantages, nearly all program production is done this way and postproduction equipment is available at all quality levels.

Postproduction techniques are highly developed, and with the arrival of digital video systems, new possibilities are being explored to even further enhance the process while making it easier at the same time.

Chapter 13

Television Receivers and Video Monitors

13.1 Introduction

The performance of today's color TV receivers is superior by orders of magnitude to those available in 1953 when color television broadcasting was first approved in the United States. They produce far better images in every measurable respect, they make larger pictures with smaller sets, and they are far more stable and reliable. Best of all, the improved performance has been achieved while offsetting manifold inflationary cost increases, and the average 1999 TV receiver is lower priced than in 1954, even in current dollars. The engineering resources of many countries have contributed to this progress, and it has truly been a superb worldwide achievement.

A remarkable feature of this performance is that it has been achieved without any revolutionary technical breakthroughs. Certainly the advent of solid-state technology has been a factor, but the same superheterodyne principle continues to be used for the radio frequency (RF) section of the receiver. The original three-gun shadow-mask principle continues to be used in most receivers. The transmission format established by the FCC in 1953 (and by the EBU in Europe) is essentially unchanged. As is often the case in war and technology, success lies in the details.

13.2 Receiver Configuration

As DTV is deployed around the world, digital TV receivers will be required. However, they will still use basically the same principles with changes caused by the different modulation methods used. This chapter describes both analog and digital TV receivers.

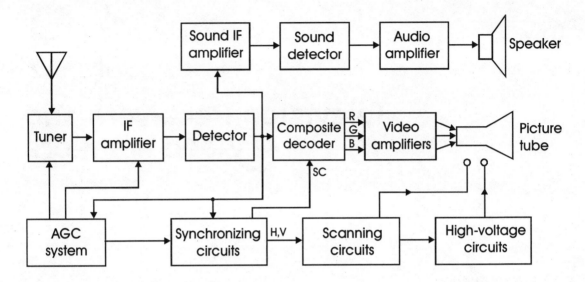

Figure 13.1 Block diagram of a typical analog TV receiver.

13.2.1 Analog TV receiver

A functional diagram of a typical analog TV receiver is shown in Figure 13.1. Its layout is conventional—a tunable RF stage for channel selection including a local oscillator and first detector, and an intermediate frequency (IF) amplifier and second detector for video and audio. The video channel uses a diode detector or synchronous detector. The audio channel uses a discriminator or other frequency-modulation (FM) detector. The output of the video channel includes the composite signal multiplexed with the timing and color synchronizing signals in the format described in Section 6.7. A decoder (NTSC or PAL) converts the video to RGB components, which are amplified and drive the color picture tube or other display device. The synchronizing circuits control the display device scanning circuits and deliver a color subcarrier (SC) signal for the decoder.

13.2.2 Digital TV receiver

Figure 13.2 shows a block diagram of a digital TV receiver. The example shown is a receiver for the ATSC DTV standards in the United States. Many features are the same, but it can be seen that the signal detection, decoding, and sound channel are entirely different. In addition, there are several new features for automatic equalization of the channel response, which is made possible within the digital standards. Digital receivers are at an early stage in their development and many improvements are sure to be made as they progress to full-scale volume manufacturing. The following sections describe each of the major areas of both receivers.

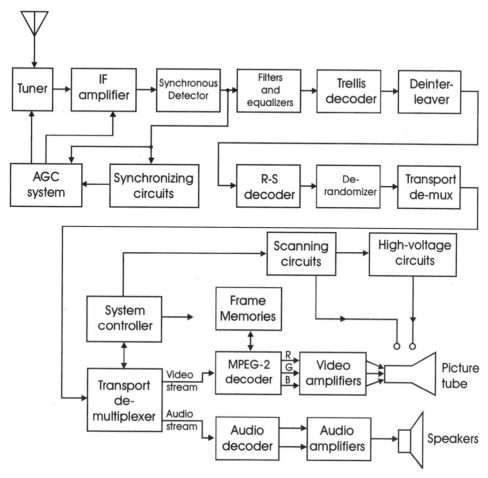

Figure 13.2 Block diagram of a DTV receiver.

13.3 Tuner Section

The purpose of a tuner is to select the desired channel and reject all others. It converts the channel frequency to a fixed IF for further amplification and detection.

Most receivers today have an automatic control system for the tuner so the viewer can select channels from a remote control device and does not have to be concerned with any adjustment of the tuner.

In earlier sets, VHF tuning was accomplished by variable capacitors with mechanical detents, while UHF tuning used a more awkward and time-consuming method of continuous tuning. The FCC then decreed that VHF and UHF receivers must have the same ease of tuning, and this requirement has been met by the use of electronically controlled systems that can have preset values for the desired channels.

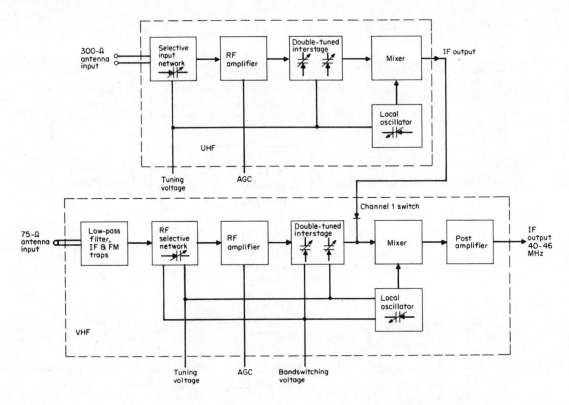

Figure 13.3 Diagram of a varactor tuner. Reproduced with permission from [1], Figure 13.3.

13.3.1 Tuner RF section

Tuners today use *varactors* for the tuning elements; these are reversed-biased solid-state diodes whose shunt capacitance varies with applied voltage [1]. Figure 13.3 is a diagram of a varactor tuner. Generally, two separate tuner circuits are used for VHF and UHF tuning. The two VHF bands are selected by bandswitching in the tuning circuits using diodes. A single tuner circuit is possible if two IF frequencies are used in a double-conversion arrangement The first IF is a high frequency, such as 900 MHz, and a single local oscillator can cover the entire frequency range needed to convert all channels to this IF. A second mixer with a fixed local oscillator drops the first IF down to the usual 40-MHz range.

The signal is amplified by an RF amplifier before mixing with the output of the local oscillator, which operates at a frequency of 45.75 MHz above the picture carrier, which is also 41.25 MHz above the sound carrier. The level of the incoming carrier signal can vary from 10 μV to 100 mV, but the automatic gain control (AGC) system maintains the amplified signal level within reasonable limits by controlling both the RF and IF amplifier gains. The maximum gain of the RF amplifier is approximately 20 dB.

Broadcast receivers can be tuned to the VHF and UHF channels in the television broadcast bands, VHF channels 2 to 13 and UHF channels 14 to 83. For reception of the additional cable channels, the receiver is tuned to channel 3 or 4 and the cable channels are heterodyned to it by means of a set-top converter. Alternatively, cable-ready receivers are capable of picking up the standard cable channels directly (see Chapter 17) without an external converter.

13.3.2 Tuner control system

The varactors are used to tune at least two circuits in each tuner to reject undesired channels and image responses. In addition, the local oscillator must be precisely set on the proper frequency to deliver the channel signal to the IF amplifier. These are controlled by one or more *tuning voltages* that are generated by the tuner control circuits. Alternatively, the local oscillator frequency may be generated by a digital *frequency synthesizer* [2].

Most receivers today have a microcomputer control system that stores tuning voltage and frequency settings in a permanent memory so they can be called up as the viewer selects channels.

13.4 IF Amplifier

The purpose of the IF amplifier is to produce the exact frequency reponse across the channel, rejecting all other frequencies to prevent adjacent-channel interference, and amplify the signal to a level high enough for detection.

13.4.1 Video channel frequency response

Tuners are designed to have essentially flat response across the selected TV channel. Thus, the video channel response of a TV receiver is determined by its IF amplifier.

13.4.1.1 Analog receiver frequency response

The frequency response of the video channel must meet rather exacting requirements (Figure 13.4). It must provide high rejection of the adjacent channel picture carrier at 39.75 MHz and the adjacent channel sound carrier at 47.25 MHz. The treatment of the channel sound carrier differs from system to system, but it must be highly attenuated at the input of the video detector. The picture carrier at 45.75 MHz should be located at the 50% (–6 dB) point on the upper slope of the response curve. This position equalizes the sum of the responses of the sidebands (at 100% in Figure 13.3) in the vestigial sideband area for baseband frequencies from 0 to 1.5 MHz. For modulating frequencies above 1.5 MHz, equalization of the video channel must be provided when the response of the IF falls below 100%. In the example in Figure 13.4, this occurs at sideband frequencies below 43.25 MHz or baseband frequencies above 2.5 MHz. Equalization is particularly important in the vicinity of the chroma subcarrier in the range 41.67 to 42.47 MHz.

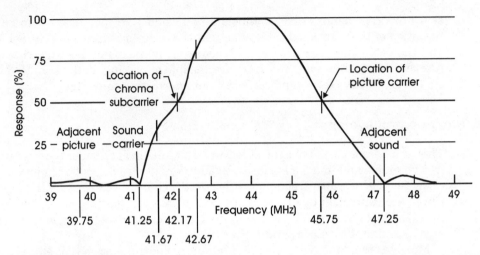

Figure 13.4 Analog video IF frequency response. (*Reproduced with permission from [1], Fig. 13.17.*)

Figure 13.5 shows the details of the equalization required.

The presence of the subcarrier imposes requirements on the performance of the IF amplifier in a color receiver that are not present with monochrome. The bandwidth must be greater to accommodate the subcarrier and its sidebands. Because the color subcarrier falls on the sloping side of the IF response, the tuning must be stable so that the correct ratio of luminance to chrominance signals is maintained. Linearity must be sufficient to avoid intermodulation between subcarrier and carrier. Envelope delay as well as the frequency response must be equalized over the channel pass band.

The frequency response is often established by a *surface acoustic wave* (SAW) filter (see Section 16.8.1.6).

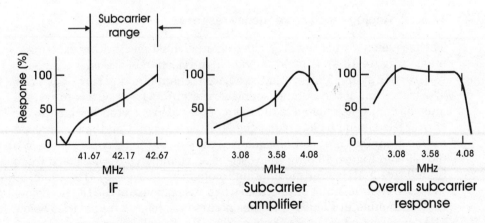

Figure 13.5 Color subcarrier equalization. (*Reproduced with permission from [1], Fig. 13.18.*)

13.4.1.2 Digital receiver frequency response

A DTV receiver is designed to have a flat IF response over the channel bandwidth with a roll-off at each end before the channel edges. This is in accord with the system (transmitter-receiver) response curve shown in Figure 7.10. Using the *training signal* feature of the ATSC standard, DTV receivers can have an automatic equalizer and ghost canceller to maintain an accurate channel response under varying transmission conditions [3].

13.4.2 Sound IF channel

In analog broadcasting, the sound is frequency modulated on a separate carrier separated by a precise distance from the video carrier. Circuits in the receiver must separate the sound carrier and demodulate it for presentation. In digital broadcasting, the sound is multiplexed in the same bit stream as the video and is separated by the digital processing that demultiplexes the bit stream. A digital decoder extracts analog sound to drive amplifiers and speakers as shown in Figure 13.2. A digital receiver has no equivalent of the analog sound IF channel.

13.4.2.1 Analog sound IF

The sound IF subcarrier operates at 41.25 MHz, 4.5 MHz below the video. It can be handled in a number of configurations. In one, it is carried through the IF stages at a level 10 to 20 dB below the picture carrier. The 4.5-MHz difference signal, including its frequency modulation, then appears at the output of the detector. In another, the sound IF is separated from the video at the output of the second IF stage and is then processed by a separate sound IF amplifier.

In both cases, the final difference between the picture and sound IF frequencies is 4.5 MHz as established at the transmitter; this is known as *intercarrier sound*. This significantly simplifies a receiver but it does have the problem that overmodulation of the video carrier can cause buzz interference in the received audio. That occurs because the video carrier level can go to zero during bright areas of the picture, causing momentary loss of the intercarrier sound carrier.

13.4.3 IF Gain

The IF amplifier gain is typically 90 dB, which is offset by 20 dB loss in the input and detector stages:

Stage	Gain (dB)
Input coupling	−10 dB
1st IF stage	+30 dB
2nd IF stage	+30 dB
3d IF stage	+30 dB
Detector	−10 dB
Total gain	+70 dB

13.4.4 AGC and AFC

Automatic gain control (AGC), and *automatic frequency control* (AFC) are standard features of most analog and digital receivers.

AGC is often supported in both the RF and IF stages of the receiver. For low- and medium-level signals, gain control is applied only in the IF section. Additional gain reduction is applied in the RF section for very strong signals.

AFC is applied to the local oscillator and maintains the IF carrier frequency at 45.75 MHz. In digital receivers, the pilot signal added to the VSB modulator at the transmitter (see Section 7.7.3.4) is detected for AFC. This signal is at the IF center frequency of 44 MHz. A *frequency-phase-lock loop* (FPLL) locks on this pilot carrier and is used for AFC and synchronous demodulation of the VSB to obtain the original trellis coded data signal.

13.5 Detection

Analog detectors can either be a diode or the more advanced synchronous and balanced type. The latter is more immune to cross-modulation between the video carrier, the color subcarrier, and the sound carrier.

Digital receiver detection of the VSB modulating signal is as described above. Following that, any automatic filtering or equalization is performed before the trellis decoding, which retrieves the error-protected data stream. This stream is then deinterleaved, R-S decoded, and derandomized to obtain the original transport stream, free of errors. This is described in more detail in [3].

13.6 Sync Separation and Scanning Generation

In analog systems, the vertical and horizontal sync pulses are transmitted at a higher modulation level than the video signal and are easily removed for driving the scanning generators.

The vertical scanning generator operates at the relatively low frequency of 60 Hz and is usually a multivibrator-type sawtooth generator that is synchronized by the vertical sync pulses as modified by the equalizing pulses (see Section 1.8.2).

The horizontal scanning generator operates at the much higher line rate, and the retrace time is much shorter. As a result, the voltage, $L(di/dt)$, developed across the deflection transformer and/or deflection yokes is very high. This voltage can be rectified and used to provide the anode voltage for the kinescope.

In a digital receiver, the scanning synchronization information for the format being broadcast is extracted from the MPEG-2 decoder. Very likely, format conversion will be required to convert the data to the native scanning format of the receiver's display section. This action will be commanded by the system controller section of the receiver.

13.7 Decoding Signal Formats to RGB

Having received the composite video in an analog receiver, or the video data stream

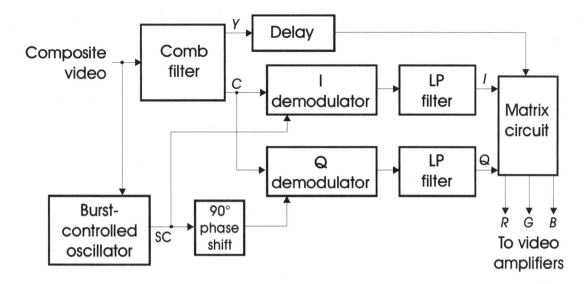

Figure 13.6 Block diagram of an NTSC composite decoder.

in a digital receiver, they must be decoded to RGB component analog video to drive the display, usually a CRT.

13.7.1 Analog decoding

A block diagram of an NTSC decoder is shown by Figure 13.6. The color burst is removed from the composite video signal received from the detector and used to synchronize a burst-controlled oscillator, which is a phase-lock loop (PLL). This reference subcarrier (SC) drives quadrature demodulators to recover the I and Q signals (see Section 5.5.2). The I and Q signals are filtered to their respective bandwidths and are matrixed with the Y luminance signals to generate the R, G, B primary signals. Video amplifiers at the output of the matrix circuits raise the video level to that necessary to drive the display device. Processing circuits for image enhancements may be added at this point.

The level of the chrominance signal is adjustable at the input of the demodulators by the "color" control to establish the saturation of the image.

The phase of the subcarrier with respect to the color burst can be adjusted in the burst-controlled oscillator to control the hue, and this is the "hue" or "tint" control.

13.7.2 DTV decoding

The decoding portion of a DTV receiver was shown in Figure 13.2. It is explained here. The output of the synchronous detector is the multilevel trellis-coded signal. This can be filtered and adjusted to correct for transmission problems such as multipath or attenuation. Then, the trellis-coding is decoded to the symbol bits

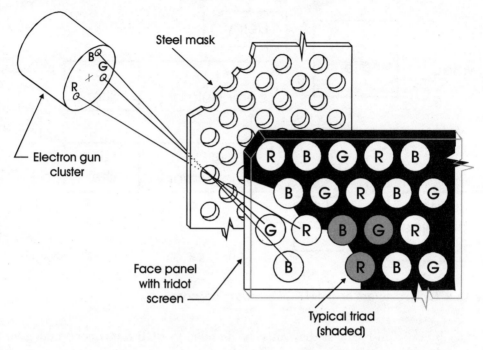

Figure 13.7 Shadow-mask kinescope, delta gun, round hole. (*Reproduced with permission from [1], Fig. 12.4.*)

(3 per symbol), and the error protection coding is undone (deinterleaving, R-S decoding, and derandomizing.) This recovers the original MPEG-2 transport stream, which can then be partitioned into its individual video, audio, and data streams.

The video stream is decoded to RGB components by an MPEG-2 decoder, the audio is decoded by the Dolby Digital decoder, and any data streams are directed to their intended destination.

A system controller microprocessor manages the viewer interface and generally directs the operation of the digital circuits. For example, it would be responsible for reading the ID information from the transport stream and directing the partitioning circuits to retrieve the components desired by the viewer.

13.8 Color Display Devices

The display device is the component that produces the visual image; this is the end point and the final key to the performance of color TV systems. No matter how perfect the performance of the signal generating and transmission equipment, the image quality at the receiver can be no better than that of the display device. Improvements in these devices, probably more than of any other single component, have been the greatest contributor to the dramatic advances in television receiver performance described in Chapter 4 and at the beginning of this chapter.

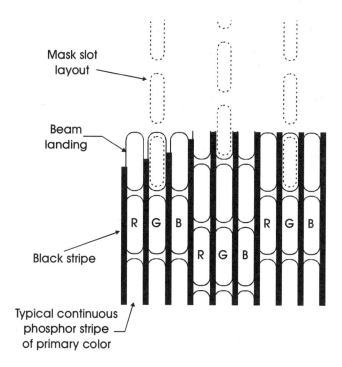

Figure 13.8 Shadow-mask kinescope, in-line guns, stripes. (*Reproduced with permission from [1], Fig. 12.3.*)

13.8.1 Shadow-mask kinescopes

The *cathode-ray tube* (CRT), sometimes called a *kinescope*, is the most popular video display device. *Shadow-mask* kinescopes were used as the display devices in the original color television receivers. This principle has been remarkably durable, and it is still in wide use 40 years later. It is produced in a number of formats, two of which are shown in Figures 13.7 and 13.8.

The format in Figure 13.7 has round holes in the shadow mask arranged in triads and a "delta gun" structure with the electron guns arranged in a triangle.

The format in Figure 13.8 has an "in-line" gun structure with the guns mounted side-by-side rather than in a triangle and with slots rather than round holes in the shadow mask. An advantage of this configuration is that there is no color registration problem in the vertical direction.

As a result of the relative positions of the electron guns, shadow-mask holes, and tricolor phosphors, the electrons from each gun impinge only on phosphors of one color. By controlling the amplitude of the signal current from each gun in accordance with the brightness of the primary component of the scene at each point, the amplitude of the image brightness for each primary color will correspond to the scene brightness for that color.

In addition to the formats shown in Figures 13.7 and 13.8, the shadow-mask principle is applicable to a variety of additional formats in the shape and location of the holes in the masks and the arrangement of the guns.

A black deposit around each phosphor dot (the surround) is an especially important feature of more recent shadow-mask designs. It absorbs the ambient light striking the faceplate so that blacks appear to be really black, thus increasing the contrast ratio (see Section 4.14). Black portions of the image can be no blacker than when the set is turned off, and this was a light gray in early TV receivers because of reflection of ambient light. The production of black blacks has been one of the major contributors to the improvement in image quality in recent years.

A trend toward larger pictures has been another important development, and kinescopes with diagonals up to 36 in (0.9 m) are now standard. The performance of larger pictures is now satisfactory because of the significant improvements in the technical quality of broadcast signals (see Chapter 4), which reduce the defects that become more visible with larger images and lower viewing ratios. Some homes are now designed with one room devoted to TV viewing—a "home theater."

The minimum screen sizes recommended by one leading manufacturer for various viewing distances are:

Viewing distance (ft)	Screen size (in)
<7	27
7–10	31
10–11	35
11–18	46
>15	52

These recommendations provide viewing ratios (see Section 4.4.2) in the range of approximately 5:1 to 6:1.

The use of large screens will be accelerated further by the introduction of HDTV—in fact, HDTV will require larger screens to be fully effective.

The transition to larger screens has been made practical by advances in design and production techniques that have steadily improved the quality and reduced the cost of shadow-mask kinescopes.

13.8.2 Projection kinescopes

Until recently, direct-view shadow-mask kinescopes have been overwhelmingly preferred by consumers for television display devices as compared with their competitor, projection kinescopes. The latter have suffered from inferior brightness, contrast ratio, resolution, and high costs. Recently, however, the same developments that have increased the demand for larger direct-view tubes have created a growing desire for screens larger than 36 in—the approximate practical limit for direct-view tubes. This has caused a modest but significant increase in the demand for projection sets, which can produce far larger images, typically with 46-in to 52-in diagonals and greater. The demand for projection sets has also been increased by impressive improvement in their performance—particularly their brightness, contrast ratio, and resolution.

Projection sets have three small but very bright tubes, one for each primary color. The images on the tubes are enlarged, converged, and focused on a screen by means of an optical system that may employ mirrors, lenses, or both.

Video projectors are now widely used to display very large pictures (2 to 8 m diagonal) for meetings and conferences. These may use three CRTs, or LCD or DMD devices (see Section 14.6).

13.8.3 Flat-panel displays

Flat panels are attractive since they would make "picture on the wall" displays possible. They are widely used for portable computer displays (see Section 14.6), but they are not presently economic in the large sizes required for TV receivers. The design and manufacturing problems for large flat-panel screens are formidable, and this must be classified as a future development, which is being spurred by the demands of HDTV.

13.9 Special Receiver Features

A number of special features are available in commercial television receivers; four of these—video inputs, closed captioning, remote control with readout status menus, and "picture-in-picture" are described below.

13.9.1 Video inputs

VCRs typically have an NTSC RF output signal, usually on channel 3 or 4, to which the receiver is tuned. This arrangement means that the VCR must modulate its video on an RF carrier and the receiver must demodulate that back to NTSC. These extra steps of processing cause significant signal degradation, which can be eliminated by providing the VCR with a baseband video output and the receiver with a baseband video input. Most VCRs and many receivers now have this capability, which is called the video monitor feature. This is also useful for interfacing a TV receiver to a video game or a computer. Another input format on some receivers is *S-Video*, which is a Y–C component format that entirely eliminates the NTSC encoding and decoding from VCR operation.

Digital receivers can use the IEEE 1394 serial standard for video input from a VCR or camcorder.

13.9.2 Closed captioning

Closed captioning is a system for displaying lines of alphanumeric text on the screen for the deaf and the hearing-impaired. This capability is now required by law in all television receivers. The text can include titles, the program dialogue, or

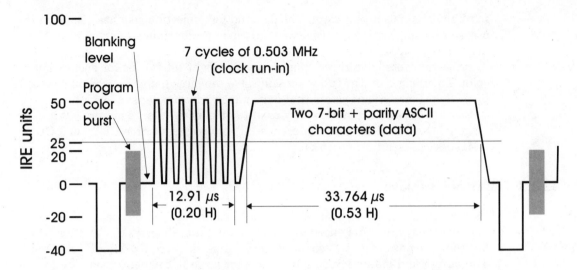

Figure 13.9 Closed captioning signal format. (Reproduced with permission from [4].)

other explanatory material. The text characters are transmitted in line 21 of odd-numbered fields by digitally encoded symbols. They are decoded at the receiver for display on the screen.

The standard format for line 21 is shown in Figure 13.9. The digital text symbols are transmitted in a 33.76-μs segment of the line at a rate of 960 bits/s. The data bit stream is synchronized with the caption generator by means of 7 cycles of a 0.503-MHz clock signal that is transmitted at the beginning of line 21. Each data block consists of two 7-bit characters coded in a nonreturn-to-zero (NRZ) ASCII format with an 8th bit added for parity. This gives a data rate of 120 characters per second, which is high enough for any captioning purpose.

Closed captioning is one example of the use of the vertical blanking interval of an analog television system for the transmission of digital data. There are a number of proposals for other uses of this capability to deliver new services such as program schedules or multilingual features.

13.9.3 Remote control and read-out menus

Nearly all modern receivers are equipped with remote control facilities, usually having infrared communication between a hand-held controller and the receiver. As a minimum, the remote unit provides capability for selecting the video channel and controlling the volume. More elaborate systems provide tone and stereo balance controls for the sound and a complete set of video controls including color, hue, brightness, contrast, and sharpness.

To make remote-controlled sets more user friendly, a menu is shown on the screen that indicates the availability and status of the controls. The format of the menu is similar to that shown on computers but it is simplified to be consistent with the capabilities of home users. When the menu is not in use, it automatically goes away.

Figure 13.10 Picture-in-picture display.

13.9.4 Picture-in-Picture

The *picture-in-picture* feature is one of the most interesting and popular in modern television receivers. It enables the viewer to display a reduced-size version of one picture while viewing a different program at full-screen. The reduced picture is displayed in a small aperture on the screen. The size and position of the aperture are adjustable by the user (see Figure 13.10).

Since the scanning standards of the two pictures may not be identical and synchronized, it is necessary to convert and synchronize the scanning of the inserted picture with the scanning of the base picture. This is accomplished with a digital standards converter (see Section 3.6.2). Geometric manipulations, such as varying the size, position, and shape of the picture, can also be introduced at this point (see Section 11.3.3). The signal is then reconverted to the analog format and inserted in the base picture by a video effects technique (see Section 11.3.3).

13.10 Receiver Performance Criteria

The performance criteria of TV receivers can be divided into two categories: the performance of its signal amplification sections, and the performance of its image display device.

The primary criteria of the signal amplification sections of an analog receiver are noise figure and selectivity (this presumes that they conform to industry practice with respect to frequency response and linearity). These criteria determine performance with weak signals and in the presence of adjacent channel interference. The primary criteria of the display device, in addition to size, are resolution, brightness, and contrast ratio.

13.10.1 Noise figure

The *noise figure* criterion was originally developed for specifying the performance of radar receivers, but it is applicable to any type of receiver, analog or digital. It is defined as the ratio of the SNR at the output of a receiver to the SNR at its input when the input is at room temperature, T_O (assumed to be 290K or 17°C).

Since the receiver inevitably adds noise, the noise figure expressed as a fraction will be less than unity, and expressed in dB, will be negative. To avoid this, it is conventional to express the noise figure in dB as a positive number as shown in Eq. (13.1):

$$N_F = -10 \log \frac{\text{SNR}_{\text{OUTPUT}}}{\text{SNR}_{\text{INPUT}}} \quad (13.1)$$

where $T_{\text{INPUT}} = T_O = 290K$. The receiver's noise figure is a critical performance criterion for receivers located in areas of weak local signals. Typical noise figures are 4 to 8 dB on VHF channels and 7 to 12 dB on UHF channels.

13.10.2 Selectivity

The *selectivity* of a receiver is a measure of its ability to reject off-channel signals. Interfering off-channel signals can be adjacent channels that are not sufficiently attenuated by the selectivity of the receiver or spurious signals such as harmonics or heterodyned cross modulation products.

The receivers's resistance to adjacent channel signals is largely determined by the response of the IF system (see Figure 13.4). Interference from cross-modulation products is minimized by the FCC's assignment plan, particularly for the UHF channels, that avoids the assignment of image frequencies and other potential off-channel sources of interference in the same area (see Section 15.2.2).

13.10.3 Picture tube resolution

The resolution of a shadow-mask picture tube is determined by the pitch or spacing between the rows or columns of dot triads. The choice of the pitch is influenced by the picture tube size, larger pitches being used for larger tubes. In part this is an economic decision and in part an assumption that a larger tube does not need any more triads per picture height. A typical pitch for a 19-in TV picture tube is 0.60 mm, which gives approximately 500 rows per picture height.

13.10.4 Picture tube brightness and contrast ratio

The typical brightness and contrast ratios for home receivers are tabulated in Table 4.2.

13.11 Professional Picture Monitors

Television production facilities require two types of monitors to view either program content (*continuity monitors*) or picture quality (*reference monitors*). Continuity monitors are similar to TV receivers without the RF sections, whereas reference monitors are high-quality professional units that are calibrated to control and maintain the quality of the images that are produced, recorded, or transmitted by the facility. In effect, they are test instruments.

The technical specifications of reference monitors differ from receivers and continuity monitors in the following respects:

Resolution

Brightness and contrast ratio

Component and composite inputs

Calibration adjustments

Stability

The resolution of shadow-mask kinescopes is determined by the pitch of the dot triads as compared with the picture height. As noted above, typical 19-inch receiver picture tubes have a pitch of 0.6 mm. The range for professional monitors of this size is approximately 0.3 to 0.4 mm giving 750 to 1000 rows of triads per picture height.

The brightness and contrast ratio of typical professional studio monitors are tabulated in Table 4.2.

Professional monitors are frequently used in system configurations where the signals are in a component format (E_R, E_G, and E_B or E_Y, E_R–E_Y, and E_B–E_Y). Inputs must be provided for both of these formats.

Since a reference monitor is used as a test device, means must be provided to adjust and calibrate contrast, brightness, and color balance by using external test signals. It is not desirable to make these adjustments constantly, and the monitor circuitry should be sufficiently stable to hold adjustment for considerable periods of time.

13.12 References

1. K. Benson and J. Whitaker (eds.) *Television Engineering Handbook,* Revised Edition, McGraw-Hill, New York, 1992, Ch. 13.

2. Ibid., Sec. 13.10.7.

3. "Guide to the Use of the ATSC Digital Television Standard," ATSC, Oct. 1995, Section 10.

4. A. F. Inglis (ed.), *Electronic Communications Handbook,* McGraw-Hill, New York, 1988, Fig. 17.12.

Chapter 14
Digital Video Display Systems

14.1 Introduction

The task of displaying an analog video signal is simplified by the fact that the incoming signal contains all the information in the order to be displayed, with constant updating. Analog display systems only have to synchronize their scanning with the signal and route the video in real time into the display device(s). Digital video display systems, used in computers and HDTV, have to do these same things, but they may need to do much more to deal with other features of a digital video signal such as packetizing, compression, scan-rate conversion, embedded commands, multiple choices, and others. These are explained in this chapter.

14.2 Computer Video Subsystem

Many of the concepts of digital video display can be introduced by describing how the video display section of a personal computer is built. Personal computer video requirements are so varied that most systems provide for them as a separate module slot that can accept a number of different devices depending on the user's needs. Even when a PC includes a built-in video subsystem, it usually can be disabled in favor of a separate plug-in device. These devices are called *video adaptors* and they provide one or more of these features:

1. Display refresh
2. Graphics acceleration
3. Motion video decompression
4. Analog video output
5. Video digitizing
6. Video compression

Each of these is discussed below.

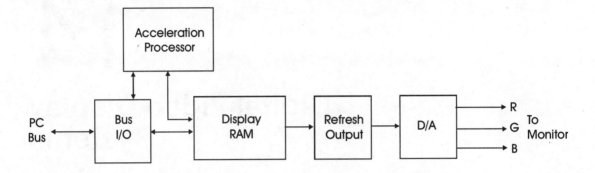

Figure 14.1 Block diagram of a video display adaptor.

14.2.1 Display refresh

Most video display devices, such as CRTs, must receive video data continuously to keep the picture on the screen. The process of doing this is called *display refresh*, and even a digital system that stores its picture in a memory must continuously read that memory out to the display.

Although it would be possible to build a display monitor as a standard PC bus device (see Section 8.2.1.1) and have the CPU send data to its address to get a picture on the screen, the amount of data transfer required would totally overwhelm the capability of most machines. The CPU wouldn't be able to do anything else. Since most of the information that shows on a computer screen remains the same for long periods of time, it makes sense to buffer the screen in separate RAM on the display adaptor and have the adaptor completely handle the screen refreshing task. That way, the CPU and main bus become involved only when screen information has to be changed. A typical hardware architecture for that is shown in Figure 14.1.

Some of the numbers for computer screens were given in Chapter 1, Table 1.2. A typical example is the super-VGA (SVGA) 640 × 480 resolution format displaying RGB-24 information and refreshing a noninterlaced screen at 70 Hz. Table 1.2 shows that requires data at 64.5 MB/s. The amount of RAM needed to store one screen on the display adaptor can also be calculated from these numbers by multiplying the horizontal and vertical resolutions by each other and by 3 (for 3 bytes per pixel). This gives 921,600 bytes. Thus, a 1-MB memory will do the job for this screen. Most SVGA display adaptors provide at least 1 MB of memory—many have more than that to support higher-resolution modes (Table 14.1).

Some display adaptors even have more memory than needed for the screen refresh to allow buffering several screens for instant recall. This is especially useful when doing animation or video because the next screen can be assembled in the buffer memory while the current one is displaying from elsewhere in memory. When the next screen is ready, the refresh circuits simply switch to a different address and begin displaying the new screen.

All of the resolution combinations in Table 14.1 except one provide *square pixels*—that is, the horizontal and vertical pixel counts are in the same ratio as

Table 14.1 Maximum SVGA resolution mode vs. color mode and display memory size

Display RAM size	Color Mode			
	4 bpp	8 bpp	16 bpp	24 bpp
1 MB	1600×1200	1024×768	800×600	640×480
2 MB	1600×1200	1280×1024	1024×768	800×600
4 MB	1600×1200	1600×1200	1600×1200	1280×1024
8 MB	1600×1200	1600×1200	1600×1200	1600×1200

the aspect ratio (4:3 for computer displays.) The 1280 × 1024 combination is not an exact 4:3 ratio, but it is close. Square pixels are advantageous on computer displays because they simplify the arithmetic involved in computing vector graphics (see Section 14.6.1.5), but any resolution combination can be accomplished with a little more software complexity. Of course, square pixels also mean equal horizontal and vertical resolutions on any display, but that is not a critical requirement.

Figure 14.1 shows that the display RAM needs two interfaces—to the PC main bus (and acceleration processor, when used) and to the display refresh circuits. Conventional *dynamic RAM* (DRAM) as used in main memory has only one interface, used for both reading and writing. To use DRAM in a display adaptor, the RAM interface must be time-shared between bus I/O and refresh tasks. That is not difficult to do, but it limits the rate at which communication with the bus can occur. Massive screen changes are therefore limited in speed. Another architecture uses special *video RAM* (VRAM) that has two ports directly on the RAM chips as shown in Figure 14.2. The shift register is loaded during horizontal blanking with a full row from the memory array. This only takes one memory cycle, and all other memory cycles are available for use by the CPU. Screen refresh is done by clocking the shift register, which is a completely independent operation from the memory cycles. VRAM is more expensive, but it speeds up screen changes. This can be important to full screen motion video applications.

14.2.2 Graphics acceleration

Screen operations that involve many pixels are often simply repeating the same operation on each pixel. Examples are: filling an area of the screen with a color, moving a block of pixels from one place to another, drawing lines or other graphic objects, and so on. If these things are done by the CPU, every pixel involved must be sent over the bus to the display adaptor. This slows the operation and ties up the bus and CPU. Most display adaptors now have their own dedicated processor directly connected to the display memory. This processor knows how to do the special graphic operations, so the system CPU only has to send commands to the display processor and it carries out the operations in parallel with whatever else

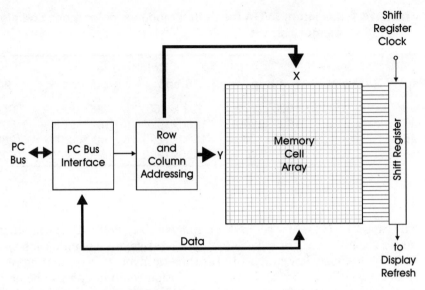

Figure 14.2 Block diagram of a VRAM memory chip.

the system is doing. This is called *graphics acceleration* and it vastly speeds up the drawing of graphic screens.

A graphics accelerator usually is a custom IC designed specifically to work in the environment of a particular adaptor. Since the accelerator capabilities to a large degree will determine the performance of the adaptor, this has become a major element of competition between adaptor manufacturers. Each manufacturer has his own proprietary accelerator design and there is no standardization between manufacturers. This is not a problem because adaptor manufacturers must supply software drivers anyway to connect their adaptors to standard operating systems such as DOS/Windows or OS/2. The proprietary accelerator design is hidden inside of the driver software; the rest of the system only sees the standard operating system interface.

The latest CPU chips also have added some of the graphics or video operations to their instruction sets. Subject to the limitation that this can tie up the CPU from other operations, this also is an effective way to speed graphics operations. Graphics and video software designers have a rich collection of processing chioces to use in developing the fastest and most efficient ways to perform their tasks.

14.2.3 Digital signal processors

Another way to accomplish graphics acceleration is to use a general-purpose *digital signal processor* (DSP) chip in the display adaptor. A DSP is a special microprocessor that is optimized for speed in doing signal processing operations. It uses a stored program just like the system CPU, so it can be very flexible in what it does. This is a more expensive approach than a hard-wired custom accelerator, but it can offer additional features and of course its features can be changed or updated by changing the software running on the DSP. The DSP software is usually handled as a part of the display adaptor driver, which can load different software into the DSP on the fly and manage its running.

14.2.4 Motion video decompression

The same kind of acceleration concepts can be applied to video processes such as decompression and new video adaptors are adding this functionality, too. With a video decompression processor or DSP on the adaptor, the CPU only has to manage the transfer of compressed data to the adaptor and all the rest of the video work occurs on the adaptor.

14.2.5 Analog video output

The output that goes from the display adaptor to the monitor is analog RGB component video. However, there is sometimes a need to deliver television-format video in either a composite or component format. This is useful, for example, if one wishes to record the computer's video on tape. It also may be used to feed a large-screen television display. (There are large-screen computer displays available, but they are more expensive than TV displays.)

Converting computer video to analog TV formats usually will introduce degradation because of TV's interlaced scanning and composite color encoding. It is satisfactory only if the video material is chosen to avoid fine detail in bright colors and to not use single-pixel horizontal lines. That means text must be significantly larger than would be used with the computer's own display. For example, a 640×480 computer display can very effectively display 80 text characters per line and 24 or more lines on the screen. An SDTV display can effectively display only half as many characters and lines.

Since the scanning parameters of computer display adaptors are usually programmable, setting up the TV scanning is easily accomplished. However, if composite encoding is required, it takes added circuitry. Fortunately, there are ICs available for TV encoding, so it still isn't expensive. The real problem is that changing the adaptor scanning for this purpose means that the computer display also must change to TV scan. That may not be desirable and some computer monitors won't even handle it.

The solution is to leave the computer display adaptor alone and intercept the SVGA RGB signals in a separate module to perform the conversion. Although that is slightly more expensive, it is then possible to have both computer and TV-scan displays at the same time. All TV display adaptors operate that way.

As HDTV comes into use, there will be a need to output that from computers, too. HDTV is more compatible with the normal contents of computer video, so there will not be the quality problems discussed above. However, there will be the problem of how to handle the different aspect ratio of HDTV. Again, the computer can be programmed to operate at HDTV scanning, but then computer displays won't work—at least not until wide-format computer displays become available. That will surely happen as HDTV display components become readily available and users become familiar with HDTV in their homes. Besides, a high-resolution, wide-format screen can display more computer data, too.

14.2.6 Video capture and compression

To bring analog video into a computer so that is can be compressed, edited, or displayed, requires digitizing and storing the digital data on the computer's hard disk. This can be done with a separate adaptor or it can be added to a display adaptor. Architectures for this are discussed in Section 8.4.3. The issues of video compression are also discussed in that section and in Sections 3.10 to 3.12.

14.3 Video Overlay

Effective audio-video-computer applications have been built for many years using a technique known as *video overlay*. The output from a computer-controllable analog video source such as a laser videodisc is displayed on the video monitor along with graphic output from the computer. The analog video is mixed with the computer graphics, but there is no attempt to input the video data into the rest of the computer. This technology has been widely used for education and training applications and it is very effective.

Video compression is not required for video overlay because the video never has to pass through the computer bus or be stored in mass storage. It is directly displayed as it is played from the laser disc. Thus, the system is simpler than a true computer digital video system. However, the overlay approach has significant disadvantages:

The laser disc player is a separate unit from the computer that must be connected with cables. That makes a complex and inconvenient setup. It also is more expensive than a computer that plays digital video.

The program (computer disk) and the video data (video disc) are stored on separate media that must be coordinated for proper use.

The video disc is not recordable except through mastering (see Section 10.20) so it is difficult to make updates.

Since the video never actually enters the computer, the computer cannot manipulate the video in any way. All the video for the application must be on the video disc in exactly the format that it will appear.

Scanning synchronization is required between the computer and the video disc. This is usually accomplished as shown in Figure 14.3 by using TV scanning on the display and synchronizing the computer graphics with the TV scan. This limits the system to interlaced scanning, which does not do a good job on the graphics. With the wide availability of PCs that can handle video, analog video overlay is being superceded, although there are hundreds of thousands of analog systems deployed that will continue to be used for many years.

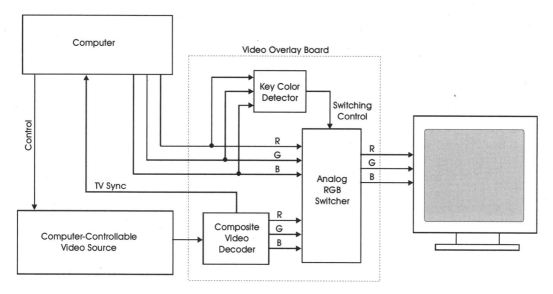

Figure 14.3 Block diagram of the use of an analog video overlay board.

14.3.1 Digital video overlay

Modern video overlay boards combine overlay with a full video digitizer and display board. Thus, the video and its overlay can be displayed on a computer monitor using progressive scanning, without the limitations of interlacing.

Digital overlay is a useful feature for video capture setups because the digitized video overlay hardware can be used for monitoring during capture without using any of the computer's resources. Such a combination video board can serve many applications:

Frame grabbing
Movie grabbing
Live overlay
Hard disk editing
Tape editing
Analog video output
TV tuner

Figure 14.4 is a block diagram of such a combination board. It differs from Figure 14.3 by performing the entire overlay process digitally. This means there must be decoding and ADC for analog video input and DAC on the output to the display. A frame memory is included with separate write and read clocks so the analog video is resynchronized to the computer's standard video. Therefore, the display scanning can be progressive as required for a high-quality computer display. By giving

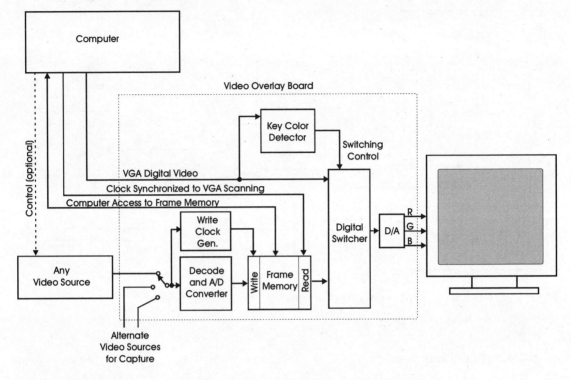

Figure 14.4 Block diagram of an all-digital video capture and overlay board.

the computer access to the overlay board's frame memory, still images or video can be compressed and stored to the computer's hard disk using the processing resources of the computer.

14.4 Digital TV Displays

The display refresh portion of a DTV display is similar to a computer display adaptor. There must be enough RAM to store the current screen and the means to continuously read that to the display at the proper frame rate. The rest of a digital TV display is concerned with processing the input signal to place it into the RAM in the correct format for display.

Note that the input processing may result in scan conversion—it is not really necessary that the display refresh scanning run at the same rate as the scanning that is implicit in the input signal. This is one of the major advantages of being digital.

14.5 Computer Display Monitors

Computer display monitors differ from analog television displays in several respects:

1. A computer display is viewed at much closer range than an equal-size TV display (see Section 6.3). That calls for significantly higher CRT resolution, higher vertical-scan rates, and a flatter screen. These features also mean that computer monitors are significantly more expensive than TV displays.
2. Interlaced scanning is not appropriate for computer display (see Section 4.16.4), so all computer displays should use progressive scanning.
3. Most computer display adaptors are capable of multiple scan rates. A display monitor must accommodate this, preferably automatically. This gives rise to such terms as "multisync" or "multiscan" when describing computer monitors.
4. Because computer display formats use progressive scanning and usually have more horizontal lines and higher vertical scan rates than analog television, horizontal scanning frequencies are significantly higher. That is another higher-cost factor.

CRT resolution is specified in terms of dot pitch, which is the distance on the screen from the center of one phosphor color dot (or line) to the next adjacent dot of the same color. Since the CRT dots are a form of sampling, there should be more than one dot for each pixel, to avoid loss of resolution. Unlike TV, the viewer will usually sit the same distance away from a computer monitor regardless of its size. A larger monitor is expected to display more information at once. Thus, the dot pitch should be the same independent of display size.

For example, a 14-in (diagonal) display has a useful screen width of about 10 in. With a dot pitch of 0.28 mm (which is typical for a good 14-in monitor), there will be 90 color dots per inch or 900 dots across the screen width. This will perform well with 640 pixels horizontally, it will also work with 800 pixels across, but at 1024 or more horizontal pixels, the dot size will severely limit display quality. For the higher resolutions, larger displays that have more dots across the screen are necessary.

14.6 Flat-Panel Displays

The "picture on the wall" display has long been a goal of TV display research. Numerous technologies have been developed but none has been commercially significant in the TV field except for small portable receivers where CRT displays are impractical. In general, flat panels do not compete with CRTs except in places where the size and power needs of a CRT display are prohibitive.

The portable computer market fits this model. It is impractical to make a portable computer with a CRT display, so this is a market that will pay the price necessary to have small, lightweight, and low-power displays. The technologies of flat-panel displays are now highly developed as a result of the impetus of this market. Compared to TV displays, portable computer displays are small—15-in diagonal displays are the largest.

Oddly enough, the entre for flat panels into the TV market may be HDTV, where large high-resolution displays are required. CRTs for this market are going to be expensive monsters in the living room and it would be an opportunity for flat

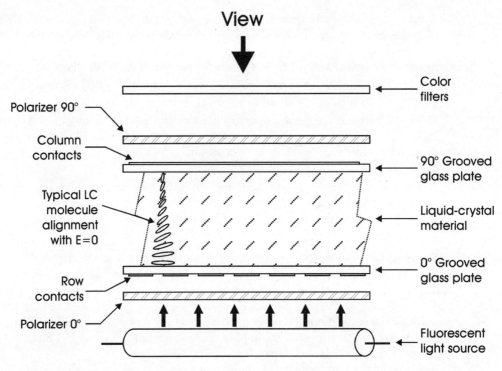

Figure 14.5 Cross-section of a twisted nematic LCD panel, transmission mode.

panels to get into the market if they can achieve the screen sizes at anywhere near reasonable costs. Many companies are working on it.

There are two versions of a single technology for flat panels that have reached mass marketability, described below.

14.6.1 Liquid-crystal displays

Nearly all portable computer displays use *liquid-crystal display* (LCD) technology. It is also used for projection displays. In these displays, the controllable element is a liquid-crystal (LC) material sandwiched between two glass plates. The LC material has the property that its oblong (*nematic*) molecules can be aligned by grooving the glass plates enclosing the material, but when an electric field is applied to the panel, the molecules will realign themselves with the field. By building a control matrix behind the glass panel that allows the electric field to be controlled locally for each pixel, a display screen can be built. This is diagrammed in Figure 14.5 for a panel that contains its own light source. (Reflective panels that use ambient light as the source are also used, but the self-illuminated panels are more popular for computer or TV displays.)

The LC material is contained between two glass plates; each plate has microscopic grooves on its inner surface, but the grooves are aligned at 90° to each other. This causes the molecule alignment to spiral by 90° between the bottom and top of the panel, which explains the name "twisted." To obtain control of light transmission in the panel, two polarizers are placed behind and above the panel, with their plane of polarization aligned with the grooves in the glass plates. As the light passes through the LC material, its polarization is rotated as the molecule alignment twists, so that full light transmission occurs even though the polarizers are at 90° to each other.

When an electric field is applied to a location on the panel by applying voltage to the control electrodes, the LC molecules begin to align with the field and the 90° twist is disrupted. The result is that light transmission is reduced—the panel gets darker where the field is applied.

Because of their use of polarized light, LCD panels in general have a limited angle of view, usually only about 30°. This is acceptable for the single-viewer application of portable computers, but it is not suitable for TV-style viewing by a group. This is one of the challenges in applying LCD panels to TV.

There are two ways to achieve control of the electric field: passive-matrix and active-matrix, explained below.

14.6.1.1 Passive-matrix LCDs

A *passive-matrix* panel is built by placing an array of row and column wires behind the panel as shown in Figure 14.5. The row and column wires are pulsed in sequence to provide a scanning action. By controlling the pulse amplitudes, the light transmission may be set for each pixel. Because the LC material changes back to its unexcited state slowly after the voltage is removed, the panel appears to be showing a continuous image that does not flicker.

Color is obtained by using a color filter pattern over the display, much like that used in single-device color cameras. Thus, each color pixel requires three LCD pixels in the matrix; they are usually arranged horizontally, so, for example, a 640 × 480 panel would actually have 1920 × 480 LC cells in its matrix.

Because of the short pulsing of each pixel and the slow decay of the LC behavior, a passive-matrix display has slow response and low contrast ratio. However, it is the least expensive form of LCD and it is used in low-end portable computers despite its shortcomings. Even so, passive-matrix LCDs are several times more expensive than CRT displays of similar size.

14.6.1.2 Active-matrix LCDs

The leading technology for portable computer displays is the *active-matrix* liquid-crystal display (AMLCD) panel. It provides the best combination of cost and performance with display resolutions of 640 × 480 in sizes up to about 14-in diagonal. It uses a LC panel that has a control transistor associated with each pixel (or subpixel for color). The transistors are integrated directly onto the bottom glass of the panel. This is why it is called "active-matrix." The control transistors set the voltage on each pixel that causes the LC molecules to scatter or transmit light. Because the control transistors are individually addressed and are much faster, the panel response time and contrast ratio are much improved. However, the addition of the control transistor array nearly doubles the cost of these panels compared to passive-matrix. Many buyers of portable computers consider that cost to be worth it.

At this writing, LCD panels are beginning to be marketed for desktop computer use. They have the advantage of saving space and power, but they are several times more expensive that same-sized CRT displays. Desktop LCD panels are on the market in sizes from 15- to 18-in diagonal.

Larger AMLCD panels have been made in sizes up to 21-in diagonal, but the production yield drops rapidly with size, making such large panels prohibitively expensive. The technique of *tiling* is being explored to assemble large panels from smaller ones. If this technique can be mastered, it promises to lower the cost of large AMLCD panels and allow HDTV-sized units to be manufactured.

14.6.1.3 LCD improvements

There are many possibilities for improving LCD panel performance and cost that apply to their use in portable computers and also the HDTV market. For example, researchers are building AMLCDs using single-crystal silicon (x-Si) for the substrate of the transistor array. This allows much higher resolutions (pixels per inch) and, because high-performance circuits can be built on x-Si (it is the substrate for nearly all integrated circuits), much of the drive hardware can be integrated onto the display panel.

Panels using an x-Si substrate are limited in size to the dimensions of integrated-circuit wafers (200 mm diameter or so), but the pixel density can be so high that the HDTV resolution of 1920 × 1080 pixels can be achieved on a device only about 38 mm across. This is ideal for a large screen projection display, and that application has a lot of potential.

14.6.2 Other promising technologies

Many other technologies exist in the laboratory but so far they are not ready to compete with the ones now on the market. However, the continued growth of the portable computer market and the emerging demand for HDTV displays will keep the laboratories working at high pitch. Several of these contenders are described below.

14.6.2.1 Field-emission displays

The field-emission display is a form of flat CRT except that there is a separate cold-cathode electron emitter for each pixel. Colored phosphors are deposited on a transparent substrate in front of the array of emitters. This is a promising technology, but it is still in the laboratory.

14.6.2.2 Micromirror displays

Another promising technology involves an array of mirrors on an IC chip. This is

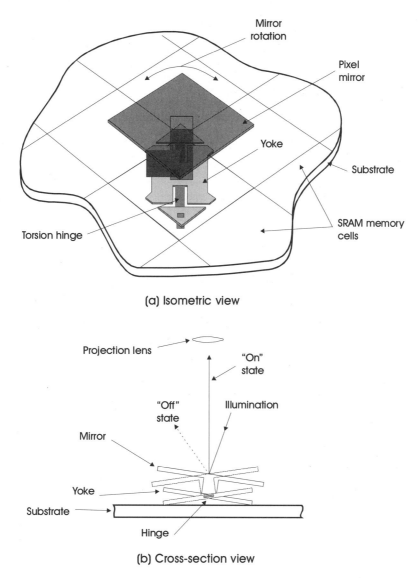

Figure 14.6 Diagram of one pixel of a DMD display. (*Reproduced with permission from [1].*)

called a *digital mirror device* (DMD). The mirrors can be physically moved electronically to control the displayed brightness by reflecting a high-intensity light source; this has been applied successfully to high-brightness projection displays. Panels up to 1.6 million pixels projected on a 12-foot screen have been demonstrated.

The DMD principle, shown in Figure 14.6, is a true digital display, in that its input signal is 8-bit parallel; the on-chip process responds to the bit values directly to produce analog amplitude modulation of the reflected light from each mirror. The micromirrors ($16 \times 16\,\mu m$) are fabricated above a static RAM array on

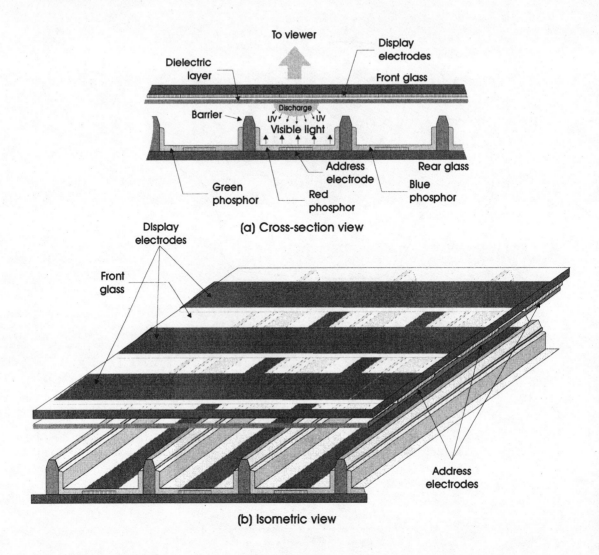

Figure 14.7 Structure of a color plasma display panel. (*Reproduced with permission from [1].*)

an IC chip; the mirrors can rotate through an angle of about ±10° under control of their input voltage. Figure 14.6(b) shows how the mirror rotation is used to modulate the amount of illumination that reaches the projection lens.

14.6.2.3 Plasma displays

For large-size, bright, direct-view displays such as are required for HDTV, the *plasma display* is an important contender. These panels produce light from a gas discharge in a vacuum. In order to produce colors, each pixel produces three controlled ultraviolet discharges that excite photoluminescent color phosphors that deliver the visible light. The design of the plasma panel includes a memory unit for each subpixel so that the pixels stay on until they are turned off. Thus, the display does not have to be continuously scanned to produce a stationary picture. Plasma panels have been demonstrated up to 40-in diagonal in HDTV resolutions.

14.6.2.4 Electroluminescent displays

These displays are an extension of the familiar red-light-emitting diode displays (LEDs) used for numerical display. *Electroluminescent* (EL) displays were used in early portable computers, but were more expensive than LCDs once that technology was developed. EL panels are now being developed for color display.

14.7 DTV Display Processing

As can be deduced from the transmission processing discussion in Chapter 7, a digital HDTV receiver has to perform a lot of processing. Custom integrated circuits will be designed for most of this processing, so it will not be expensive in the long run. Figure 14.8 is a simplified block diagram of a typical digital HDTV receiver (actually it is the GA prototype receiver). Since only the transmission format will be standardized, there will likely be many variations in receiver design.

14.7.1 Tuner, IF, and detector

The first step of reception is to tune the desired channel and down-convert to an intermediate frequency suitable for detection of the VSB amplitude modulation. This is done by a double-conversion heterodyne system in the GA prototype to obtain the desired selectivity, ending at an IF of 44 MHz. A *frequency and phase-lock loop* (FPLL) synchronizes to the pilot carrier of the signal to drive a synchronous detector that converts the 44-MHz VSB amplitude-modulated IF signal to the baseband 8-level symbol sequence.

A comb filter is available in the signal path for rejection of NTSC co-channel interference when it exists. (Because this filter slightly degrades the system noise performance, it is automatically switched out of the signal path if no NTSC interference is present.) This filter provides seven comb nulls in the 6-MHz channel, three of which reduce the energy at the NTSC visual carrier, the NTSC color subcarrier, and the NTSC aural carrier frequencies. The comb nulls and digital processing rates are coordinated so that the NTSC rejection filter has no effect on the digital data.

14.7.2 Segment sync and symbol clock recovery

The 10.76-MHz symbol clock is recovered by a phase-lock loop that locks to the random symbol pattern by looking for the 4-symbol segment sync intervals. This

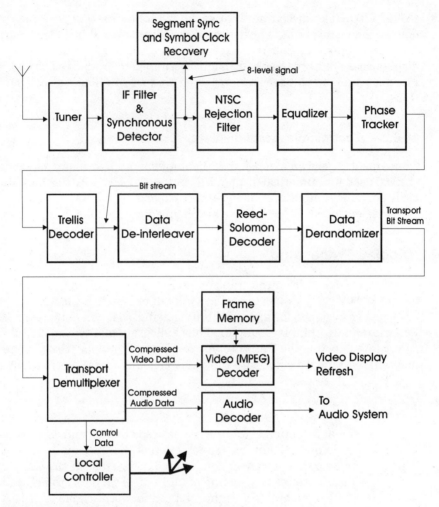

Figure 14.8 Block diagram of a typical DTV receiver.

simultaneously synchronizes the symbol clock and the segment sync. Data field synchronization is also obtained by looking for the data field sync pattern in the 8-level data.

14.7.3 Equalizer and phase tracking

The channel equalizer compensates for linear frequency response distortions and ghosts due to multiple reflections in transmission. It contains a 64-tap transversal filter that is adjusted by examining the training signal that is contained in the field sync pattern.

The phase tracking loop is an additional feedback loop that further reduces phase noise that may not have been removed by the IF FPLL operating from the pilot signal. It operates on the 8-level data signal to identify and track out residual phase errors.

14.7.4 Trellis decoder

Because of the 12-symbol interleaving at the transmitter (see Section 6.7.3.3), the receiver contains 12 trellis decoders in parallel, where each decoder sees every 12th symbol. The 8-level symbols are converted back to the original 2-bits per symbol. The output of this is a serial bit stream in the same order it existed at the transmitter before trellis encoding.

When the NTSC rejection filter is used, the trellis decoder is automatically modified to provide optimum performance with the different system response caused by the NTSC filter.

14.7.5 Deinterleaving, Reed-Solomon decoding, and derandomizing

The data interleaving applied at the transmitter must be removed before Reed-Solomon (R-S) decoding. This ensures that long burst errors will not overload the R-S decoder. The R-S decoder is capable of correcting up to 10 bytes of errors per data segment. The result of the combination of this and the data interleaving is that bursts of noise or interference up to 193 μs can be handled.

The randomizing pattern introduced at the transmitter is now removed using the same randomizing code as at the transmitter. This restores the signal to the transport packet format.

14.7.6 Transport demultiplexing

Since there is a 1:1 relationship between data segments and transport packets, the segment sync already extracted is used to decode the packet headers. Packet IDs are used to separate the packets into individual bit streams for audio, video, or data. This process results in separate bit streams feeding audio and video compression decoders. When control data is present in the packets, that is sent to the system controller (CPU) that identifies it and acts appropriately.

14.7.7 Audio and video decompression

The final step of digital processing is to decompress the audio and video bit streams. These processes are described in Section 3.12 and Section 6.5.

14.8 Reference

1. A. C. Luther, *Principles of Digital Audio and Video*, Artech House, Norwood, MA, 1997.

Chapter 15

Interactive Video Systems

15.1 Introduction

Television viewing is ordinarily considered to be passive—the user sits back and watches the program but has no control except to change the program or turn it off. The other side of the coin is the personal computer style of viewing, where the user carries on a continuous dialog with the machine. This is known as interactive viewing or simply *interactivity*. Interactivity is everywhere—any product that is controlled by its user is interactive—this even includes mundane products like the faucet on the kitchen sink. Other important interactive products (Figure 15.1) are automobiles, photographic cameras, and musical instruments.

Interactivity is creeping more and more into video systems. Even broadcast television receivers will acquire interactive features as the capabilities of digital transmission are applied to achieve richer capabilities for entertainment, information delivery, education, and play. This chapter discusses interactivity in video systems of all types.

Figure 15.1 Interactive products.

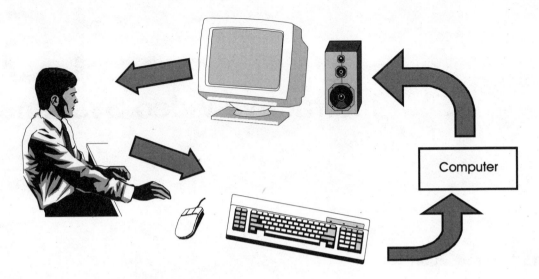

Figure 15.2 The interactivity feedback loop.

15.2 Ingredients of Interactivity

Any interactive situation requires three main ingredients (Figure 15.2):

1. Two-way communication—in a personal computer, the machine talks to the user through its video display and sound capability. The user talks to the machine with keyboard, pointing device (mouse or trackball), sound (sometimes), or other more specialized controller devices. In an automobile, the driver communicates with the vehicle through the steering wheel, the pedals, and other controls. The vehicle communicates with the driver through its motion, the speedometer, and other instruments or displays. In each of these examples, a continuously operating feedback loop is made possible by the two communication channels.

2. User interface—this is the protocol of the communication involved in interactivity. It is ordinarily something that the user has to learn, which may be easy or difficult. Often, a degree of skill or dexterity is involved in operating a user interface. A good example of a difficult interface is the violin. Years of learning and practice are required to master that interface, and only a few of those who attempt it succeed. Yet the result is worth the effort and no proposal to redesign the violin has ever been taken seriously. The user interface of an automobile is much easier and can be learned by almost anyone. Most of the learning with an automobile is concerned with the rules of the road rather than the physical dexterity needed to operate the controls.

3. Computing capability—when the user issues a command, the system must know how to respond—that is, to close the loop. This is most effectively handled with a programmable computing device (an embedded processor; see Section 15.5)

or an actual computer. Most video devices already contain computing capability for this and other purposes.

Each of these ingredients of interactivity will be discussed in more detail, specifically for video-based environments.

15.3 Communicating with the User

The communication devices available in video systems are the video display or sound (machine to user), and pointing devices, keyboards, touch screens, remote controls, or sound (user to machine). Even more advanced devices are possible for user-to-machine communication based on forms of body sensors such as gloves or eye tracking.

Although in the examples given above the user operates the interface devices hands-on, interactivity can also occur remotely, where one or both communication channels involve radiofrequency, optical (infrared), or cabled communication. A simple example of this is the programming of a VCR using an infrared remote device and on-screen menus. The user sits across the room from the machine but still communicates with it for interactive control.

Good performance of the user-machine feedback loop depends on an adequate speed of response around the loop. If the loop becomes too slow, the user has to wait and he/she may perform superfluous actions that could even cause the loop to become unstable. Whenever the user performs an action, the system should respond somehow in (preferably) less than a second. If it doesn't, the user may think the action was not noticed by the system and perform it again. Even if the system action requested will take longer than a second, an acknowledgment of the request should be given in less than a second. An example of this is when a system responds with a click sound when a button is pressed. The user knows from this that the system got the message, even though the requested action may take much longer.

15.4 Video-Based User Interfaces

In a video environment, there are many ways to construct user interfaces [1]. The video screen is such a flexible device that there is usually no need to have any other means for the system to communicate with the user. The one exception to this is sound, which has the advantages that the user doesn't have to be able to read from the screen (maybe he/she doesn't always look at the screen or maybe he/she can't read), so a sound channel generally operates in parallel with the visual channel. The parallel nature of sound and vision means that together they can convey more information.

These communication channels provide rich capabilities that challenge designers to effectively use them. However, that same richness offers so many different ways to accomplish the same task that there can be serious problems in making interfaces easy to learn and consistent between different products. Standards are really valuable but in most fields they have been difficult to establish.

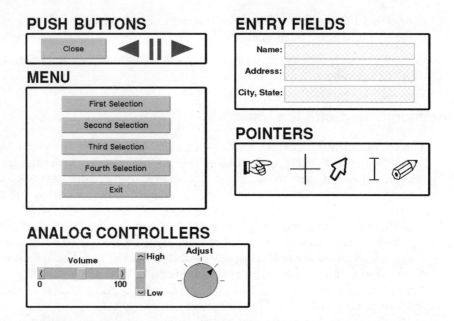

Figure 15.3 Visual user interface objects.

15.4.1 Types of user interfaces

Video-based interfaces usually employ metaphors that help the user remember how the interface should work. Typical metaphors are shown in Figure 15.3:

Push buttons—these are objects that the user *touches* to cause a specified action. ("Touching" means just that, if a touch-screen is employed, or it can mean moving a pointer object over the push button object and clicking a button.) Push buttons usually are labeled to indicate the action to be invoked when they are touched.

Menus—these are lists from which the user makes a selection by touching. The menu can be text or it may be actual pictures or other graphical representations of things that show the choices available in the menu. Another form of menu is the *check box* list, where the user touches one or more objects to choose several things at once. In this kind of list, the selected objects must highlight some way so the user can distinguish between checked and unchecked items.

Continuous controllers—these are knobs or scroll boxes (sliders) to set an analog parameter. They must show the setting of the parameter within its range by using position or rotation.

Entry fields—these are used when the user is required to type in something, such as a name. This ordinarily involves the use of a keyboard for typing, although in a very simple situation other kinds of controllers might be used.

Pointers—a pointer is an on-screen object that is movable by the user. It usually functions to operate other controls or to identify specific locations or

Interactive Video Systems 373

Figure 15.4 Icons.

objects on the screen (called selection). In complex situations, the shape of a pointer can change to indicate the type of action it currently represents. For example, the normal pointer might be an arrow, but if it is moved over an entry field, it changes to an I-beam shape and now indicates the text insertion point.

Metaphorical devices shown on a video screen should look and behave like the physical device they represent; this tells the user how they work. Such on-screen representations are called *icons* (Figure 15.4) and they can be realistic or abstract. The important thing about icons is that they should be used consistently so that once a user learns the meaning of an icon and its use, it will behave the same way every time it is encountered in the interface.

15.4.2 Typical interfaces

Depending on the application, video-based user interfaces range from simple to very elaborate with multiple dimensions and levels. Examples of both extremes are given below.

15.4.2.1 A simple interface

Controlling a tape deck is a simple task; usually no more than five buttons are needed for Play, Pause, Stop, Fast Forward, and Rewind. A CD player is a little more complicated because of the possibility of choosing a track (skipping). Figure 15.5 is an example of a very simple video-based interface for a computer CD-audio player. The buttons provide all the possibilities listed above plus a display of the

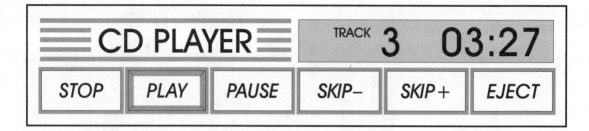

Figure 15.5 A simple interface for a CD player.

player status. The selected button (PLAY) is highlighted. This is a good interface whose use is almost obvious.

15.4.2.2 An elaborate interface

Microsoft Windows is a good example of a video-based interface that can handle tasks of almost any complexity or depth. Figure 15.6 shows the Windows video screen used for the formatting of this book. The large window contains this page open for editing. That window is surrounded by other windows that offer choice of special characters, page, format tags, file selection, etc. It is clear from this screen how Windows got its name. The screen also shows icons at the bottom for programs that are running but their windows are not open.

15.4.2.3 TV-based interface

Both examples shown above are from personal computers. However, video-based interfaces aren't only on computers. Figure 15.7 is a screen from a television receiver showing the menu setup used to adjust color parameters. In this case, the user is responding by means of an infrared remote unit. This actually proves awkward because the user has to learn how the buttons on the remote controller relate to what is on the screen. The computer idea of using a pointing device is really better because once the user has learned how to use the pointing device, all other instructions about the specific task at hand can be presented on the screen.

15.5 Computing Capability for Interactivity

Responses must be constructed for the user's actions in an interactive environment. This is defined here as "computing capability," although it may range all the way from a little hard-wired logic in a simple situation to a full-scale personal computer in other cases. Intermediate situations can be handled by an *embedded microprocessor*, which is a special microprocessor designed to be built in to a product. Embedded processors are available as CPUs only, or they come in the form of a computer-on-a-chip, which contains all the functionality of a complete computer.

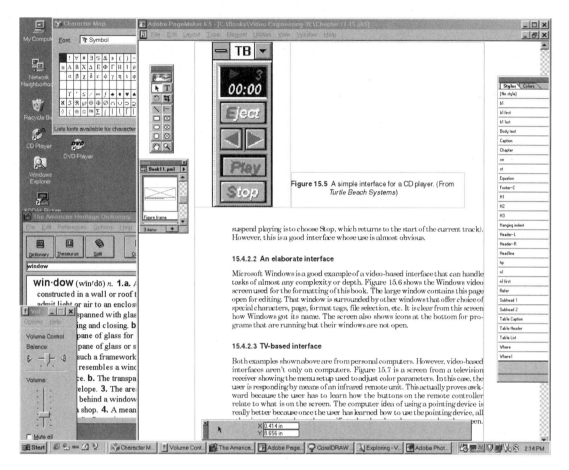

Figure 15.6 The Microsoft Windows user interface.

Typically, this includes the CPU, RAM, ROM (for programs), and I/O capability. The embedded processor approach is usually applied in "smart" appliances or in electronic products such as VCRs, CD players, or tape decks.

As video products get more and more interactive features, which will require greater computing capability, the trend will be to emulate more of the features of personal computers in video products. This could include such things as loadable programs, standard keyboards, and internal mass storage.

This latter feature is especially important in order to provide rich interactivity in a situation where the video device is receiving broadcasted or cable signals. Although two-way communication with the broadcast station or cable head-end is possible to set up interactivity, the return path is likely to have slow response (seconds, or longer). That may be all right for a task such as program selection, but it certainly is too slow for exploring a database or playing a game. The problem can be solved by having enough mass storage in the local video product that data and programs for close interactivity can be downloaded from the head end.

Figure 15.7 A user interface for adjusting a TV receiver.

Then the user at the local product can interact with the data at his/her leisure, without the need for constant communication with the head end—see Section 15.7. (As digital communications to individual users becomes faster, this will be less important.)

15.6 Multimedia

As PCs have become capable of presenting realistic audio and motion video, the word *multimedia* has been used to refer to applications that use these new capabilities. All new PCs today have audio, video, and CD-ROM capabilities, so there is no longer anything special about a "multimedia PC." Most applications now combine audio and video with other PC data types. These are typically distributed on CD-ROM or DVD discs because of the large amounts of data involved in audio and video.

Probably the most widespread example of multimedia today is the *World Wide Web* on the Internet. This is described in detail in Section 20.3.

15.6.1 Ingredients of multimedia

Computer multimedia is unique in its ability to combine many different data types into one application and under one user interface. The sections below explain each of these and their significance to multimedia. Figure 15.8 shows on-screen examples of each of the visual data types.

Interactive Video Systems

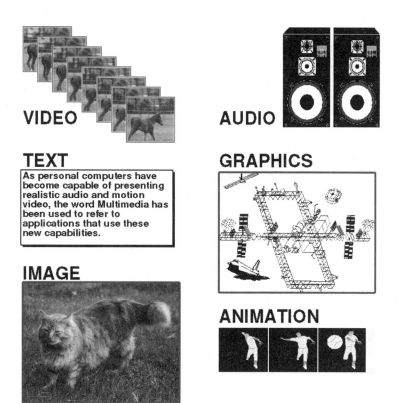

Figure 15.8 Multimedia data types.

15.6.1.1 Motion video

PC motion video uses the video compression technologies discussed in Chapter 3, and special algorithms have been developed for video playback using software without extra hardware except for an accelerated display adaptor. This performs well on late-model PCs. There is also special hardware for higher-performance algorithms, such as MPEG-2, but the market still favors the software approach.

To speed up software-based video or allow larger video windows, some display adaptor vendors have added video capabilities to their boards, which can be accessed by driver software provided with the boards. This is the best solution until all machines are as fast as the best today. The advantage of not requiring dedicated video hardware that may quickly become obsolete is overwhelming.

15.6.1.2 Audio

Multimedia audio mostly uses simple compression or none at all. PC audio hardware supports PCM audio at both 8 bits/sample and 16 bits/sample, with very little load on the system CPU. Using compressed audio would increase the CPU loading and since the CPU is already heavily loaded by video (that often plays at the same time), there isn't much left for audio processing. Dedicated hardware could be used, but the arguments against it are the same as for video.

PC audio formats provide a wide range of performance choices from speech quality up to full audio CD-quality (see Table 1.3). In general, applications should use the lowest quality level that serves their purpose, because higher audio quality costs data space.

Uncompressed audio takes considerably less data than compressed video, so it tends to be more heavily used than video. A single CD-ROM disc can hold only an hour of video, but it can hold 10 or more hours of medium-quality audio. The CD-ROM capacity can only be used once; space taken for audio data subtracts from that available for video. Trading between different data types based on the capacity of a CD-ROM disc is one of the challenges of designing multimedia applications.

15.6.1.3 Text

The original computer interface medium was text and, even in these days of video interfaces, text is still an inescapable medium. It also is very efficient in terms of the data used. Consider that a CD-ROM disc devoted to text can hold upwards of 200,000 pages! Many existing databases are in the form of text and a lot of design effort in the multimedia field has gone into ways to use the video capabilities of computers to give easier access to text.

One very important multimedia text format is known as *hypertext*. With hypertext, words or phrases in a block of text can be made into push buttons (see Figure 15.9). If the user touches one of these words, which are highlighted so he/she can tell they are special, the program branches to a new screen or window that offers additional or related information about the subject of the word touched. For example, when reading a biography of a person, the person's birthplace may be mentioned. If the birthplace name is hypertext, the user can touch it and get more information about that place. When the user is done with the birthplace information, he/she can instantly return to the biography and continue reading. It's just like what is done in a book with "see... references" except that the reader doesn't have to look up the reference—it pops up when he/she simply touches the hypertext button. This is a valuable user interface technique.

15.6.1.4 Graphics

Graphics is the capability of a computer to draw pictures from a table of instructions—the technique is called *vector graphics*. The process of drawing a vector graphic picture is called *rendering*, and it can take into account the parameters of the current environment and draw the picture to suit the screen resolution and amount of screen space available for the picture. It is also possible to implement zooming or shrinking the image to show more or less detail. Further, vector graphics is very data-efficient compared to storing pictures in pixel form (see below).

The disadvantages of vector graphics is that the pictures are drawings and may

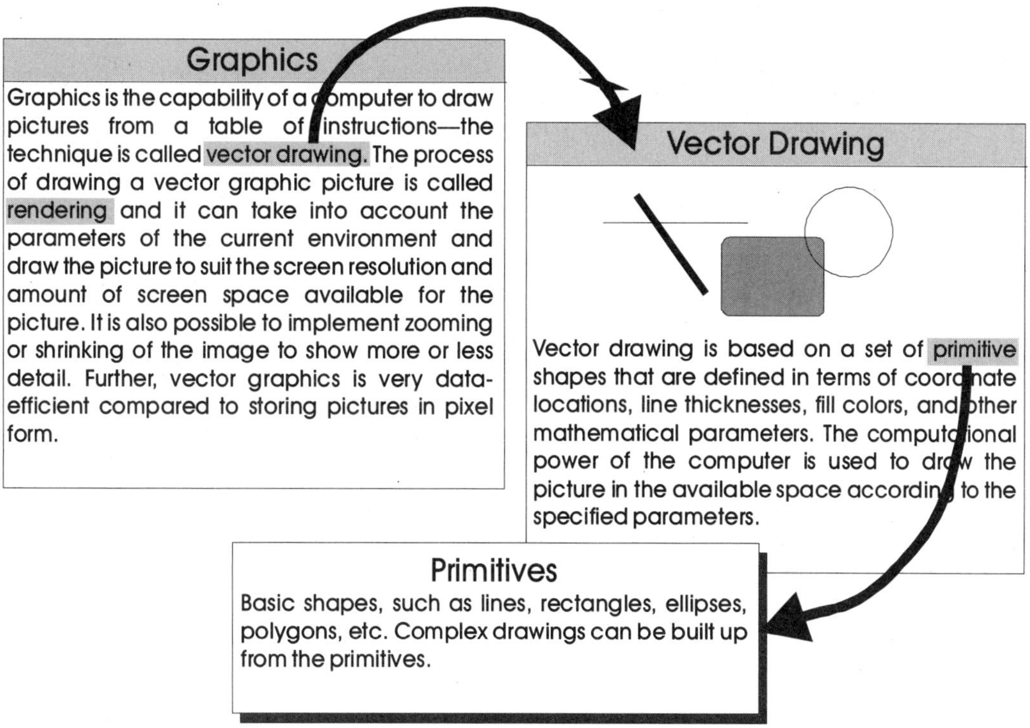

Graphics

Graphics is the capability of a computer to draw pictures from a table of instructions—the technique is called vector drawing. The process of drawing a vector graphic picture is called rendering and it can take into account the parameters of the current environment and draw the picture to suit the screen resolution and amount of screen space available for the picture. It is also possible to implement zooming or shrinking of the image to show more or less detail. Further, vector graphics is very data-efficient compared to storing pictures in pixel form.

Vector Drawing

Vector drawing is based on a set of primitive shapes that are defined in terms of coordinate locations, line thicknesses, fill colors, and other mathematical parameters. The computational power of the computer is used to draw the picture in the available space according to the specified parameters.

Primitives

Basic shapes, such as lines, rectangles, ellipses, polygons, etc. Complex drawings can be built up from the primitives.

Figure 15.9 Hypertext.

not be realistic. Also with drawings, the result depends on the skill of the artist who did the original work.

15.6.1.5 Images

Pictures can also be displayed on a video screen by digitizing photographs or capturing stills from motion video. Both of these produce *bit-mapped* images, where the actual pixels to display are stored (of course, they can be compressed using JPEG or another still-compression technique—see Section 3.12). Bit-mapped images are very realistic as long as the image has enough pixels and enough colors.

Displaying bit-mapped images consists of decompressing if necessary, and copying the pixels to the desired location on the screen. Since the number of pixels in the image is fixed, the size of the image displayed depends on the total resolution of the screen being used. For example, a 640 × 480 image shows full-screen on a 640 × 480 display, but it will only be quarter-screen on a 1280 × 960 display. If a bit-mapped image is enlarged (zooming in), it will become blocky because the pixels become larger and easier to see.

15.6.1.6 Animation

The techniques of vector graphics can be extended to render a series of frames to produce motion video. This is computer *animation*. It is very effective and data-efficient, but it has the same dependence on artistic skill as drawing. Animation is very difficult to do, an animator has to devote his/her life to it. If animation skill is available for an application, it can be very valuable.

15.6.1.7 User interface

The preceding sections show how many different kinds of information and presentation techniques are available for multimedia applications. Making use of them all can present a user with a dazzling array of choices and that can mean a real user interface problem. The process of the user moving about in a multimedia application is called *navigation*. Many approaches are available for navigation and many user interfaces have been designed. The challenge is to find the one that suits the intended user's skills and the application at hand.

15.6.1.8 Operating system

Multimedia also places demands on the computer's operating system (see Section 8.3.2). Lots of data has to be moved quickly from mass storage to display or sound system, sometimes with several things going on in parallel. This calls for a multitasking operating system (see Section 8.3.4.1). Multimedia is one of the things that are moving users to the more advanced operating systems.

Multimedia could, of course, be done on any equipment that had the same capabilities as a personal computer. Non-PC products for multimedia (multimedia players) have been proposed and some are even on the market. They work for their intended audience, but they offer little cost advantage compared to a standard PC and they lack the access to the vast software resources of the standard PC marketplace. They may or may not succeed in the long run.

15.6.2 Integration of multimedia applications

Upon learning about the capabilities of PC multimedia, many people in industry, education, military, or the general public get ideas of applications they would like to do. But the production of multimedia is complex, expensive, and just plain difficult. This has not been lost on the software industry, and there is a large market for multimedia *development tools*—software programs that assist the multimedia development process.

The most important tools are those known as *authoring tools*. These are programs that support the assembly of all the content material into a single application and provide the user interface and navigation features needed. That, of course, is something that any computer programmer could do using his/her favorite language, but most of the people who want to create multimedia are not programmers. Some authoring tools are specifically designed to allow nonprogrammers to create multimedia; other authoring tools still require programmers, but they make the job easier. Authoring is beyond the scope of this book—there are whole books about it [2]—but a few important considerations will be covered here.

An authoring tool is always associated with a program that will be used to run the authored application. These are called *runtime packages*, or *players*. The authoring tool packages the application in a form that will be understood by the player, so that the end user does not have to have access to the full authoring package, only the player. Often the application content (audio, video, text, etc.) is kept in its native format and is simply accessed from mass storage by the player. However, this poses a security issue if the application developer wants to prevent others using their content material for different purposes. Some professional-level authoring systems employ data encryption techniques to prevent unauthorized use of the application or its materials.

The simplest and lowest-cost authoring systems may not have a player at all. The user is required to have the full authoring package to play the application. This is fine for individual users who create their own applications for their own use, but it is unacceptable for professional distribution of applications. This is just one example of why there are authoring packages available for less than $100 and other packages that cost thousands of dollars. The range of features matches or exceeds the range of prices.

15.6.3 A multimedia example

Multimedia applications are usually dynamic and thus, they are difficult to convey in book illustrations. There are also hundreds of different kinds of applications. Figure 15.10 is an example of a hypothetical multimedia encyclopedia. The screen shows an entry for the city of San Francisco and offers controls for navigating that entry or the rest of the encyclopedia. It shows how text, images, graphics, and video are combined on the screen to provide a rich presentation of any subject that the user chooses.

15.7 Interactive Television

Many experiments have been done over the years to add interactive features to TV. Most of these were conducted on cable systems using reverse transmission on the cable for the user to head end communication. Generally, the interactivity was concerned with subscriber control of access to programs that were already on the cable but scrambled for limited access. By communicating with the head end, a subscriber could request access to a particular program and the head end would charge for the service. This has become known as *pay-per-view*. It can also be implemented using the telephone for the reverse channel and that has proven to be a better means of communication for this purpose.

With DTV on the horizon and with digital capabilities already appearing in receivers, there are some new technical possibilities for interactive TV. It hasn't yet been shown that any of these will prove valuable enough to consumers for them to succeed, but that is being explored. One example was described above in

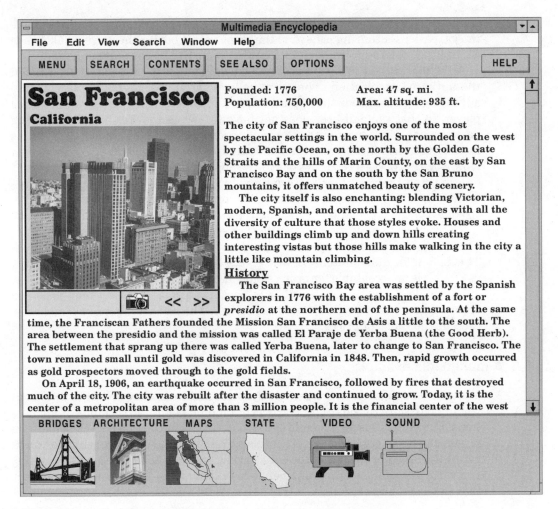

Figure 15.10 An interactive application—a multimedia encyclopedia.

Section 15.5; it will be covered more here.

Consider a television commercial for an automobile. The commercial shows glitzy attention-getting pictures of the car and an announcer describes one or two key features of it. That is all that there is time for in a 30-second or even a 1-minute commercial. However, an interested viewer would like to know a lot more about the product. This can be supported by having the digital transmitter download a database and access program to the receiver during the commercial. After the commercial is over, the consumer can activate the downloaded program and explore specifications, pictures, even prices for the car that was advertised in the commercial.

With the ATSC DTV transmission system that sends data at 19.3 Mb/s, it is possible to send several megabytes of programs and data during a 30-second commercial simply by turning up the screws on the compression of the commercial to reduce the data rate below the 19.3 Mb/s. For example, downloading a 3-MB interactive package during a 30-second commercial requires a data rate of 3,000,000 × 8 /30 = 0.8 Mb/s. This is only a little more than 4% of the channel capacity—compressing the commercial's video by another 4% to handle the downloading would have a negligible effect on picture quality.

The transmission part of this scenario seems very reasonable, but what is more questionable is the capability required at the receiver to store the data and then run the interactive program. That will take much of the functionality of a PC and so far, it does not seem reasonable to build that into a TV receiver because it would probably double the cost of the TV. Another solution might be to add an interface feature in the TV that would allow connection of a standard PC to the TV. Since a high percentage of homes already have PCs, the incremental cost to the consumer would be much less.

Further ideas about interactive TV are discussed in Chapter 21.

15.8 References

1. Brenda Laurel, *The Art of Human-Computer Interface Design*, Addison-Wesley, Reading, MA, 1990.

2. Arch C. Luther, *Authoring Interactive Multimedia*, AP Professional, Cambridge, MA, 1994.

Chapter 16

Terrestrial Broadcasting Systems

16.1 A Brief History and the Future Outlook

By way of example, this chapter describes the terrestrial TV broadcasting situation and technology in the United States. Similar scenarios have occurred in the development of TV broadcasting eleswhere in the world, except that the technical solutions are slightly different. Only some of these possibilities are discussed here; the reader may consult the references for further details of TV broadcasting outside the United States.

16.1.1 History in the United States

Commercial television broadcasting began in the United States in 1946. After a slow start, during which there were real questions as to its financial viability, growth accelerated, and by year-end 1950, there were 107 stations on the air.

Two serious questions about previous Federal Communication Commission regulatory policies were raised at this time:

1. What technical standards should be adopted for the transmission of color?
2. Were the technical restrictions that had been adopted for the assignment of channels adequate to avoid excessive co-channel interference?

To resolve these questions, the FCC convened a rule-making proceeding in September 1948. While the hearing was in progress, it imposed a "freeze" on the granting of construction permits for new stations. The hearing was lengthy and contentious, and it lasted nearly four years until April 1952. During this time the profitability of television broadcasting was demonstrated by the 107 on-air stations.

The FCC decision at the conclusion of the hearing was far-reaching, and it included the following provisions:

1. The system for transmitting color, the *field sequential system*, proposed by CBS was approved.
2. More stringent requirements for mileage separations between VHF co-channel stations were imposed. The result was a major reduction in the number of stations that could be assigned VHF channels.
3. To make more channels available, a large section of the UHF spectrum, 470 to 806 MHz, was allocated for broadcasting.

The decision for CBS color did not survive the test of time, primarily because its transmission standards were not compatible with the existing monochrome standards, i.e., color broadcasts could not be received on monochrome sets, even in black and white. In December 1953 the FCC approved a compatible color system developed by the *National Television Systems Committee*, which was an outgrowth of a system originally proposed by RCA. It has since been known as the *NTSC system*.

Except for CBS color, the standards and rules adopted by the FCC at the conclusion of the hearing, which specified analog video signals and *vestigial sideband* transmission, have been remarkably durable and they are largely in place more thn 45 years later. The technical rules for the assignment of channels are virtually unchanged, and UHF channels are widely used for broadcasting, although their coverage with analog transmission is inferior to VHF. The NTSC color system and its variations (PAL) have been adopted by most of the major countries of the world (France and Russia excepted).

But technology continues to progress, and a shift from analog to digital transmission is now underway, a shift that is now incorporated in the FCC's rules. The immediate impetus for this change is the increased information transmission rate required for HDTV. With analog transmission, HDTV requires greatly increased video bandwidth and scarce or unavailable spectrum space. But digital systems, making use of data compression technologies, can transmit HDTV signals in the same bandwidth as analog NTSC. Digital transmission also has other important advantages as described in Chapters 4 and 7.

As of this writing (1999), digital transmissions has been authorized as a short-term (15-year) supplement to analog and as a long-term replacement. The added capacity can be used (technically) not only for HDTV but also for additional SDTV signals and as a part of the "information highway" now being advertised widely.

The first sections of this chapter are devoted to a description of the existing analog transmitting system. That is followed by a description of the digital broadcasting system specified by the ATSC DTV standards.

16.2 Allocations and Assignments

The cognizant regulatory agency in each country allocates bands of frequencies to specific services, subject to international agreement. Within these bands, specific channels are assigned to individual licensees.

Table 16.1 TV broadcast channels in the United States

Channel	Frequency (MHz)	Channel	Frequency (MHz)
Low-band VHF		UHF (cont'd)	
2	54–60	35	596–602
3	60–66	36	602–608
4	66–72	37	608–614
5	76–82	38	614–620
6	82–88	39	620–626
High-band VHF		40	626–632
7	174–180	41	632–638
8	180–186	42	638–644
9	186–192	43	644–650
10	192–198	44	650–656
11	198–204	45	656–662
12	204–210	46	662–668
13	210–216	47	668–674
UHF		48	674–680
14	470–476	49	680–686
15	476–482	50	686–692
16	482–488	51	692–698
17	488–494	52	698–704
18	494–500	53	704–710
19	500–506	54	710–716
20	506–512	55	716–722
21	512–518	56	722–728
22	518–524	57	728–734
23	524–530	58	734–740
24	530–536	59	740–746
25	536–542	60	746–752
26	542–548	61	752–758
27	548–554	62	758–764
28	554–560	63	764–770
29	560–566	64	770–776
30	566–572	65	776–782
31	572–578	66	782–788
32	578–584	67	788–794
33	584–590	68	794–800
34	590–596		

16.2.1 Current U.S. allocations

In the United States, the VHF bands 54–72, 76–88, and 174–216 MHz and the UHF band 470–806 MHz are allocated for television broadcasting. (The entire

Table 16.2 Minimum co-channel separation in miles

Zone	Channels 2–13	Channels 14–69
I	170	155
II	190	175
III	220	205

VHF portion of the spectrum extends from 30 to 300 MHz while the UHF extends from 300 to 3000 MHz.) These bands are divided into channels, 6 MHz wide, for assignment. Each channel is identified by a number as shown in Table 16.1.

The VHF channels are divided into two segments, low-band—channels 2 to 6, and high-band—channels 7 to 13. The small gap between channels 4 and 5 is reserved for aeronautical use.

16.2.2 Current United States assignments

AM radio channels were assigned on the basis of demand. The FCC established co-channel and adjacent-channel interference criteria, and if the applicant for a new station could show that his or her proposal would meet these criteria, a construction permit and ultimately a license could be granted. But stations were often authorized that did not meet these criteria because of the intense commercial pressure for additional AM stations, and this led to overpopulation of the AM band and excessive mutual interference.

The FCC was determined not to repeat this mistake with TV. Stations were authorized on a demand basis, but only in accordance with a predetermined table of assignments (which was developed during the "freeze") that assigned specific channels to each city. The table could be changed but only as the result of a cumbersome rule-making procedure.

At the end of the freeze, a new table of assignments was issued, which was based on increased spacing between co-channel and adjacent channel stations as shown in Table 16.2. This resulted in a major reduction in the number of VHF stations, particularly in the larger cities.

Zone I in Table 16.2 encompasses the northeast and middle west states. Closer spacing is permitted in this zone in recognition of the denser population.

Zone III includes the Gulf Coast states. Greater spacing in this zone is required because of the prevalence of inversions in the troposphere and greater tropospheric propagation.

Zone II covers the remainder of the United States.

The table also includes the effect of additional criteria or "taboos" for UHF channels based on the characteristics of superheterodyne receivers. For example, no channel is assigned on the receiver image frequency of another channel in the same city.

The application of these criteria has resulted in many unassigned channels in

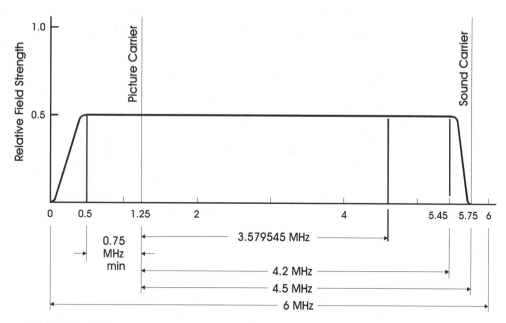

Figure 16.1 United States standard TV broadcast channel.

each city, for example, channels adjacent to assigned channels. This has provided an important opportunity for the assignment of additional channels for DTV.

This table of assignments has generally been successful in minimizing both co-channel and adjacent channel interference.

16.2.3 DTV channel assignments

The channel assignments for DTV in the present VHF and UHF broadcast bands are established so that these bands will be shared by NTSC and DTV stations. The criteria for assigning DTV channels include the following principles:

1. The DTV channels will occupy the same 6-MHz bandwidth as NTSC.
2. Restrictions on the assignment of adjacent channels and taboo UHF channels will be greatly reduced for DTV channels; adjacent NTSC and DTV channels will be assigned in the same city.
3. Co-channel mileage spacing requirements for DTV stations, whether to other DTV stations or to NTSC stations, will be equally strict or perhaps stricter.

Active programs of laboratory and field testing were undertaken to establish firm standards for the assignment of DTV channels. These include tests for the robustness of the signals—their insensitivity to interference, and their benignity—their lack of interference potential to other signals. These tests are necesary because the characteristics of digital television broadcasting are radically different from analog broadcasting.

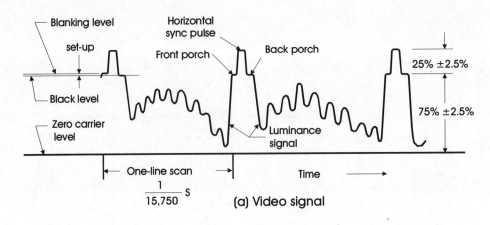
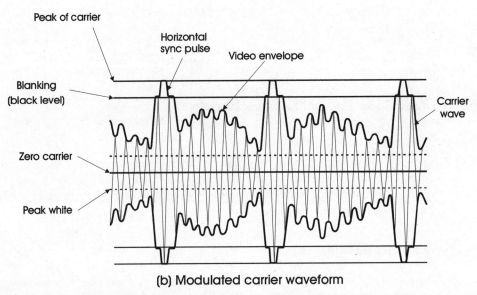

Figure 16.2 Amplitude modulation—video signal.

The resulting DTV assignment standards assign a DTV channel as a companion to every NTSC channel [1]. The added channels in this standard are overwhelmingly UHF—45 VHF and 1645 UHF.

16.3 The Standard Analog Broadcast Television Channel

The amplitude/frequency characteristic of the 6-MHz television broadcast channel as specified by the FCC is shown in Figure 16.1. It has two carriers, the visual carrier 1.25 MHz above the lower channel edge and the aural carrier 4.5 MHz above the visual carrier.

16.4 The Visual Carrier

Most of the TV channel bandwidth is occupied by the visual carrier and its sidebands.

16.4.1 Modulation method and power rating

The visual carrier is amplitude-modulated by the video signal (Figure 16.2). Unlike the audio signal, the video signal has a dc component. It is transmitted by *clamping* (fixing) the carrier power at its maximum rated value (100%), at the peak of sync and adjusting the gain of the system so that its value at the reference white level is 7.25%. The rated transmitted power is the value at the peak of sync. The average transmitted power is less and will depend on the picture content, being least for an all-white picture and greatest for all black. This is known as *negative modulation* because the transmitter power decreases with increasing picture brightness.

Although the sync and blanking waveforms are very effective in fulfilling their intended functions, they are extravagant in their use of transmitter power. On a voltage scale with the peak of sync at 1.4 and the blanking pulses at 1.0, the rms voltage of the blanking intervals over a complete frame is about 1.25. The rms voltage during the active picture intervals depends on the picture content, but it typically would be around 0.2. In this example, the average power level during blanking intervals would be proportional to 1.25^2 or 1.56, while the average picture power would be proportional to 0.2^2 or only 0.04. Although the blanking intervals occupy only 25% of the time, they utilize 97.5% of the transmitter power averaged over an entire frame. With a transmitting system having an effective radiated power of 100 kW, only 2.5 kW is used for transmitting the picture information. The result of this is that receiver synchronization continues to work even when the picture has almost disappeared into the noise.

Co-channel interference is also a problem with the high-powered sync and blanking pulses. They are the major source of interference because the lower-level picture signal from the desired station must override the higher-power sync and blanking pulses from adjacent interfering stations.

Both the specing between co-channel stations and the average transmitted power is reduced substantially by using the United States DTV system as described later in this chapter.

16.4.2 Vestigial sideband format

The spectrum of the visual carrier and its sidebands was shown in Figure 16.1. Both the upper and lower sidebands are transmitted for video signal frequency components up to 0.75 MHz. The lower sideband is attenuated for frequency components above 0.75 MHz, and only the upper sideband is transmitted for signal components above 1.25 MHz. This is known as a *vestigial sideband* format, and its purpose is to reduce the amount of spectrum space required for the channel.

The effect of transmitting only a single sideband is shown on the vector

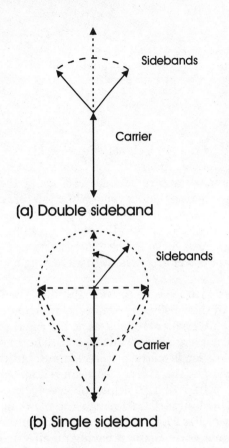

(a) Double sideband

(b) Single sideband

Figure 16.3 Single-sideband operation.

diagrams in Figure 16.3. Double sideband operation is shown in Figure 16.3(a). The sidebands are represented by two vectors rotating in opposite directions around the carrier vector. When both sidebands are present, the amplitude of the vector sum varies by ± twice the sideband amplitude and its phase is constant. If one sideband is removed, the amplitude variation of the resultant is reduced by one-half and phase modulation known as *quadrature phase distortion* is introduced. [Quadrature phase distortion is not the same as *incidental phase distortion*, the variation of the phase of the carrier with level. Incidental phase distortion can be corrected (see Section 16.10.3.3).]

The loss in amplitude can be compensated by the amplitude response of the IF amplifier in the receiver. Typically it is tuned so that the response at the carrier is 6 dB below the maximum (see Section 13.4.1). Quadrature distortion cannot be easily corrected, but this is not necessary because of the relatively low amplitudes of the high-frequency components in a video waveform. For most picture contents, the maximum instantaneous high-frequency sideband amplitude occurs for the color burst signal.

16.5 The Aural Carrier

The aural carrier is frequency modulated with a peak deviation of ±25 kHz as compared with ±75 kHz for FM radio broadcasting. The lower deviation is specified to make as much spectrum space as possible available to the visual signal, and higher deviation of the aural carrier is unnecessary because the visual signal establishes the limiting signal-to-noise ratio parameter for the system.

The power of the aural transmitter is normally 20% of the peak visual power. This is usually more than adequate to produce an acceptable aural SNR in any location with a satisfactory visual SNR.

16.6 Elements of a TV Transmitting System

A TV transmitting system includes a transmitter that generates the modulated visual and aural carriers, a filter that removes a portion of the lower sideband, a diplexer that combines the visual and aural carriers, and an antenna that radiates the signal.

In older systems, the modulated visual carrier, after amplification, passed through a sharply tuned *vestigial sideband filter* that removed a portion of the lower sideband and produced the vestigial sideband visual carrier. This wasted a lot of power. In current designs the lower sidebands are removed from the carrier at an intermediate frequency where the power level is low by a *surface acoustic wave* (SAW) filter (see Section 16.8.5).

The aural and visual carriers are combined in a multiplexer, usually a *notch diplexer* that has high attenuation of the lower subcarrier sideband, for transmission to the antenna in a single transmission line. Some antennas require two feeds separated in phase by 90°. The phase shifter for this follows the multiplexer. In most cases the line is two coaxial copper tubes with the ratio of their diameters chosen to give it a characteristic impedance of 51.5 or 75 ohms.

The antenna is invariably directional in the vertical plane to concentrate the radiated energy toward the horizon and along the earth rather than wasting it into space.

Some stations, especially in the UHF band, use antennas that are directional in the horizontal plane as well. The FCC restrictions on this mode are severe, and antenna directivity cannot be used to permit closer station spacing. It can be used, however, to increase the station's coverage in a desired direction, and this provision of the FCC rules is sometimes used by UHF stations.

The *gain* of the antenna is the ratio of the maximum radiated energy density to the radiated energy density from a reference antenna with equal power input. The reference antenna used for broadcast systems is the half-wave dipole. A half-wave dipole radiating 1 kW in free space produces a field intensity of 137.6 mV/m at a distance of one mile.

The half-wave dipole reference differs from the *isotropic radiator*, the reference for microwave and satellite systems that radiates equally in all directions. An isotropic radiator is a more fundamental reference and it simplifies calculations,

Table 16.3 FCC power/antenna height limits for analog broadcasting

Channels	Maximum ERP	Maximum Height
2–6	100 kW	305 m
7–13	316 kW	610 m
14–68	5000 kW	610 m

The power limitations apply to the horizontally polarized component of the radiation. Stations using circular polarization are permitted to radiate an added equal amount of vertically polarized energy.
See Section 16.15.2 for definition of antenna height.
See Figures 16.4 and 16.5 for power limitations if antenna height exceeds the maximum.

but it cannot be physically constructed. The gain of a half-wave dipole is 2.15 dB with respect to an isotropic radiator.

The FCC rules specify limitations on the *effective radiated power* (ERP), which is the visual carrier power input to a half-wave dipole that would result in the same radiated energy density. It is given by the equation:

$$\text{ERP} = P_T \, G_A \, E_{\text{EFF}} \tag{16.1}$$

where ERP is the effective radiated power, P_T is the output power of the visual transmitter, G_A is the gain of the antenna with respect to a half-wave dipole, and E_{EFF} is the efficiency of the transmission lines, diplexer, and other system elements between the output of the transmitter and the input to the antenna.

16.7 FCC ERP/Antenna Height Limitations

The FCC limits the ERP and antenna height of television stations to provide the basis for an orderly plan for assigning channels to cities in the United States. These limits are higher in high-band VHF, Channels 7 to 13, than at low-band, and higher at UHF than at VHF. This reflects the less favorable propagation characteristics of the higher frequencies. In consideration of these limitations, the FCC established criteria for mileage separation between co-channel and adjacent channel stations to minimize interference while permitting a reasonable number of assignments. The radiated power limitations for antenna heights up to indicated maximums are shown in Table 16.3. Additional power limitations apply for heights above these maximums as shown in Figures 16.4 and 16.5. The zones on which the limitations are based are defined in Part 73.699 of the FCC rules. Zone I includes the northeast and midwest. Zone III includes the Gulf Coast states while zone II includes the remainder of the country.

16.8 Television Transmitters

The designs of TV transmitters differ widely depending on their frequency band and power rating. Other differences result from proprietary choices of manufac-

Terrestrial Broadcasting Systems 395

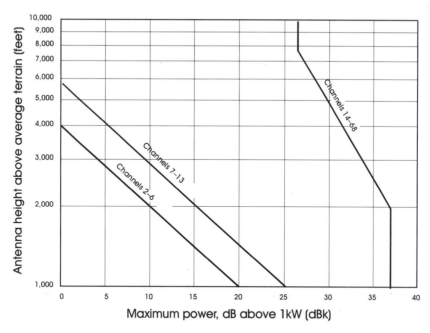

Figure 16.4 FCC power/antenna height maximums (zone 1).

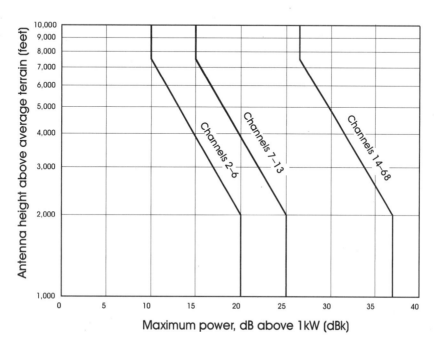

Figure 16.5 FCC power/antenna height maximums (zones 2 and 3).

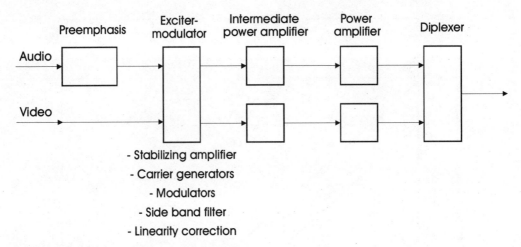

Figure 16.6 Functional diagram of an analog TV transmitter.

turers for the basic system arrangement and components. All transmitters, however, must include the following components:

1. Aural preemphasis to preemphasize the high-frequency components in accordance with FCC requirements.
2. Processing amplifiers to correct deficiencies such as improper sync level in the input visual signal and to provide DC restoration.
3. Visual and aural carrier generators with the frequency tolerances specified by the FCC.
4. Visual and aural modulators.
5. Filters to remove the lower sidebands of the modulated visual carriers and to remove spurious frequency components from both carriers.
6. Linearity correction devices in the visual chain to compensate for incidental phase modulation and differential gain. Ideally, these corrections should be made in both the video and RF domains.
7. Intermediate and final power amplifiers. Figure 16.6 is a functional diagram of a representative analog television transmitter that shows the locations of these functions.

16.8.1 Exciter modulator

The exciter modulator is the heart of the transmitter and includes functions 1 to 6 in Section 16.8. Its outputs are the modulated visual and aural carriers at low power levels.

16.8.1.1 Aural preemphasis

The aural preemphasis must be in accordance with FCC requirements. The time constant of the preemphasis circuit is 75 μs.

Table 16.4 Carrier frequency tolerances

	Visual	Aural
FCC requirements	±1,000 Hz	(Visual + 5.5 MHz) ±1,000 Hz
Industry objective	±200 Hz	(Visual + 4.5 MHz) ±200 Hz

16.8.1.2 Stabilizing amplifier

The stabilizing amplifier regenerates sync pulses and stabilizes their amplitude. The signal is dc restored by clamping the signal at black level. White peaks are clipped is applied to avoid overloading the modulator.

16.8.1.3 Linearity and frequency response correction

Linearity and frequency response correction is made in both the video and RF domains.

16.8.1.4 Aural and visual carrier generation

The aural and visual carriers are generated at IF frequencies. Their frequency tolerances are small (see Table 16.4), in part because of their use of offset carriers (see Section 3.17.2), which require the difference in carrier frequencies between adjacent co-channel stations to be stabilized at ±10 kHz. Stations are assigned frequencies on this basis and the technique would not be effective if the difference in carrier frequencies greatly exceeded this tolerance.

It is equally important that the frequency difference between aural and visual carriers be maintained within close tolerances. This is accomplished by maintaining their frequency difference at 4.5 MHz so that the carrier frequencies drift together.

16.8.1.5 Visual modulator

The engineering profession was originally divided on the issue of low-level vs. high-level modulation, i.e, whether the carrier should be modulated at a low or high power level. With the passage of time, technical advances have favored low-level modulation, and most new transmitters use this modulation mode.

The advantages of low-level modulation are numerous. They include the elimination of high-power modulators and vestigial sideband filters, and the ability to correct linearity and frequency response distortions with compensation in both the video and radio frequency domains. The disadvantage is the requirement for linear power amplifiers.

16.8.1.6 SAW filter and sideband removal

The device that finally tipped the balance for low-level modulation was the surface acoustic wave (SAW) filter. SAW devices are tuned filters that employ piezoelectric semiconductors. SAW filters can produce amplitude and envelope delay responses that are unachievable with conventional RLC filters. They have steep skirts at the band edges, but they do not introduce the envelope delay distortion that is inherent in RLC filters. Transmitter designers have taken full advantage of this ability to provide the vestigial-sideband amplitude-frequency response required by the FCC for the visual carrier and to make frequency and linearity corrections in the RF domain. The insertion loss of the SAW filter is high, typically 25 dB, and low-noise amplifiers must be used with it to achieve a satisfactory SNR.

16.8.1.7 Up-converter

The up-converter is a conventional heterodyne circuit that mixes the outputs of the visual and aural modulators with that of a local oscillator operating at the sum of the IF and final carrier frequencies. The up-converter outputs at the final carrier frequencies are then sent to the visual and aural intermediate power amplifiers.

16.8.2 Power amplifier drivers

Power amplifier drivers are linear amplifiers that provide the input to the final power amplifier (PA). They must have sufficient output power to drive the PA to its full rated power, and their frequency response and linearity must be consistent with the performance specifications of the transmitter.

The power requirements for the driver vary from less than 25W for klystron UHF power amplifiers to nearly 2 kW for some tetrode VHF power amplifiers. In common with the trend for all electronic devices, electron tubes are being replaced by solid-state devices as soon as the latter can meet the power level and other performance specifications at an economic price. The technology of solid-state devices has now progressed sufficiently to meet these conditions, and most new commercial transmitters use solid-state drivers.

16.8.3 Power amplifiers

In common with all TV technologies, transmitter PAs have benefitted from continuous technical improvement. Early VHF transmitters used tetrodes that were grid-modulated when high-level modulation was used. Different models of early UHF transmitters employed tetrodes and klystrons, a tube that employs velocity modulation of an electron beam to amplify the carrier. The features of klystrons include stability, high output power, reasonable efficiency (about 40%), and high gain so that very little driving power is required. This characteristic eliminated the need for a long chain of preamplifiers with their linearity problems and was

Table 16.5 Representative analog TV transmitter power ratings

VHF low band	VHF high band	UHF
5 kW	15 kW	30 kW
15 kW	25 kW	60 kW
25 kW	45 kW	120 kW

one of the factors that hastened the trend to low-level modulation. Unlike tetrodes, whose performance deteriorates with increasing frequency, klystrons work better at higher frequencies. This led to their early choice as the power amplifier for UHF transmitters.

The *klystrode*, a member of a class known as *inductive output tubes* (IOTs), has been introduced recently that employs both a control grid and velocity modulation. This has extended the use of the klystron principle to VHF transmitters and provides higher efficiency than the klystron at UHF.

In the meantime, solid-state technology continues to advance, and in time solid-state devices will probably supersede electron tubes of all types. They are already being used in PA drivers and in low-power PAs, and their power and frequency limits are continuously being expanded.

Solid-state devices as used in transmitters are not particularly efficient, and a considerable amount of heat must be dissipated, which is difficult because of the small size of the devices. The *heat pipe* is one solution. A liquid in a wick transfers heat from the small area of the solid-state device to larger cooling fins that can be fan-cooled if necessary.

16.9 Representative Transmitter Power Ratings

Although the FCC does not specify power ratings for transmitters, certain power ranges have been established in commercial transmitters that are strongly influenced by the maximum ERP permitted by the FCC and practical antenna gain values.

Table 16.5 shows typical power ratings for commercial TV transmitters in the United States. Note that they become progressively higher in the higher-frequency bands. This is not because of equipment limitations but because the propagation characteristics of higher frequencies are not as favorable—a fact that is recognized in the power limitations specified by the FCC.

16.10 Transmitter Performance Standards

It is important to make a distinction between equipment performance and system performance in interpreting performance standards. Degradations and distortions are cumulative in analog systems, and for many criteria, e.g., SNR, the performance of each element in the system must be better than that of the total system.

Table 16.6 Transmitter amplitude-frequency response tolerances

Frequency Range (MHz)	Tolerance
0.5 – 4.1	±1.0 dB
4.1 – 4.18 (VHF)	+ 0.0, -2.0 dB
4.1 – 4.18 (UHF)	+ 0.0, -4.0 dB

The upper sidebands between 4.75 and 7.75 MHz and the lower sidebands between 1.25 and 4.25 MHz should be attenuated at least 20 dB below the visual carrier, using the 200-kHz sideband as reference.

The standards for transmitter performance in this section are for the transmitter only [2]. The standards for the radiated signal, which includes the effect of other components of the system, would be somewhat lower. The transmitter standards should be compared with the EIA standards for point-to-point transmission systems (such as microwave) described in Chapter 18.

The following paragraphs summarize the most significant performance standards for television transmitters as contained in two sources: the FCC rules, Section 73, and informal industry equipment goals3 that have been established based on the actual performance of high quality transmitters. The industry goals are more complete and generally more stringent than the FCC rules.

16.10.1 Frequency-related standards

16.10.1.1 Carrier frequency tolerance

See Table 16.4 for FCC requirements and industry goals for frequency tolerances.

16.10.1.2 Amplitude-frequency response

The amplitude-frequency response of the transmitter as specified by the FCC is shown in Figure 16.1. The goals for the maximum deviation from this response (as measured with a sideband analyzer) are shown in Table 16.6.

16.10.1.3 Envelope delay vs. frequency

The standard envelope delay (see Chapter 4) and its tolerances are shown in Table 16.7.

16.10.1.4 Frequency response vs. brightness

Measurements of frequency response are made with a modulating sine wave having a peak-to-peak amplitude of 25% of reference black to reference white and

Table 16.7 Standard transmitter envelope delay vs. frequency

Frequency (MHz)	Delay
0.05 – 3.0	0 ns
3.0 – 4.18	increases linearly with frequency with 0 at 3.0 MHz and –170 ns at 3.58 MHz.

The tolerance goals for these values are:

Frequency (MHz)	FCC, ns	Goal, ns
0.05 – 2.1	±100	±50
2.1 – 3.0	± 70	±35
3.0 – 3.58	± 50	±25
3.58 – 4.18	±100	±60

with axes on the brightness scale of 25, 45, and 65%. The reference is the 45% axis.

Tolerance: FCC—not specified; goal—±1 dB

16.10.2 Signal-to-noise ratio, visual signal

The visual SNR is defined as:
S[blanking to reference white] N[weighted rms] ratio (dB)
(See Section 4.15.3 for definition of weighted noise.)
Tolerance: FCC—not specified; goal—55 dB

16.10.3 Linearity standards

Linearity standards specify the variations in gain or phase delay with brightness or other signal parameter.

16.10.3.1 Differential gain and low-frequency nonlinearity

Differential gain is the measure of variations in the transmitter's gain at the color subcarrier frequency for different luminance levels of the video signal (see Section 6.10). It is measured by superimposing a 3.58-MHz sine wave, having a peak-to-peak amplitude of 20% of blanking-to-reference white, on a stairstep luminance signal and measuring the differences in the sine wave amplitude at the transmitter output on each of the steps.

Low-frequency nonlinearity is the measure of the variation in the transmitter's gain with brightness for lower video frequencies. It is also measured with a stairstep waveform, and the nonlinearity is indicated by variations in the height of the steps at the transmitter's output.

Tolerance: FCC—not specified; goal—5%

16.10.3.2 Differential phase

Differential phase is the complement of differential gain and is the variation in the color subcarrier phase with brightness (see Section 5.10.4). The same waveform is used for measurement.

Tolerance: FCC—not specified; goal—±3°

16.10.3.3 Incidental phase modulation

Incidental phase modulation differs from differential phase distortion in that differential phase refers to the phase shift of the color subcarrier envelope whereas incidental phase modulation refers to the phase of the carrier. It is usually larger than the differential phase.

Tolerance: FCC—not specified; goal—±7°

16.11 Transmission Lines

The output energy of the transmitter is conveyed to the antenna by means of a *transmission line*. It is usually two coaxial copper tubes that are insulated from each other, although wave guides are occasionally used for high-power UHF stations (see Section 16.12).

16.11.1 Coaxial line transmission modes

When the RF output of the transmitter is applied to the input of a coaxial line with a gaseous dielectric and a diameter that is small compared with the wavelength, it generates a voltage and current wave that travels at approximately the velocity of radiation in free space. (A small deviation results from the fact that the transmission line is not a perfect conductor. In most practical cases it is small enough to be ignored.) If the dielectric is solid, the velocity and the wavelength are reduced by the square root of the dielectric constant.

In this primary mode, the electric field is perpendicular to the axis of the line. If the wavelength is equal to or less than the circumference of the line, a secondary mode can occur in which the electric field is circular and coaxial with the line. This is known as *multimoding*. Multimoding is undesirable because it increases the losses and reduces the power handling capability of the line. In practice, this limits the usefulness of coaxial lines to frequencies with wavelengths greater than the circumference of the line, a factor that must be considered at the upper end of the UHF band.

The voltage and the current waves are in phase and their ratio, E to I, is Z_C, the *characteristic impedance* of the line. If the impedance of the load at the end of the

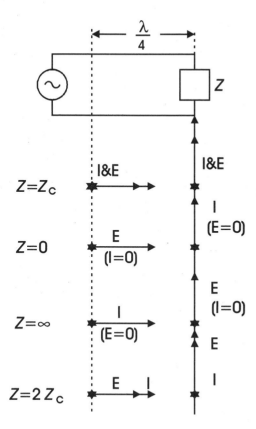

Figure 16.7 Transmission line voltages and currents.

line equals the characteristic impedance, all of the energy in the wave is absorbed. If the load impedance does not equal the characteristic impedance, some of the energy is not absorbed, and voltage and current waves are reflected back to the transmitter. The magnitude and phase of the voltage and current at each point on the line is the vector sum of the direct and reflected waves, and a stationary standing wave is formed with maxima and minima separated by one-half wavelength.

16.11.2 Voltage and current relationships

Figure 16.7 shows the relationship between voltage and current vectors at the load and ¼ wavelength toward the source for four load impedances: characteristic impedance, short circuit, open circuit, and double the characteristic impedance.

The ratio of the voltage maximum to the voltage minimum in the standing wave is called the *voltage standing wave ratio* (VSWR). It is a fundamental measurement of transmission line performance.

VSWR is calculated by Eq. (16.2):

$$\text{VSWR} = \frac{1 + |C_R|}{1 - |C_R|} \tag{16.2}$$

where $|C_R|$ is the magnitude of the reflection coefficient. C_R is calculated by Eq.

(16.3):

$$C_R = \frac{Z_{\text{LOAD}} - Z_C}{Z_{\text{LOAD}} + Z_C} \tag{16.3}$$

where Z_{LOAD} and Z_C are the load and characteristic impedances. Z_C has a small reactive component as the result of the ohmic resistance of the transmission line conductors. This is usually so small that it can be ignored.

If Z_{LOAD} equals Z_C, the standing wave pattern disappears, and the voltage and current are constant except for a gradual diminution toward the load owing to losses in the line.

If Z_{LOAD} is a pure resistance and greater than Z_C, a voltage maximum will appear at the load and at 180° intervals toward the load on the line. Similarly, if Z_{LOAD} is a resistance less than Z_C, a voltage minimum will appear at the load and at 180° intervals along the line. If Z_{LOAD} has a reactive component, the maxima and minima will be shifted so that the standing wave at the load will be neither a maximum nor a minimum in the standing wave pattern.

The input impedance undergoes an inversion every quarter wavelength. For example, an open-circuit quarter-wavelength line will have zero input impedance, and a short circuited quarter-wavelength line will appear to be an open circuit at its input.

The relationships between load impedance, VSWR, and input impedance in the general case can be determined by the use of a *Smith chart*.

16.11.3 Characteristic impedance calculation

The characteristic impedance of transmission lines is $(L/C)^{1/2}$, where L and C are the inductance and capacitance of the line per unit length. It is determined by the ratio of the diameters of the conductors and the dielectric constant of the insulating material between them:

$$Z_C = \frac{138 \log_{10}\left(\frac{b}{a}\right)}{\sqrt{\mu}} \tag{16.4}$$

where b and a are the diameters of the outer and inner conductors, and μ is the dielectric constant of the insulator between them.

The ratio of outer and inner conductor diameters for lines with air dielectric, ($\mu=1$), is 3.52 for a Z_C of 75 ohms and 2.30 for a Z_C of 50 ohms.

16.11.4 Standard transmission line diameters

The de facto standard dimensions that have been developed for rigid coaxial transmission lines are tabulated in Table 16.8. Rigid transmission lines normally come in 20-ft (6.1-m) lengths that are joined by flanges welded to their ends. A wide variety of elbows, T-connections, and other fittings is available.

Table 16.8 Transmission line diameters

Type	Z_c (ohms)	Outer diameter (in.)
560	75	7/8
561	75	1-5/8
562A	75	3-1/8
563	75	6-1/8
573	50	6-1/8

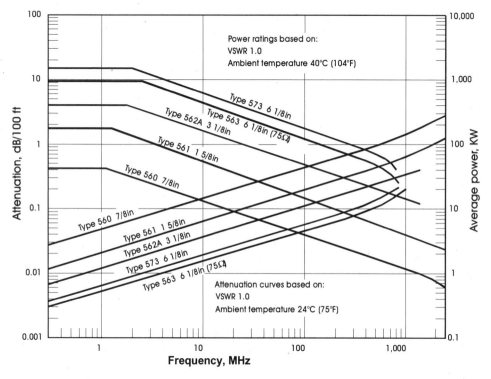

Figure 16.8 Coaxial transmission line power ratings and attenuation.

16.11.5 Power ratings and attenuation

The power ratings and attenuation of transmission lines are normally established for a VSWR of 1.0. If the VSWR exceeds 1.0, the higher currents and voltages at their maxima will cause greater attenuation and a lower power rating. Figure 16.8 shows the power rating and attenuation as a function of frequency for the standard transmission line sizes in the VHF and UHF television frequency ranges, all with a VSWR of 1.0. The rapid increase in attenuation for 6⅛-in line at the upper end of the UHF band is caused by multimoding.

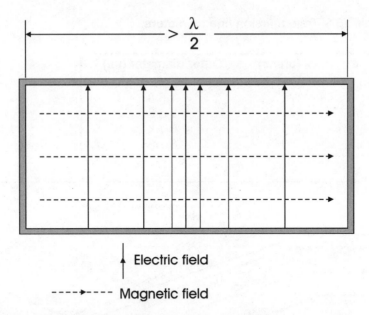

Figure 16.9 Cross-section of a waveguide operating in the TE_{10} mode.

Note that the power ratings are based on average transmitter power. The peak-power rating for television transmissions would be higher, but on the basis of an all-black picture and after allowing for the aural carrier, the difference is small.

16.11.6 Mechanical features

A number of mechanical features of coaxial transmission lines should be noted.

The inner and outer conductors are subjected to different temperatures so that differential expansion occurs. On a long run the difference can be large, and provision must be made to compensate for it. One solution is to join the inner conductors with slotted plugs that fit inside the two ends. This allows the conductors to expand or contract without breaking the electrical connection.

Provision must also be made for expansion of the outer conductor, and the line is usually hung from the tower with spring hangers rather than being rigidly attached to it.

Moist air inside the line can cause condensation and leakage across the inner conductor spacers. To prevent this, most lines are pressurized with dry air. This requires the joints to be sealed so that they are airtight.

16.12 Waveguides

It is unfortunate that the trends of the power capacity and attenuation vs. frequency curves of coaxial transmission lines are in the wrong direction for UHF transmissions. More power is required for UHF channels, but the power rating of

Table 16.9 Waveguide power ratings and attenuation

Type	Width Inside (in)	Freq. Range (MHz)	Power Rating (MW)	Attenuation (dB/100 ft)
WR 1800	18.0 – 9.00	410 – 625	93.4 – 131.9	0.056 – 0.038
WR 1500	15.0 – 7.00	490 – 750	67.6 – 93.3	0.069 – 0.050
WR 1150	11.5 – 5.75	640 – 960	35.0 – 53.8	0.125 – 0.075

coaxial lines is lower at the higher frequencies and their attenuation is greater. The use of larger-diameter lines at the upper end of the UHF band to reduce losses and increase the power rating is limited by multimoding. As a result, waveguides, which are better suited for higher frequencies, are sometimes used for the upper UHF channels.

Rectangular waveguides operating in the TE_{10} mode (Figure 16.9) are the most commonly used. In this mode, the width of the waveguide must be more than one-half wavelength at the transmission frequency. The frequency at which the wavelength equals twice the width of the waveguide is its *cutoff frequency*, and the waveguide will not transmit lower frequencies.

The losses in waveguides are much lower than transmission lines at these frequencies. The dimensions, frequency range, power rating, and attenuation of aluminum waveguides operating in the TE_{10} mode and in the standard sizes that cover the UHF broadcast band are shown in Table 16.9. Note that the attenuation decreases and the power rating increases as the frequency increases.

Like coaxial transmission lines, waveguides are subject to moisture condensation as they cool. This can be avoided by pressurization, but the amount of pressure is limited by the poor mechanical strength of the rectangular structure. Higher pressures can be used with *double truncated* waveguides. The corners of a rectangular waveguide are rounded, and it is fitted inside a circular tube. The circular structure not only permits greater pressurization but also reduces the wind loading.

16.13 Properties of VHF and UHF Radiation

This section provides a brief review of the properties of VHF and UHF electromagnetic radiation as an introduction to a description of television broadcast antennas and the coverage areas of broadcast stations.

16.13.1 Electric and magnetic fields

As its name implies, electromagnetic radiation is a moving wave of electric and magnetic fields that are at right angles to each other and to the direction of travel (see Figure 16.10). The path of the wave is a straight line unless altered by refraction, diffraction, or reflection.

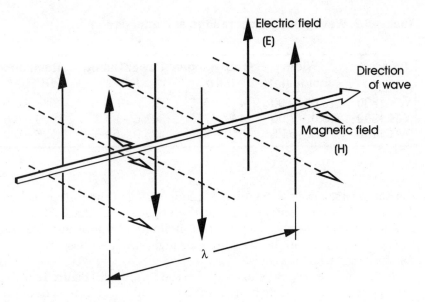

Figure 16.10 Electromagnetic wave.

16.13.2 Frequency and wavelength

The frequency and wavelength of electromagnetic radiation in free space are related by Eq. (16.5):

$$\lambda = \frac{c}{f} \tag{16.5}$$

where λ is the wavelength in meters, c is the wave velocity in m/s, and f is the frequency in hertz.

16.13.3 Velocity and index of refraction

The velocity of electromagnetic radiation in free space, c, is 3×10^8 m/s. Its velocity through gases, liquids, or solids is c/n, where n is the *index of refraction* of the transmission medium.

The index of refraction of the atmosphere varies in accordance with altitude, barometric pressure, and humidity. Typically it is 1.003 at sea level and decreases exponentially toward exactly 1.000 with altitude. The deviation from unity is small, but is sufficient to have a major effect (normally an increase) on the coverage of TV broadcast stations because of refraction.

16.13.4 Refraction

Refraction occurs when the radiation path passes obliquely across a boundary between two media having different refractive indices. The path is bent toward the denser medium that has a higher index of refraction.

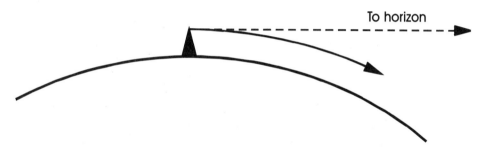

Figure 16.11 Tropospheric propagation.

The path of a broadcast television signal is directed tangentially along the earth's surface at the transmitter and passes through progressively less dense layers of the troposphere, the lower portion of the atmosphere. Because of refraction, it is bent downward and passes beyond the geometric horizon; this is known as *tropospheric transmission* or "tropo" (Figure 16.11).

Under normal atmospheric conditions the effect of refraction can be approximated by assuming that the earth's radius is 4/3 of its actual value, thus increasing the distance to the calculated radio horizon. Atmospheric abnormalities sometimes exist in which the atmospheric density gradient deviates from its normal pattern. The most common is an *inversion* in which the temperature of the air increases with height, thus causing the density to decrease more rapidly than normal and the distance to the radio horizon to increase correspondingly.

16.13.5 Diffraction and reflection

Electromagnetic radiation travels in a straight line in free space, but if a part of the wave front is blocked by an intervening object the path will bend, and some radiant energy will appear behind the geometric shadow of the object. This is one of the effects of a phenomenon known as *diffraction*.

Radiant energy is reflected, and, as the result of the combined effects of reflection, refraction, and diffraction, television signals can be received at most points in a station's service area even though not all of them have a line-of-sight path to the transmitting antenna.

16.13.6 Multipath transmission

The signal at the receiving antenna is the vector sum of a direct signal and one or more reflected signals. If the difference in path lengths is small, the signal strength will be increased or decreased depending on the relative phase of the incoming signals. If the difference in path lengths is large, e.g., as the result of a reflected signal from a large distant object, the time difference will be great enough to cause a ghost. This effect is called *multipath*, and it can be serious in mountainous or congested urban areas.

16.13.7 Polarization

The *polarization* of a wave describes the orientation of its field vectors, and by convention it is expressed as the orientation of the electric field. Two polarization modes are permitted for TV broadcasting signals, horizontal and elliptical. With elliptical polarization, the electrical vector rotates as the wave passes a fixed point. Circular polarization is a special case of elliptical in which the vector has an equal value in all orientations.

FCC rules permit the same ERP for the horizontal component of circularly polarized radiation (or the component on the major axis for elliptical polarization) as for horizontal polarization, so that the total radiated power can be twice as great.

Circular polarization has a number of advantages. Since the total radiated power is twice as great, the received signal is usually greater with "rabbit ears" or other indoor antennas that are not optimized for horizontal polarization. Improvements of up to 3 dB in received signal strength can be achieved by using circularly polarized receiving antennas that derive energy from both the horizontal and vertical components.

Circularly polarized receiving antennas also reduce the magnitude of ghosts since the direction of rotation is inverted upon reflection, and the reflected wave is not received.

As a result of these advantages, there has been an increasing trend toward the use of circular polarization.

16.13.8 Radiated field strength and energy

The intensity of electromagnetic radiation is usually specified by the strength of its electric field. The usual units for the magnitude of the fields encountered in broadcasting are millivolts/meter or microvolts/meter.

The field strength of an electromagnetic wave in free space is inversely proportional to the distance from the antenna. For radiation from a ½-wave dipole, it is:

$$E = \frac{137 \, P^{1/2}}{D} \tag{16.6}$$

where E is the field strength in mV/m, P is the radiated power in kW, and D is the distance in miles.

The strength of a wave can also be measured by the power passing through a unit area, usually a square meter, at right angles to the wave's path. Field strength and energy in space or in the atmosphere are related by Eq. (16.7):

$$\text{Power/m}^2 = \frac{E^2}{377} \tag{16.7}$$

where: power is measured in μW, E is the rms value of the field strength measured in mV/m, and the constant, 377 (120π), has the dimension of ohms, and it is sometimes described as the impedance of free space.

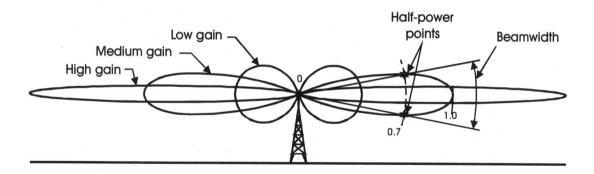

Figure 16.12 Vertical antenna directivity and gain.

16.14 Antennas

The role of the antenna is to couple the transmitter to space by converting its electrical output energy to electromagnetic radiation. The primary antenna specifications are vertical directivity and gain, horizontal directivity, polarization, input impedance and bandwidth, and power-handling capacity.

16.14.1 Vertical directivity and gain

Television transmitting antennas increase the signal strength on the earth's surface by minimizing the power that is radiated into space and concentrating it at angles toward or just below the horizon. This property, the antenna's *directivity*, is indicated by a *radiation pattern*, a plot of the radiation intensity at different elevations.

Figure 16.12 shows the vertical radiation patterns for three representative antennas, a low-band, low-gain VHF antenna (see Section 16.6 for definition of "gain"), a high-band, medium-gain VHF antenna, and a high-gain UHF antenna.

The points on the pattern where the field strength is 70% of the maximum are known as the *half-power points*, and the angular separation of these points is the *beamwidth* (Figure 16.12).

The gain of the antenna is increased by narrowing the beamwidth to concentrate the radiated power in the desired directions, and the product of the gain and beamwidth tends to be constant.

The directivity of a television broadcast antenna can be calculated by defining its *aperture*, an imaginary cylinder in space surrounding the radiating elements. The cylindrical aperture is *illuminated* by the radiating elements, and for analytic purposes each point on the aperture can be considered to be a source of radiant energy.

The antenna's directivity is determined by the dimensions of the aperture and by the amplitude and phase of the radiant energy at each point on its surface. In the simplest case, all the points on the aperture are illuminated by a field of the same phase and of equal amplitude. The beamwidth is inversely proportional to the ratio of the aperture height to the wavelength. If the illumination of the aperture is uniform, the beamwidth in degrees is:

$$\text{Beamwidth} = 48 \frac{\lambda}{h} \tag{16.8}$$

where λ is the wavelength, and h is the height of the aperture measured in the same units as λ.

If the illumination of the aperture is not uniform, Eq. (16.8) does not apply precisely, but in general the longer the aperture relative to the wavelength, i.e., the taller the antenna, the narrower the beam.

16.14.2 Horizontal directivity

The FCC rules make a distinction between antennas that are intentionally designed to be directional and those that are directional by virtue of inevitable imperfections in the design. There are no firm rules with respect to the circularity of the horizontal pattern of "nondirectional" antennas, but most of them maintain it within ±3 dB or less.

As the result of its unfortunate experience with AM directional antennas, which were overused to increase the number of stations on each frequency, the FCC ruled that directional antennas cannot be used in TV to reduce the spacing between stations. Also, the maximum ERP from directional antennas cannot exceed the maximum allowed for nondirectional antennas. Their major purpose of horizontal directivity is to avoid wasting energy over sparsely populated areas and to provide the maximum allowed ERP to densely populated areas with lower total power.

16.14.3 Bandwidth, VSWR, and echoes

Most antenna components are frequency sensitive, and achieving the proper input impedance over the bandwidth of a single channel was one of the early challenges of TV technology. In subsequent years, techniques have been developed to make it possible to achieve satisfactory response over more than one channel.

The conventional indicator of antenna bandwidth is the range of frequencies at which its input impedance approximates the characteristic impedance of the transmission line. This is measured by the VSWR of the transmission line feeding the antenna, and it is desirable that this be less than 1.05 for frequencies from 0.75 MHz below to 5.45 MHz above the visual carrier. The VSWR at the color subcarrier 3.58 MHz above the carrier is particularly important.

A low VSWR is not an absolute guarantee of a satisfactory match between antenna and transmission line, and reflections from the antenna may be great enough to produce visible echoes or ghosts without causing the VSWR to exceed otherwise acceptable limits. The visibility of the echoes depends on their amplitude and their separation from the primary signal. For a transmission line length of 1000

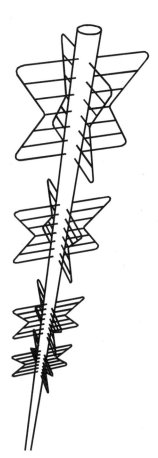

Figure 16.13 Superturnstile antenna.

ft, echoes having an amplitude of 1.5% or more will be visible. For a 500-ft length, visibility begins for echoes having an amplitude of 3.0% or greater [4].

16.14.4 Radiating systems

Antennas can be classified by the type of radiators that illuminate the aperture, and these can be divided into two categories, dipoles and slots.

16.14.4.1 Dipole radiators

This section describes representative dipole radiators.

1. Superturnstile antenna. The *superturnstile* antenna was one of the first to be introduced in the early years of TV broadcasting, and it is still in wide use for VHF stations. The dipole is a pair of radiators, sometimes called *bat wings* because of their shape, mounted on opposite sides of a supporting pole. The radiator consists of two dipoles mounted at right angles, and the antenna consists of two to twelve layers of dipoles (Figure 16.13). The spacing between

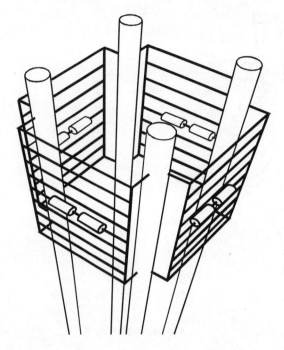

Figure 16.14 Panel antenna.

layers is approximately one wavelength, and the gain is approximately equal to the number of layers. The dipoles are plane surfaces rather than cylinders to increase their bandwidth. The dipole pairs are fed in quadrature and generate a reasonably circular pattern in the horizontal plane.

2. Panel antenna. The superturnstile antenna requires a cylindrical mounting pole of limited diameter so that it is not well suited for the lower element of a stacked array where two or more antennas are mounted above one another. The panel antenna (Figure 16.14) solves this problem by mounting the dipole in front of a reflector and mounting a dipole/reflector assembly on each face of a tower that can be either 4- or 3-sided. Panel antennas are used for both high- and low-band VHF stations.

3. Circularly polarized antennas. In one form of circularly polarized antenna, the single horizontally polarized dipole in the panel antenna is replaced by crossed dipoles that are fed in quadrature, thus producing a circularly polarized wave.

16.14.4.2 Slot radiators

Slot radiators are most frequently used for UHF stations but are also used for VHF. They consist of a hollow cylindrical column in which vertical slots are cut. A conductor in the center of the column, which is connected to the transmitter output, is capacitively coupled to the slots by radial pins mounted on their edges (Figure 16.15).

Figure 16.15 Slot radiator.

An electric field is produced across the slots with its amplitude and phase determined by the length and position of the pins. The electric field across the slots and the current around the cylinder generate the radiated energy.

There are two types of slot antennas, *standing wave*, commonly used for UHF and *traveling wave*, commonly used for VHF. The inner conductor of the standing wave antenna is connected to the outer cylinder at the end opposite the feed point, and a standing wave pattern is produced. The inner conductor of the traveling wave antenna is terminated by a capacitive connection to the end slots that absorbs any remaining energy and provides a characteristic impedance load. The amplitude of the voltage in the standing wave antenna is a standing wave pattern with diminishing maxima amplitudes. The amplitude of the voltage in the traveling wave antenna diminishes continuously from feed point to termination.

16.14.4.3 Helical radiators

Helical antennas are occasionally used. The radiating elements are two helixes, one above the other, wrapped around but separated from a supporting pole. The two are wrapped in opposite directions with the result that the vertical components of the radiation from the two halves cancel each other while the horizontal components are reinforced.

16.14.5 High-gain UHF antennas

The gain of VHF antennas is lower than UHF because, first, the longer wavelengths would lead to impractical dimensions and, second, because the lower power limit for VHF stations reduces the necessity for high gain. A gain of 12 is common for high-band VHF antennas as compared with a gain of 50 or more for UHF antennas.

UHF antenna gains and their accompanying narrow vertical radiation patterns create a number of problems.

High-gain antennas must be adjusted with a high degree of precision because of the leverage involved in the geometry of the radiation path.

Practical antenna radiators can only approximate the theoretical illumination patterns and the antenna patterns must be verified before the antenna is erected.

If the pattern is adjusted so that its main lobe is horizontal, the pattern maximum will be above the surface of the earth and miss it because of the antenna height, even at great distances. This problem can be solved or alleviated by beam tilt.

Nearby areas will receive a weak signal since they will be in the null of the pattern. This problem can be solved by null fill.

16.14.5.1 Beam tilt

The beam is tilted below the horizontal so that it is aimed toward the horizon or slightly below it. The depression angles and distances to the horizon for different antenna heights are:

Height (ft)	Angle (deg)	Distance (mi)
500	0.34	31.6
1,000	0.49	44.7
2,000	0.69	63.2

The beam is often pointed below the horizon, and beam tilts of 1.0° or more are common. Figure 16.16 shows a high-gain pattern with 2° of beam tilt. This pattern is based on uniform illumination of the aperture, but the phase varies linearly from +75° at the top of the aperture to −75° at the bottom.

16.14.5.2 Null fill

The second problem of high-gain antennas, close-in coverage, is solved by null fill, which deliberately increases the signal in the nulls of the pattern to increase ERP at the angles for close-in points. By careful adjustment, the pattern can be shaped

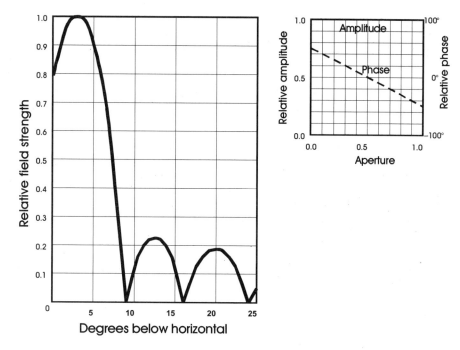

Figure 16.16 Vertical pattern with beam tilt.

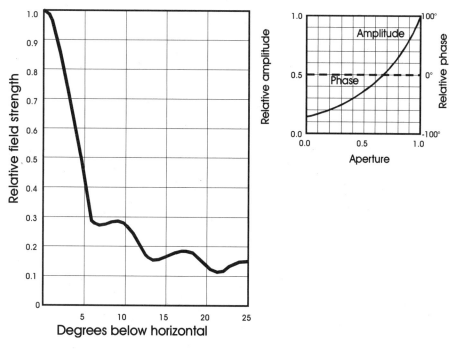

Figure 16.17 Vertical pattern with null fill.

so that the product of distance and $(\text{ERP})^{1/2}$, the indicator of received field strength, remains fairly constant out to a considerable distance. Figure 16.17 shows the radiation pattern of an antenna with null fill. Here the phase is constant, but the relative amplitude rises exponentially from 0.15 at the bottom to 1.0 at the top.

16.14.6 Power ratings

The power-handling limitations of television antennas are established by corona and voltage breakdown and by heating due to ohmic resistance. Both must be taken into account in the design and construction of the antenna, and the manufacturer's published specifications should be carefully followed.

16.15 TV Station Service Areas

A TV station's service area includes the locations, usually contiguous, in which the signal level is adequate to produce a satisfactory SNR in the receiver, and that are not subject to an unsatisfactory level of co-channel or adjacent channel interference.

16.15.1 FCC definitions

The FCC has established standards for both parameters, and they are generally accepted by the industry.

The specification of the signal level required for a service area must recognize that signal levels vary widely and randomly within relatively short distances because of differences in absorption, reflection, diffraction, and refraction at adjacent points on the highly irregular surface of the earth. At greater distances they also vary with time because the refractive index of the atmosphere changes with the weather.

As the result of these variations, it is necessary to define service areas on a statistical basis, and signal levels are usually defined by the term F(50,50). A level of F(50,50) means that the field strength, F, is exceeded at 50% of the potential receiver locations for 50% of the time within a limited area. A receiving antenna height of 30 ft is assumed. F is usually defined in dBμ, the number of decibels above a field strength of 1 μV/m.

The FCC defines three grades of television service: Principal Community, Grade A, and Grade B. These grades are defined in Table 16.10. As shown in the table, a different probability of service is defined for each of these grades, but all are specified in terms of the corresponding F(50,50) values.

The specified values are based on the signal levels that were required to overcome both thermal and man-made noise with the monochrome receivers available in the early 1950s [5]. Modern receivers produce less thermal noise, but viewers have become accustomed to lower noise levels and manmade noise is not affected by receiver performance. Accordingly, as a subjective standard, the signal levels specified in the 1950s are still reasonable.

Table 16.10 Grades of TV service

Service	F(50, 50) (dBµ) Channels		
	2 – 6	7 – 13	14 – 68
Principal community 90% of locations 90% of time	74	77	80
Grade A 70% of locations 90% of time	68	71	74
Grade B 50% of locations	47	56	64

Table 16.10 indicates important differences between the propagation characteristics of the three bands. For example, higher values of F(50,50) are required at the higher-frequency bands because the smaller receiving antennas require greater field strength to produce the same voltage at their terminals. Note that the difference between the F(50,50) value for the principal community and Grade B services is 27 dB for channels 2 to 6 but only 16 dB at UHF. In part this is because the level of manmade noise is higher in the low-band VHF channels than in UHF, particularly in urban areas.

16.15.2 FCC field-intensity curves

The FCC developed and published families of curves on the basis of measurement data that show the values of F(50,50) for a range of distances and transmitting antenna heights. One set of curves is for high-band VHF, channels 7 to 13 and the other, somewhat paradoxically, for low-band VHF and UHF.

The FCC also published F(50,10) curves that show the values of field strength that are occasionally exceeded. These are mainly of value in calculating interference between stations since a signal level that exists only 10% of the time has little value in providing service but can be very annoying as interference. These curves were used in constructing the FCC Table of Assignments.

It must be emphasized that these curves show statistical results only and do not make allowance for specific terrain features. They have little value, therefore, in forecasting the field strength at a particular point at a particular time. With an understanding of this limitation, however, they fill a useful role in estimating the gross features of television coverage.

Figures 16.18 and 16.19 show the FCC F(50,50) curves for high-band VHF and for low-band VHF and UHF. The abscissa is the antenna height, which is defined as its height above the "average terrain," the average elevation in the 2- to 10-mile distance from the antenna.

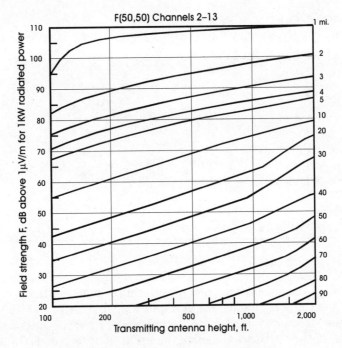

Figure 16.18 FCC F(50,50) field-intensity curves, high-band VHF.

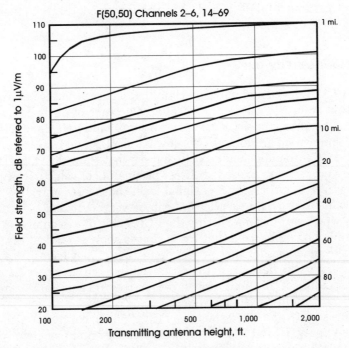

Figure 16.19 FCC F(50,50) curves, low-band VHF and UHF.

The antenna height has a major effect on the received field strength, in part because it increases the distance to the horizon but also because it results in a more favorable phase relationship between the direct waves and those reflected from the earth. As an example, according to the curves for UHF coverage, increasing the antenna height from 500 to 1,000 feet increases the field strength by nearly 8 dB or 8 times.

The predicted distance to the principal community, Grade A, and Grade B contours for antenna heights of 500 and 1,000 feet and the maximum permitted ERP are shown in Table 16.11. Achieving this coverage for UHF stations requires a costly transmitting system. Also, UHF transmissions are more adversely affected in rough terrain or built-up areas because of their shorter wavelength. For the same reason, variations in signal strength within short distances are greater and more rapid. (These problems may be greatly reduced with digital modulation, however; see below.)

16.16 Digital Broadcasting

The digital TV standards adopted around the world call for DTV terrestrial broadcasting. This section describes some of the important considerations.

16.16.1 Current status

In the United States, DTV standards were developed by the Grand Alliance and the ATSC and were adopted by the FCC on December 24, 1996 and placed in the FCC Rules as Section 73.682(d). A phased plan is now being followed for the deployment of digital transmitters throughout the country. First installations are in major cities and began going on-air in November, 1998.

In Europe, the DVB consortium has published its standards for DTV and they are being implemented. The rest of the world is studying the two standards mentioned above and are making decisions about DTV also.

16.16.2 DTV channels

The FCC has ordered that a second set of channels will be assigned in the same areas of the VHF and UHF spectra that are now allocated for NTSC broadcasting.

These channels are assigned to existing broadcasters and are to be used exclusively for DTV transmissions during a 15-year transition period. During this interval broadcasters would provide programming on both their existing NTSC channels and the new DTV channels. At the end of the 15-year period, broadcasting would become all DTV and broadcasters will be required to surrender their NTSC channels.

The DTV channels may be used for any form of digital broadcasting, including SDTV (multiple channels, if desired), HDTV (required to do some of this), interactive TV, and possibly other digital services. It is recognized that the owners of the DTV chanels are primarily TV broadcasters, and are therefore to always provide basic broadcasting service.

16.16.3 Digital format advantages

The use of a digital rather than an analog format for broadcast signals as proposed by the Grand Alliance has a number of major advantages that are summarized here and described more fully elsewhere in this book.

16.16.3.1 Bandwidth utilization

Perhaps the most obvious advantage of digital formats is the major improvement in bandwidth utilization by the use of data compression algorithms. For example, the HDTV display formats included in the ATSC DTV standards would require a video bandwidth of nearly 24 MHz for analog transmission. With digital transmission, the modulated carrier, including video and high fidelity stereo sound signals, can be accommodated in a 6-MHz channel (see Section 3.11). Similarly, a single 6-MHz digital channel can carry four or more SDTV broadcasts simultaneously.

16.16.3.2 Interference immunity

Digital signals are both more benign and more robust than analog, i.e., they both cause less interference to other signals and are less susceptible to interference from other signals. Sufficient tests of these signals have been made to give reasonable assurance that the added DTV channels can be fitted into the spectrum gaps in the NTSC assignment table that were created by adjacent channel and UHF taboo restrictions (see Section 16.2.3).

16.16.3.3 Noise performance

Digital transmission systems have the characteristic (see Section 3.8) that they add no noise to the signal provided the BER remains low enough to be within the range of the error-correction circuits. Above this threshold, a BER of approximately 3×10^{-6} for the encoding formats proposed for broadcasting, the correction circuit loses control and the picture is lost. This is sometimes known as the cliff effect. The result is that signals transmitted digitally are either excellent or they disappear. This is an advantage to most viewers; but fringe area viewers, who receive noisy reception with analog signals, may receive no pictures with digital.

16.16.3.4 Signal distortion

Analog transmission inevitably causes at least some nonlinear and frequency distortion of the signal. Digital transmission should not introduce any beyond that in the signal at the point of ADC.

16.16.3.5 Display format flexibility

In the analog systems that are presently in use, the transmission and display formats are "coupled," which means the display format is rigidly controlled by the transmitted signal. Digital broadcasting is a major change in this policy, and the formats are decoupled. The display format is chosen by the receiver, and it can differ considerably from the transmission format. Digital receivers are designed to accept any of the transmitted formats, but they may display all formats on their internal display format.

16.16.3.6 Support of interactive TV systems

A number of interactive TV systems are now under consideration (see Chapter 15). Any of them can be broadcast over DTV channels.

16.17 DTV Transmission Standards

The ATSC DTV transmission standards are described in this section.

16.17.1 Channel bandwidth

A bandwidth of 6 MHz, the same as for analog NTSC channels, is specified for the DTV channels. Among other advantages of this choice is the simplification of the transition from NTSC to DTV.

16.17.2 Bit rate

A bit rate of approximately 20 Mb/s is specified for DTV transmission. This is high enough (with data compression) to provide the picture definition required for DTV systems. It is low enough to be accommodated in a 6-MHz channel provided a suitable modulation method is used.

16.17.3 Modulation type

8-VSB modulation (see Section 3.7) has been chosen as the modulation type to be used for the DTV channels. The factors included in the choice were the signal speed (bs/Hz), the power per bit (E_B/N_O)[1] required to achieve a satisfactory bit error rate, and the benignity and robustness of the desired signal in the presence of other signals [6].

16.17.4 Channel occupancy

The amplitude/frequency characteristic of a 6-MHz DTV channel was shown in Figure 7.10. It can be compared with the amplitude/frequency characteristic of an NTSC channel as shown in Figure 16.1. The main difference is the absence of the

---

1. E_B = energy per bit; N_O = noise spectral density.

Table 16.12 Aspect ratios

Designation	Frame size (mm)	Aspect ratio
16 mm	10.26 × 7.49	1.37
Super-16	12.52 × 7.42	1.69
35 mm	21.95 × 16.00	1.37
Super-35	24.89 × 18.67	1.33
Cinemascope	(anamorphic)*	2.35
70 mm	48.56 × 22.10	2.20

* Image is expanded 2× horizontally during projection

separate aural and visual carriers. The aural carrier is unnecessary because the audio signal is included in the digital bit stream along with the video. The 8-VSB modulation is transmitted with the carrier suppressed. A small in-phase pilot signal having an amplitude 11.3 dB below the level of the digital bit stream provides the reference for recovering the carrier at the receiver in the presence of noise and interference.

16.18 Display Formats for HDTV (DTV) Systems

The decoupling of the transmission and display formats gives the system designer wide latitude in the choice of display formats. In practice, because of economic factors and the desire to limit the number of receiver models, the ATSC standards limit the transmission format choices to variations of three numbers of active scanning lines (480, 720, and 1,080—see Section 7.4). A choice of progressive and interlaced scanning is offered for 480 and 1,080-line systems, while 720-line system choices all use progressive scanning.

16.18.1 Aspect Ratio

As noted in Chapter 1, the standard aspect ratio for NTSC broadcasts is 4:3 or 1.33:1. The DTV standards has added a wide screen format, 16:9 or 1.78:1, available in any of the line number formats.

A number of aspect ratios are used for films produced for direct projection. They are shown in Table 16.12. All of the wide-screen film formats on this list have an aspect ratio that is much greater than the NTSC TV standard and also, although to a lesser extent, greater than the 1.78 HDTV ratio. These incompatibilities can be handled in one of three ways:

1. A TV image is produced with the same aspect ratio as the film by leaving blank areas on the top or bottom of the TV frame. This is called *letterbox* transmission. This conversion can be accomplished photographically by producing a film print for TV reproduction.

2. The full film frame height is reproduced and the sides of the image are cropped

Table 16.13 HDTV receiver scanning standards

Aspect ratio	16:9
Progressive scanning alternative	
Active scanning lines	720
Pixels per scanning line	1280
Field/frame rate	60/s
Interlaced scanning alternative	
Active scanning lines	1080
Pixels per scanning line	1920
Field rate	60/s
Frame rate	30/s

to produce the correct aspect ratio. This can also be done photographically.

3. The horizontal location of an aperture on the film image having the TV aspect ratio is moved dynamically so that the portion of the image with the greatest interest is reproduced, a process called *pan and scan*. This can be done while making a film-to-tape transfer. This method produces the best results, but it is laborious and it requires excellent editorial judgment.

None of these methods is ideal, and it is expected that films made specifically for HDTV will be produced with a 16:9 aspect ratio.

16.18.2 Scanning standards

The decoupling of the DTV transmission and display formats in the ATSC standards is reflected in the flexibility of the receiver's scanning format. For example, any transmission format permits the receiver to choose either progressive or interlaced scanning for the display. Two receiver scanning choices that represent the range available are tabulated in Table 16.13. The transmission standards in this example could be either progressive or interlaced, and the conversion, if necessary, is made at the receiver.

16.19 DTV Channel Assignments

In January 1995, the *Association for Maximum Service Television* (AMSTV) published a proposal for the assignment of 1691 DTV channels to existing broadcasters and license holders, although final criteria for these assignments had not been established. They were chosen with the aid of a computer program that had three goals:

1. to provide a DTV channel for each current NTSC channel
2. to provide a DTV service area (see Section 16.20) that was at least comparable with that of the NTSC station

Table 16.14 NTSC/DTV comparative service

	NTSC		HDTV	
	VHF	UHF	VHF	UHF
Transmitter ERP	10 kW*	500 kW*	0.63 kW**	31.6 kW**
Sites measured	169	199	169	199
Satisfactory reception	39.6%	76.3%	81.7%	91.5%

* Peak power
** Average power

3. to minimize interference to existing NTSC service

The proposed assignments are overwhelmingly, although not quite totally, in the UHF portion of the spectrum.

16.20 DTV Station Coverage

The most complete data currently available for calculating the coverage of DTV stations are included in a set of measurements made in Charlotte, North Carolina, in September 1994 under the sponsorship of the Grand Alliance [7]. It compared the coverage of 8-VSB digital broadcast signals and NTSC signals in urban and rural areas in the vicinity of Charlotte. Measurements were made both at VHF (Channel 6) and UHF (Channel 53).

The results of the measurements are summarized in Table 16.14. The HDTV average power was adjusted to be 12 dB below the NTSC peak power, which is the ratio of the average to the peak power of an NTSC signal with typical picture content. This ratio was estimated to provide "equivalent" service with the two modulation methods.

The criterion for "satisfactory" reception for DTV transmissions in Table 16.14 was a bit error rate of less than 3×10^{-6}, the point at which it is expected that the error-correction circuits would become ineffective. For NTSC transmission, the criterion was subjective, and was chosen to be the point at which expert observers judged noise and interference to be "slightly annoying."

The measurements were not sufficiently extensive to be totally conclusive, for example they did not cover extremely densely populated urban areas such as Manhattan, but they give a strong indication that the coverage with digital transmission will be as good or better as with analog. (They do not reveal, however, that digital signals will either be excellent or absent while analog signals suffer "graceful degradation"; some degree of analog reception might be achieved at locations having a SNR where digital signals might disappear.)

The comparatively poor performance of the VHF NTSC signals was attributed to a number of local factors including interference from a cable system and an educational FM station.

16.21 DTV Channel Utilization Alternatives

The 19.3 Mb/s signal speed in the HDTV channel, together with digital data compression, opens up a wide variety of possibilities in addition to HDTV for the utilization of this channel. They have not been fully explored at the present time, but a few of the options are:

1. With additional compression, a SDTV channel in addition to HDTV
2. Two or more SDTV channels
3. A multiplicity of digital data channels

The decision as to the uses of these extra channels involves issues of economics and public policy that are beyond the scope of this book.

16.22 DTV Transmitters

Since the DTV channels are in the same portions of the VHF and UHF spectrum as the NTSC channels, there are some similarities in the designs of NTSC and DTV transmitters, but there are also major differences as described in the following sections.

16.22.1 Visual and aural channels

NTSC transmitters have separate aural and visual power amplifiers. In a DTV transmitter, the aural and visual bit streams are combined at the Transport layer (see Section 7.6) and only a single power amplifier is required.

16.22.2 Signal processing and carrier modulation

The processing of analog signals for transmission and the modulation of carriers in NTSC transmitters is accomplished in the exciter-modulator (see Section 16.8.1). The performance of equivalent functions in DTV transmitters is described in Section 7.7.3 and is considerably more complex, although it is amenable to the design of digital ICs. Thus, a digital exciter-modulator might seem simpler than an NTSC one, because the complexity is hidden inside of the ICs.

16.22.3 Transmitter power ratings and levels

Digital signals are characterized by a high ratio of peak-to-average power. The power rating of digital transmitters, therefore, must specify both peak and average power.

16.22.4 Transmitter performance standards

The need for good linearity in amplitude-modulated analog transmitters is obvious, since distortion of the signal waveform causes direct distortion of the gray scale. The importance of linearity in digital signals is not so obvious, but for the ATSC digital modulation system, it is equally important.

With 8-VSB signals, the amplitude as well as the presence or absence of pulses is significant in transmitting information and, like noise, amplitude nonlinearity can cause bit errors. Nonlinearity can also create cross-modulation products that are seen by the system as noise. The overall effect of nonlinearities, therefore, is to create noise in the signal that is added to the ordinary transmission noise. This causes the threshold bit error rate to be reached at a lower level of transmission noise, which in turn results in a reduction of the range of the transmitting system.

16.23 References

1. *Broadcasting and Cable*, January 16, 1995.
2. Donald G. Fink and Donald Christiansen (eds.), *Electronics Engineers' Handbook*, 3d edition, McGraw-Hill, New York, 1989, Chap. 21, para. 41.
3. Electronics Industry Association, EIA/TIA Standard, Electrical Performance Standards for Television Transmission Systems, EIA/TIA-250-C (1990).
4. K. Blair Benson (ed.), *Television Engineering Handbook*, McGraw-Hill, New York, 1986, Chap. 8 (for a more complete description of antenna echoes).
5. Federal Communications Commission, "Sixth Report and Order," Docket Nos. 8736, 8975, 8976, 9175, April 11, 1952.
6. Grand Alliance, "Transmission Subsystem Comparison Testing (January 13–February 11, 1994)," FCC Advisory Committee on Advanced Television Service.
7. Jules Cohen and Victor Tawil, "Testing of the Grand Alliance HDTV Subsystem," *J. SMPTE*, May, 1995.

Chapter 17
CATV Systems

17.1 A Brief History of CATV

Cable television (CATV) began in the very late 1940s as community TV, a system for bringing television service to small communities that did not receive adequate off-the-air signals in individual homes, either as the result of their distance from broadcasting stations or because of intervening hills or other terrain features. Local TV dealers or other entrepreneurs erected high-gain antennas on mountain tops or towers to provide usable signals to cable systems that delivered it to local homes.

Within a few years the variety of program services available to community television systems was increased by importing signals from distant stations by microwave, and service was delivered to communities far outside the normal service areas of the originating stations.

In the late 1970s, communication satellites made it possible for cable systems to receive programs from all over the world. The variety of available programming became great enough to be attractive to viewers in large cities, even though they had several off-the-air sources available. At this time community TV was no longer descriptive of the service, and cable TV or CATV became the accepted nomenclature. CATV is now a large business, and more than 50 million homes in the United States are receiving programs by cable.

Looking to the future, the introduction of digital technology will improve the performance of CATV systems, increase capacity, and extend functions to include interactive multimedia services.

17.2 Cable System Elements

Figure 17.1 shows the elements of an analog CATV system. The incoming signals—off-air, microwave, satellite, and locally originated—are received at the *head end*. Except for the off-air signals, they are in the baseband mode and modulate

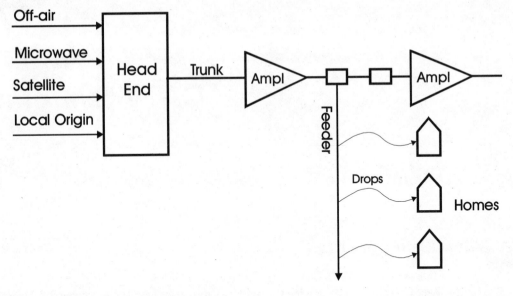

Figure 17.1 Cable system elements

carriers in accordance with broadcast standards. The carrier frequencies of the off-air signals are usually shifted before multiplexing with the other signals.

The cable distribution system is a combination of trunks, feeders, and drops. *Trunks* are cables that feed major portions of the service areas; *line amplifiers* are used to extend the length of trunks. *Feeders* are connected to trunks to serve individual streets, and *drops* make the connections to individual homes. At the present time, with the exception of drops that may employ flat twin lead, these system components use coaxial cable or fiber optics.

Some systems have *super trunks* to interconnect more distant parts of the system or even different systems. Super trunks may use coaxial cable, fiber-optic cable (see Chapter 19), or a type of microwave known as an *amplitude modulated link* (AML).

17.3 Cable Television Channels

The first community antenna systems employed the five low-band VHF channels, 2 to 6, because of the low attenuation of cable at these frequencies (see Section 17.4.2.2). As the demand for channels grew, systems expanded into the high-band VHF, channels 7 to 13. The spacing between line amplifiers had to be reduced, but unmodified standard receivers could be used.

As the demand for channels grew still further, the industry had the choice of adding channels in the UHF spectrum or utilizing the channels between low-band and high-band VHF and between high-band and UHF. The latter course was chosen because of extremely high cable attenuation in the UHF band.

A standard CATV configuration uses frequency division multiplex with fifty-

Table 17.1 CATV channels

Channel Designation	Visual Carrier MHz	Channel Designation	Visual Carrier MHz
Low-band VHF		*Superband*	
2	55.25	J	217.25
3	61.25	K	223.25
4	67.25	L	229.25
5	77.25	M	235.25
6	83.25	N	241.25
		O	247.25
Mid-band VHF		P	253.25
A-2	109.25	Q	259.25
A-1	115.25	R	265.25
A	121.25	S	271.25
B	127.25	T	277.25
C	133.25	U	283.25
D	139.25	V	289.25
E	145.25	W	295.25
F	151.25	*Hyperband*	
G	157.25	AA	301.25
H	163.25	BB	307.25
I	169.25	CC	313.25
High Band VHF		DD	319.25
7	175.25	EE	325.25
8	181.25	FF	331.25
9	187.25	GG	337.25
10	193.25	HH	343.25
11	199.25	II	349.25
12	205.25	JJ	355.25
13	211.25	KK	361.25
		LL	367.25
		MM	373.25
		NN	379.25
		OO	385.25
		PP	391.25
		QQ	397.25
		RR	403.25

five adjacent channels in the NTSC format, the highest located just below the UHF band. More channels can be added by expansion into the UHF band (Table 17.1).

In addition, four 6-MHz channels are available in the 5- to 30-MHz region that are sometimes used for upstream transmissions to provide two-way capability. The region from 30 to 55 MHz is reserved as a guard band to separate the upstream and downstream channels.

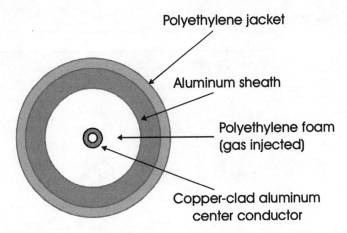

Figure 17.2 Coaxial cable construction. (*From E. R. Bartlett,* Cable Television Technology and Operation, *McGraw-Hill, New York, 1990.*)

Adjacent channels are not assigned to the same geographic location in off-the-air broadcast practice because receivers may not be able to discriminate between a weak desired signal and a strong undesired one on an adjacent channel. In CATV systems, however, adjacent-channel signals can be maintained at approximately equal levels and adjacent-channel reception is possible. It requires very careful control of signal levels to avoid adjacent-channel interference and of system linearity to avoid cross-modulation between channels. This is an inherent weakness of analog systems that will be solved by the introduction of digital transmission.

Most of the channels in the CATV configuration cannot be received directly by a standard TV set designed for broadcast reception. Set-top converters (see Section 17.11) heterodyne all incoming channels to a single output channel, usually channel 3 or 4. Alternatively, *cable-ready* receivers that can tune to the cable channels are also available.

17.4 Coaxial Cable

Coaxial cable is most commonly used for the distribution of CATV signals.

17.4.1 Physical construction

The physical construction of a CATV coaxial cable is shown in Figure 17.2. The inner conductor is copper-clad aluminum. Copper, with its low resistivity, is used on the outside of the inner conductor where the current density is highest. Aluminum is used for the outer conductor and core of the inner conductor because of its light weight.

The insulating material between conductors is polyethylene foam. Gas is sometimes injected to prevent water from migrating through the insulation. A flexible steel tubing is sometimes placed over the polyethylene jacket for mechanical protection. Standard cable diameters are 0.50, 0.75, 0.875, and 1.00 in. Cables, particularly in the larger diameters, are only semiflexible, and the bending radius is an important specification.

Cable expands and contracts with temperature change, approximately 23 parts per million per degree Celsius, and provision must be made to allow for this in the installation of the cable. The problem is complicated by the fact that the steel cable (called a messenger) that supports an overhead cable expands at a different rate, and allowance also must be made for that.

17.4.2 Electrical characteristics

The important electrical characteristics for CATV cables are characteristic impedance, attenuation, return loss, and loop resistance. These are discussed below.

17.4.2.1 Characteristic impedance

The characteristic impedance of flexible transmission line has the same electrical significance as for the solid-tube lines used to connect transmitters to antennas (see Section 16.6). The standard characteristic impedance for CATV transmission line is 75 ohms.

17.4.2.2 Attenuation

The signal is attenuated as it passes down the cable owing to ohmic losses. The amount of loss depends on the diameter of the cable, the frequency, the standing wave ratio, and the temperature. The attenuation with a VSWR of 1.0 and a temperature of 68°F is shown in Table 17.2. Note that the attenuation (in dB) is approximately proportional to the square root of the frequency. The attenuation increases with temperature at the approximate rate of 1.8% per 10°C change.

17.4.2.3 Return loss

The *return loss* is a measure of the deviation of the actual value of the cable's characteristic impedance from its nominal value. It is equal to the ratio of the incident power, P_I, to the reflected power, P_R, in decibels:

$$L_R = 10 \log \frac{P_I}{P_R} \qquad (17.1)$$

The closer the actual characteristic impedance of the line matches its nominal value, the smaller the P_R and the larger the return loss. For a perfect match, $P_R = 0$ and $L_R = $ infinity. No line is perfect, and practical values of L_R vary from 28 to 32 dB. If the return loss is too low, reflections occur along the line causing variations in voltage or signal ripple.

Table 17.2 Cable attenuation (dB/100 ft)

Frequency (MHz)	Cable Diameter (in)			
	0.500	0.750	0.875	1.000
5	0.16	0.11	0.09	0.09
30	0.40	0.26	0.24	0.23
50	0.52	0.35	0.32	0.30
110	0.75	0.52	0.47	0.44
150	0.90	0.62	0.55	0.52
180	1.00	0.68	0.60	0.59
250	1.20	0.81	0.72	0.68
300	1.31	0.90	0.79	0.75
400	1.53	1.05	0.91	0.87
450	1.63	1.12	0.98	0.92
500	1.70	1.17	1.04	0.97
550	1.80	1.23	1.09	1.02
600	1.87	1.29	1.15	1.06

17.4.2.4 Loop resistance

The primary power for the line amplifiers that are installed to compensate for cable attenuation is usually supplied by low-voltage ac or dc transmitted along the cable with the RF signal. Since the amplifiers operate at relatively low voltage, typically 45 V, and high current, the loop resistance, the combined resistance of the inner and outer conductors to dc or power-frequency ac, is an important cable specification.

The loop resistance of coaxial cable with copper-clad aluminum inner conductors is slightly greater than all-copper conductors as shown in Table 17.3.

17.5 Line and Bridging Amplifiers and Equalizers

Line amplifiers are inserted in the cable at intervals to compensate for attenuation, and their specifications are a key element in determining the performance of

Table 17.3 Loop resistance, coaxial cable, ohms/1000 ft, 20°C

Outer diameter (in)	Center conductor	
	Solid copper	Copper-clad aluminum
0.412	1.75	1.93
0.500	1.23	1.68
0.750	0.56	0.76

a system. Each amplifier includes an equalizer that compensates for the increase of cable attenuation with frequency.

Bridging amplifiers have a high-impedance input and are used to tap the signal from a trunk without disturbing its performance.

17.5.1 Technical requirements

The technical requirements for amplifier/equalizer combinations are demanding because of the cumulative degradation of a large number of amplifiers in series:

1. They must operate over a wide frequency range, and their gain must be high enough at the top of the frequency range to enable the amplifiers to be separated by a reasonable distance.
2. The equalizers must compensate precisely for the variations in cable attenuation with frequency.
3. The linearity of the amplifiers must be good to minimize crosstalk and the generation of beat frequencies between channels.
4. Automatic gain and frequency response, (slope correction) must be provided to compensate for variations in these parameters with temperature or equipment aging (see Section 17.5.3).
5. The SNR of individual amplifiers must be high enough so that the cascaded noise level of all the amplifiers in a trunk meets industry standards (see Section 17.8).

17.5.2 Gain-frequency response

The gain of the amplifiers in typical amplifier/equalizer combinations is essentially constant at 20 dB from 50 to 450 MHz. Assuming that the cable loss between amplifiers increases from 6.9 dB at 50 MHz to 20 dB at 450 MHz, the equalizer is designed to reduce the gain of the combination to 6.9 dB at 50 MHz so that the frequency response of the transmission line/amplifier-equalizer combination is approximately constant over the entire frequency band.

17.5.3 Automatic gain and slope control

If a system has many amplifiers in series, a small change in the gain or frequency response of each amplifier resulting from a common cause such as a temperature change will produce a major change in the gain and frequency response of the system because of the compounding effect. To maintain these parameters within close tolerances, inverse feedback is introduced by the addition of pilot tones near the bottom and top of the frequency band. Using these tones as references, the gain and frequency response of each amplifier is automatically maintained at a stable level.

17.5.4 Signal-to-noise ratio

The system noise amplitude is the root sum square of the noise amplitudes of the system components. The SNR and carrier-to-noise ratio (CNR) are usually expressed in decibels, and the SNR for n identical cascaded amplifiers, each with a SNR of S_O/N_O is:

$$\text{SNR}n = \text{SNR}_O - 10 \log n \tag{17.2}$$

17.5.5 Distortion and cross-modulation

Nonlinearities in system amplifiers can cause distortion, both of the amplitude-modulated carrier envelope and of the carrier itself. The most serious effects of envelope distortion are differential gain and differential phase distortion (see Sections 5.10.3 and 5.10.4). Distortion of the carrier causes "beats" and other cross-modulation effects between the closely spaced channels. The beat frequencies retain some of the modulation characteristics of the original signals. With 50 or more channels, the possibilities for the production of spurious signals by cross-modulation are almost endless. A common type is a triple beat caused by the interaction of the signals on three channels.

A number of rules, some empirical and others based on theory, have been developed for estimating the cumulative distortion of a number of amplifiers in series. If $d_1, d_2, \cdots$ are the distortions of successive elements in a system, the total distortion, D, may be the linear sum:

$$D = d_1 + d_2 + d_3 + \cdots \tag{17.3}$$

or follow a 3/2 power rule:

$$D = \left(d_1^{\frac{3}{2}} + d_2^{\frac{3}{2}} + \ldots\right)^{\frac{2}{3}} \tag{17.4}$$

or the root-sum-square rule:

$$\sqrt{d_1^2 + d_2^2 + \ldots} \tag{17.5}$$

Differential gain and phase distortion increase in accordance with the 3/2 rule.

17.6 Trunk Circuits

The frequency response of a trunk circuit is the product of the responses of its individual components.

Line amplifiers divide trunk circuits into segments, each having the length at which the attenuation at the highest transmission frequency is equal to the gain of the amplifiers. Equalizers are added that reduce the gain at lower frequencies so that the gain of each cable-segment amplifier/equalizer combination is 0 dB. As an example, assume the system bandwidth is 50 MHz to 450 MHz, the gain of each amplifier is 20 dB, and the cable attenuation rises from 3.1 dB/1,000 ft. at 50 MHz to 10 dB/1000 ft at 450 MHz. The spacing between amplifiers would be (1,000

Table 17.4 CATV system performance standards

	FCC 76.605	NCTA	EIA RS250B
Frequency response	±2 dB	—	±0.7 dB
Differential gain	—	10%	4%
Differential phase	—	5°	1.5°
Cross modulation (C/XMO)	46 dB	49 dB	—
Composite triple beat (C/CTB)	—	52 dB	—
Second-order beat (C/2ndOB)	—	60 dB	—
Co-channel interference	36 dB	—	—
CNR (unweighted)	36 dB	44 dB	—
SNR (weighted)	—	55 dB	56 dB

ft.) × (20 dB/10 dB) or 2000 feet. An equalizer is added to each amplifier with an attenuation varying from 6.9 dB at 50 MHz to 0 dB at 450 MHz. If the slopes of the equalizers are properly designed, the combined gain of the cable, amplifier, and equalizer will be 0 dB over the entire band.

17.7 Reverse Path Circuits

It is sometimes desired to bring signals from remote program sources back to the head end in a reverse direction. This can be accomplished on the frequencies below channel 2 (6.0 to 30.0 MHz) where there is space for four 6-MHz channels. These channels have the advantage of lower attenuation but the disadvantage of higher ambient noise levels.

Special care must be exercised in the design of two-way systems. Every drop to a receiver is a noise source, and every return path must be blocked except the one carrying the desired signal. Otherwise the SNR of the system drops to a level that makes it unusable.

17.8 Video Performance Standards

As the cable TV industry has matured, various industry and governmental organizations have issued performance standards for CATV systems. Table 17.4 summarizes the basic performance standards that have been issued by the FCC, the NCTA, and the EIA for the transmission of NTSC signals over cable. It illustrates the fact that there is not complete industry agreement on performance standards or on parameters that should be specified.

The standards for carrier-to-cross-modulation products and co-channel interference in part 76.605 of the FCC rules are particularly important specifications for CATV systems because there are no guard bands between channels.

Like many industry standards, Table 17.4 is a compromise between what is desired and what is realistically possible in the current state of the art. Meeting them requires meticulous adjustment and maintenance of the system with very little "headroom" for error. The introduction of fiber-optic cables and/or digital transmission will make it possible to bring about significant improvements in CATV system performance (see Section 17.14) and also to make it more insensitive to minor misadjustments.

17.9 Head Ends

The head end of a CATV system collects the signals from all its program sources, processes them for transmission, generates and modulates channel carriers for signals that are in baseband form, shifts the channel frequency where necessary, and combines the resulting channels for transmission on the system.

17.9.1 Head-End Signal Sources

The enormous expansion of CATV service has been made possible by satellite-transmitted signals, but the distribution of off-air broadcast signals continues to be essential to its popularity. A CATV system head end, therefore, requires high-quality receiver-antenna combinations for off-air pickup.

17.9.1.1 Broadcast channel antennas and receivers

The antenna must be located at a sufficient height to receive strong signals from the desired broadcast stations. It may be a commercial version of an all-channel home antenna, although the all-channel feature is often not satisfactory—one problem is that the off-air stations may be located in different directions—and, therefore, most CATV systems use single-channel, high-gain antennas or antennas that receive only a few channels.

The most common antenna formats are the familiar YAGI, which is frequently used in home antennas, and the *log periodic* in which each element is driven, thus giving better control over the antenna's directivity and bandwidth.

The principal components of the receiver are the preamplifier and signal processor. If the field strength is weak at the antenna location, the SNR of the system can be improved by placing the preamplifier at the antenna location. The signal at the output of the preamplifier is converted to the IF frequency for further amplification and signal processing. The IF signal is then heterodyned to the desired RF channel where it is mixed with other signals for transmission and distribution over the system.

17.9.1.2 Video, microwave, and satellite signals

The signals from local program sources, microwave systems, and satellite systems are in the video and audio baseband formats, and after processing they modulate channel carriers, using a separate modulator unit for each channel. The modulated signals should meet FCC standards for broadcast signals.

17.10 Set-Top Converters

Set-top converters provide an interface between the cable system and a standard television receiver.

17.10.1 Converter functions

A converter's original function was to shift the frequencies of the incoming cable channels on a tunable basis to a single output channel. This is done by a heterodyning process and it means that cable channel selection is done by controlling the converter while the TV receiver remains on the specified single channel. Subsequently, the functions of set-top converters were expanded as described below.

17.10.1.1 Increased system capacity

Because of the converter's heterodyne frequency shifting described above, cable systems are not limited to broadcast channels for signal distribution. This allows 55 channels in the 450-MHz cable bandwidth as listed in Table 17.1.

17.10.1.2 Off-air signal interference reduction

Cable signals are subject to interference from off-air signals picked up by imperfectly shielded receivers. Since cable signals travel more slowly than off-air signals, there can be a time difference of a few microseconds between the cable signal and the off-air signal at the receiver. This may cause ghosts or, in less severe cases, a smearing of horizontal detail in the picture. With a converter in use, the receiver is tuned to a different channel than the original broadcast channel and off-air interference is greatly reduced or eliminated.

Further protection from off-air signal interference (as well as interference from other cable signals) is achieved by using a double heterodyning process in the converter (Figure 17.3).

17.10.1.3 Design objective conflicts

There is a major conflict in the sensitivity design objectives between off-air and cable operation of receivers. Off-air reception needs high sensitivity since signals may be weak, but the FCC channel assignment policies protect an off-air receiver from adjacent-channel interference. By contrast, cable reception does not require high sensitivity because cable signals are strong, but with 55 channels located side by side, a cable receiver must have excellent selectivity. The converter's design is optimized for cable reception, thus eliminating this requirement for the receiver.

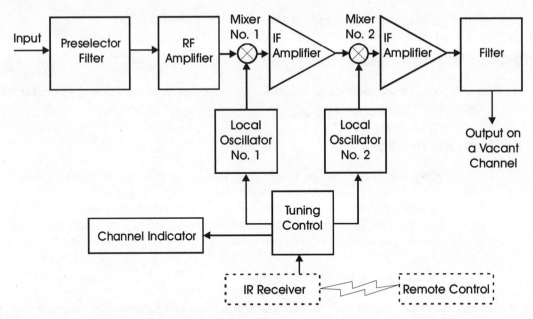

Figure 17.3 Block diagram of a set-top converter.

17.10.1.4 Pay-TV services

Subscriber and pay-per-view services require a means for controlling access to signals. Set-top converters with controlled descramblers provide this capability.

17.10.1.5 Digital signal distribution

It can be expected that cable systems will eventually support digital distribution because of its many advantages (see Section 17.15). The conversion will be costly, but it will occur as the public buys digital broadcast receivers to receive broadcast DTV. Of course, a digital cable system can be used with analog receivers by use of set-top converters equipped with digital decompression and DAC circuits. This approach limits the performance to that of the analog receiver, so there is incentive for all-digital receivers that can offer the higher picture and sound quality inherent in DTV systems.

17.10.1.6 Interactive and other special services

A number of interactive services for CATV are under consideration (see Chapter 15 and Section 17.15). This will become even more important with DTV because the new standards provide capability for any type of interactive content to be included in the transport bit stream. Interactivity requires an interface at the receiver, which may become a function of set-top converters.

CATV Systems

17.10.1.7 Set-top converter summary

Since the introduction of set-top converters in the early history of CATV, their importance has grown enormously. Once considered to be a temporary expedient to increase system capacity and reduce interference from off-air signals, they are now an essential component of cable systems because of their many additional useful purposes. This is indicated by the fact that there are 25 to 30 million converters in use (not including VCRs, which also have many converter functions—see Chapter 10).

17.10.2 Converters vs. cable-ready receivers

All of the functions of set-top converters can be incorporated in receivers, and many recent receiver designs include at least the basic converter features that allow the nonbroadcast cable channels to be received. Such receivers are known as *cable-ready*. However, many of the additional features listed above are specific to particular cable systems, so it has so far been difficult to include them in standard TV receivers. Efforts are being made in the development of digital systems to include features such as access control in a standardized way, so that they may be included in future digital receivers and eliminate the need for the set-top converter.

17.10.3 Converter operation

Figure 17.3 shows a block diagram for a set-top converter of moderate complexity. The input from the cable system is first passed through a preselector filter that removes signals outside the CATV passband. The signal is then amplified and passes through two mixers that perform double heterodyning to provide image channel rejection and increased selectivity. The output of the second mixer passes to the receiver through a second filter that is tuned to the output channel. The infrared (IR) remote control shown in the figure is an optional feature.

Three optional features available on some converters are also not shown in the figure. The first is a connection for an upstream signal at the converter input (see Section 17.3). The second is a baseband video output obtained by detection of the signal from the second mixer, and the third is a descrambler for pay-TV services.

17.11 Pay-TV Systems

Most cable systems offer a choice of pay-TV services, usually known as basic, premium subscription, and pay-per-view. Basic service is offered for a fixed monthly base rate and comprises a group of channels, ordinarily including off-the-air broadcast signals, for which there is no extra charge. Premium subscription service is optional and includes channels that carry special programs for an additional monthly charge. As its title indicates, pay-per-view service is offered for individual programs for which customers are billed.

All pay-TV services must provide a means for limiting the availability of service to subscribers who have paid for it. This can be done rather simply for premium service, which is billed on a monthly basis, by *trapping*—installing a filter on the subscriber's drop that blocks premium channels for nonsubscribers or by *jamming*, adding a signal between the sound and picture carrier that is removed by a trap for desired signals. A serious disadvantage of these techniques is that a technician must pay a costly visit to the subscriber's premises to add or disconnect service.

More sophisticated systems eliminate the requirement for visiting subscribers' homes by *scrambling* the signal, i.e., making it unusable unless it is descrambled. The set-top converters are provided with addressable decoders that descramble the signal upon command from the system head end. This system can be used either for premium channels or for pay-for-view.

A variety of scrambling techniques have been developed that differ in security and cost. An example is sync attenuation, in which the amplitude of the horizontal sync pulses is reduced for transmission sufficiently to make the signal unusable until it is restored by the descrambler.

With the new DTV signal standards, scrambling systems of even greater sophistication can be introduced with an increase in flexibility and security.

17.12 Wireless Cable

Wireless cable is an oxymoron that is used to describe cable-like television distribution systems that are transmitted by radio rather than cable [1].

17.12.1 Wireless cable services

Wireless cable services include four point-to-multipoint media for program distribution—(*Instructional Television Fixed Service* (ITFS), *Multipoint Distribution Service* (MDS), *Multipoint Multichannel Distribution Service* (MMDS), and (*Operational Fixed Service*) (OFS)—and one multichannel point-to-point service, *Cable Television Relay Service* (CARS). All of these services use single-sideband amplitude modulation with 6-MHz bandwidths. They differ in their spectrum allocations and usage authorizations.

17.12.1.1 ITFS

The ITFS was first authorized in 1963 for the distribution of instructional and educational programs to schools and colleges. Twenty channels are allocated for ITFS, but they are not all used full-time for instructional purposes. FCC rules permit them to be leased for commercial use by their licensees with the proviso that they be used part-time for the distribution of a minimum amount of instructional and educational programs (20 hours per week per channel two years after they begin service).

The requirement for part-time use of these channels for education means that a commercial lessor cannot have full-time use of a channel. This problem is solved by channel sharing. A channel map is prepared that indicates the time periods during which each channel is in use for educational programming. Full-time schedules can be developed for commercial services by combining available time periods from two or more channels. The program being transmitted is switched to an unused channel and the subscribers' receivers are automatically retuned to it. Several channels may be used in the course of a single program. With this technique, as many as three full-time commercial services can be distributed by four ITFS channels.

17.12.1.2 MDS

MDS was first authorized in 1969 with two channels, MDS-1 that was available in the entire United States and MDS-2 that was available in the fifty largest metropolitan areas. It was used primarily for the distribution of pay-TV programming. It was so successful that there were strong pressures for the authorization of additional channels for commercial programs. The FCC made three responses to these pressures:

1. The part-time use of ITFS channels for commercial programs was authorized (see above);
2. Eight ITFS channels were reallocated to the MMDS service (see below); and
3. Three channels from the operational fixed service portion of the spectrum were authorized for commercial program distribution.

MDS is allocated 10 MHz of spectrum space from 2150 to 2160 MHz.

17.12.1.3 MMDS

Eight channels are allocated for MMDS full-time commercial use. They were made available for commercial use in 1983 by the reallocation of ITFS channels. They are very much in demand, and obtaining a new assignment is difficult if not impossible in many areas.

MMDS is allocated 48 MHz (eight amplitude-modulated TV channels) from 2596 to 2644 MHz. It is used for the distribution of television programs in sparsely populated areas where cable is uneconomic.

17.12.1.4 OFS

Three channels were allocated in 1983 for commercial point-to-multipoint distribution of television programs in the operational fixed service portion of the spectrum. Like the MMDS channels, they are very much in demand.

17.12.1.5 CARS

CARS is an inexpensive microwave service for transmitting a large number of CATV program signals from one site to another, e.g., from head end to head end. The amplitude modulated CATV channels from 112 to 300 MHz are heterodyned

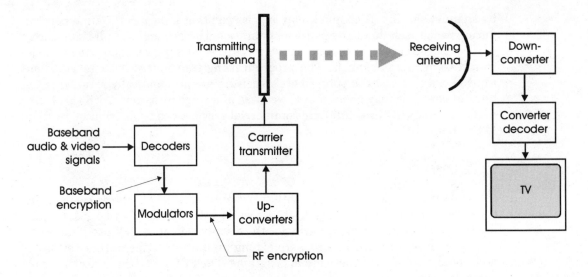

Figure 17.4 Configuration of a wireless cable system.

upward to 12,758.5 to 12,946.5 MHz and transmitted as microwave channels. This is called an *amplitude modulated link* (AML). For relatively short distances, an AML is usually more economic than cable or conventional microwave service.

17.12.2 Spectrum allocations

The spectrum allocations of wireless cable distribution services all have channel bandwidths of 6 MHz, and their modulation format—vestigial sideband for video and FM for audio—is identical with that in CATV systems except that the position of the video and aural subcarriers is inverted in the two MDS channels where the aural subcarrier is below the visual.

17.12.3 Wireless cable distribution system configuration

The configuration of wireless cable distribution systems is shown in Figure 17.4. At the head end, it is identical with most CATV systems until the signals of the combined channels are delivered to the carrier transmitter. This is typically a highly linear broad-band device that amplifies eight channels simultaneously, rather than a cable network. The linearity is required to avoid the production of intermodulation products by the simultaneous amplification of multiple channels.

The transmitter power level is usually 10 to 100W, typically generated by a GaAsFET (gallium arsenide, field effect transistor) amplifier.

The connection from the transmitter to transmitting antenna may be by wave guide or cable.

Since propagation of signals at the MMDS/IFTS frequencies is essentially line-of-sight, the transmitting antenna must be high enough to clear major terrain and man-made obstacles. Either horizontal or vertical polarization may be used.

Like a broadcast antenna, a wireless cable transmitting antenna is usually highly directional in the vertical plane (which requires that it have a large vertical dimension). Its vertical beamwidth may be so narrow that beam tilt is required to cover close-in subscribers as with high-gain UHF broadcast antennas (see Chapter 16). The antenna's directivity in the horizontal plane depends on the angular sector coverage; it may be from 90° to 360°. The antenna gain is determined by the combined vertical and horizontal directivity and is typically 15 dBi.

The receiving antenna located at the subscriber's premises should be high enough to clear local obstructions. It usually consists of a dipole-reflector combination, but unlike the transmitting antenna it is usually directional in the horizontal plane since it receives signals from a fixed direction. Typical receiving antenna gains are 18 to 24 dBi.

The receiver antenna feeds a block downconverter, which should be mounted near the antenna to minimize losses in the cable to the receiver. It translates the carrier frequencies to the standard CATV channel configuration (see Section 17.3). At this point the signal is functionally identical to a subscriber feed from a CATV system. Pay-TV signals pass through a decoder at this point.

17.12.4 System coverage

In the absence of terrestrial obstacles, the theoretical distances (miles) to the 45, 50, and 55 dB SNR contours with fresnel zone (see below) path clearance and typical system parameters are shown in Table 17.5. Note that these distances are theoretical maxima, and they assume a path that is far enough above the line-of-sight to provide fresnel zone clearance. This is given by Eq. (17.6):

$$F = 72.1 \sqrt{\frac{d(D-d)}{fD}} \qquad (17.6)$$

where F = fresnel zone in feet, D = path length in miles, f = frequency in gigahertz, and d = distance from the point of minimum path clearance to the nearest path end. As an example, the height of the fresnel zone on a five-mile path with the minimum path clearance at the path mid-point and a frequency of 2.5 GHz is 51 ft. With a 20-mi path it would be 102 ft.

If, as is often the case, the path clearance is less than the fresnel zone, the signal will be substantially attenuated [2]. If there is a major obstruction in a path, it sometimes can be cleared by a *beam bender*. This is a low power repeater that picks up the signal at the top of the obstruction and retransmits it into the shadowed area.

17.13 Digital Transmission and Fiber-Optic Cable

The principal technologies used by current CATV systems—analog signal formats, FDM modulation, and coaxial cable distribution—have served the industry well

Table 17.5 Distance (mi) to SNR contours with fresnel zone path clearance

	Transmitter power (W)			
	100		50	
Receive antenna gain (dBi)	24	18	24	18
SNR contour (dB)				
45	60	30	36	19
50	33	17	20	10
55	18	9	12	6

Transmit antenna gain   15 dBi
Antenna/downconverter noise figure   7.5 dB
Operating frequency   2.6 GHz

for more than 40 years. They have limitations, however, that are being overcome by newer technologies that will replace or supplement them in the years ahead. The most significant of these are digital transmission and fiber-optic cable.

As with broadcast systems, the use of digital transmission on cable is almost a necessity for HDTV and it will be equally important for interactive pay-per-view systems (see Section 17.15.4). The new ATSC DTV standards (see Chapter 7) have recognized this and includes standards for cable digital transmission.

17.13.1 Digital transmission

Formats and standards for the encoding, compression, and modulation of digital signals on CATV systems are included in the ATSC DTV standards and the DVB standards. In the case of the ATSC DTV standards, they are a variant of the terrestrial broadcast standards. Like the broadcast standards, they specify MPEG-2 compression and VSB modulation, but they are modified somewhat to take advantage of the higher signal levels on cable systems.

The system parameters of the broadcast and CATV DTV formats both require channel bandwidths of 6 MHz. The most important differences are:

	Broadcast	CATV
Modulation mode	8-VSB	16-VSB
Data rate (Mb/s)	19.29	38.58
CNR threshold (dB)	14.9	28.3

The result is that the digital data rate of a 6-MHz cable channel is twice that of a broadcast channel. The additional data rate can be used in a number of ways— one example is to deliver two complete HDTV programs in each 6-MHz cable channel. Another example is to transmit multiple SDTV programs on one cable channel—possibly as many as eight. This would multiply the capacity of the cable system without any increase in its bandwidth.

17.13.2 Fiber-optic cables

Fiber-optic cables (see Chapter 19) offers additional potential for increasing system capacity and performance improvement. These enhancements will be essential for the distribution of interactive video-on-demand programs or for the merging of TV program and telephone service on the same transmission facilities. Fiber optics provides substantially more transmission bandwidth and greater distances between line amplifiers, but it is more expensive. Because of this, most systems using fiber optics will probably be a hybrid, using both fiber and cable transmission.

17.13.3 Hybrid cable systems

Hybrid cable systems' territories are divided into serving areas, each containing perhaps 500 subscribers within a limited geographical area. Signals are distributed within the serving area by coaxial cable from nodes that are close enough to the subscribers to be within the range of cable transmission. The nodes are connected to the head end by fiber optics. The conversion from the optical to the electrical domains is then necessary only at the nodes.

As will be described later, this division of the system's territory into small serving areas is a basic technique for providing individual pay-per-view service to all subscribers.

17.14 Future Cable Systems

The remainder of this chapter is devoted to a forecast of some of the most important features of future CATV systems. It includes a description of the services they may offer, their utilization of the cable bandwidth, and their functional design.

17.14.1 Service offerings

It seems likely that future cable systems will offer the following services:

SDTV programs that may be offered as basic or premium service

HDTV programs also may be offered as basic or premium service

Pay-per-view programming in which subscribers pay fees for receiving individual programs

Interactivity—the ability of subscribers to communicate directly with the head end. This service might be used, for example, for ordering programs on the pay-per-view channel (video-on-demand) or for purchasing items from a catalog

Limited data and telephony service

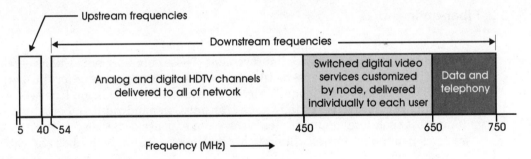

Figure 17.5 Future cable system bandwidth allocation.

17.14.2 Bandwidth utilization

There are numerous possibilities for the future utilization of the bandwidth available in cable systems. One of them, based on a 5- to 750-MHz passband, is shown in Figure 17.5.

In this example, the band from 5 to 40 MHz is used for upstream signals from subscribers to head end (see Section 17.7). The 40- to 50-MHz band is unused for transmission and it serves as a buffer between the downstream and upstrean channels.

The standard analog NTSC cable channels, 2 to RR, are transmitted in the 50- to 450-MHz band. As in the broadcast plan, digital HDTV channels will be interspersed with the NTSC channels in this band.

The 200-MHz band from 450 to 650 MHz is used for digital pay-per-view signals. As an option, the band from 650 to 750 MHz could be used for telephony and data transmission.

17.14.3 System functional diagram

Figure 17.6 is a functional diagram of a CATV system that includes the services described above. The output of the video server includes basic and premium analog NTSC and DTV signals, together with the switched pay-per-view signals. These are transmitted by fiber-optic cable to the nodes from which they are distributed by coaxial cable to individual subscribers.

17.14.4 Pay-per-view services

A computerized control center receives orders for specified programs from individual subscribers. When the center receives an order, it directs the server, which may have 500 or more titles in storage, to feed the specified title to the fiber-optic trunk to the node for the service area where the subscriber is located.

To completely avoid the equivalent of a busy signal on a telephone line, the fiber trunk must have the capacity to carry a separate program for each subscriber connected to its node. As with telephony, however, it is possible to limit the capacity of the trunk to a lower value based on statistical analysis of subscriber

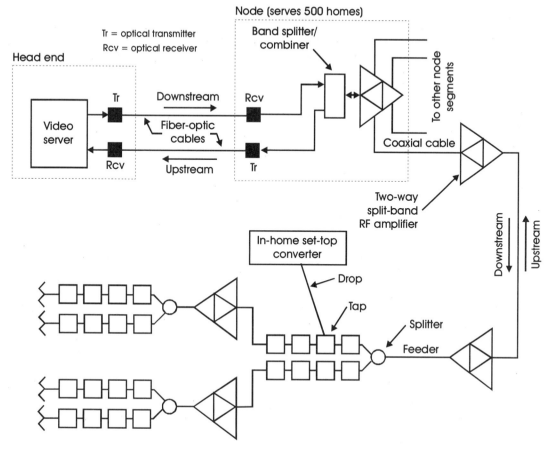

Figure 17.6 Diagram of a future CATV system.

usage. A standard is established that specifies the maximum acceptable percent of time that the entire trunk or a particular service area network may be fully loaded. The necessary number of circuits can then be calculated by standard statistical methods. A typical requirement for a 500-subscriber service area is 110 circuits, the assumption being that 500 subscribers will very rarely wish to see more than 110 different programs at any given time.

Note that the capacity of the 200-MHz band above 450 MHz for digital SDTV is as much as 260 simultaneous programs, considering that 8 programs of digital SDTV can be carried in one 6-MHz band.

17.15 References

1. Frank Baylin and Steve Berkoff, *Wireless Cable and SMATV*, Baylin Publications, Boulder, CO, 1992.

2. Andrew F. Inglis (ed.), *Electronic Communications Handbook*, Chap. 17, McGraw-Hill, New York, 1988.

Chapter 18

Satellite Video Communications

18.1 Satellites and Television

The use of satellites for television broadcasting and CATV program transmission and distribution began in the mid-1970s, and its subsequent growth has been phenomenal. Satellites are now indispensable to both media and their utilization continues to grow.

The value of satellites for transmission and distribution results from four unique properties:

1. The cost of a satellite circuit (an end-to-end transmission path via satellite) is independent of its length. It costs no more to send a signal across the country than across the street. This property generally makes satellites cost effective for long-haul, point-to-point service.

2. Except for a short segment at the beginning and end of each circuit, the path of the satellite signal is far above the surface of the earth and is unaffected by the nature or accessibility of the intervening terrain. This property makes satellites especially useful for paths over oceans, jungles, mountains, and arctic regions.

3. Once the satellite is in place, a new circuit can be established very quickly. This makes satellites valuable for *satellite news gathering* (SNG), where circuits must be set up quickly and with little preplanning. In this role satellites are of equal value for local and international news events.

4. Finally, satellites are especially adapted for point-to-multipoint service. When the satellite is in place, service can be supplied to an indefinite number of receiving points for the relatively low cost of receive-only earth stations (see Section 18.6). This capability is used for the distribution of programs to CATV systems and for direct-to-home broadcasting. It is also used by the television broadcast networks for distribution of programs to their affiliates.

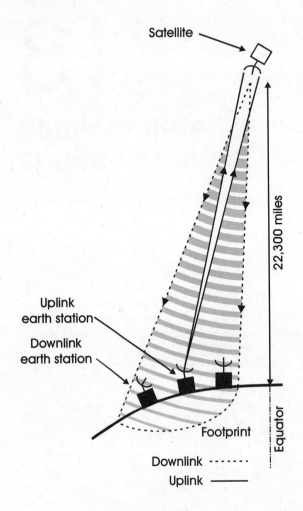

Figure 18.1 Satellite communication system.

Point-to-multipoint services for broadcasting directly from satellites to homes are called *direct broadcast services* (DBS). The most rapidly growing type of DBS uses digital transmission. With digital transmission and high-power satellites, it is possible to receive a satisfactory signal with a very small (c. 45 cm) receiving antenna. The digital transmission mode uses video compression technologies (see Chapters 3 and 7) that greatly increase the channel capacity of the satellite. More than 1,000 video program services are now being distributed by the digital satellite service and millions of receivers are in use.

Part-time channels in the broadcast service are leased on an occasional use basis, primarily for SNG and sporting events.

18.2 Satellite Communication Systems Elements

The elements of a satellite communication system are shown in Figure 18.1. An *uplink earth station* transmits a signal to the satellite in a very narrow beam. There it is received and amplified, and its frequency shifted to the downlink fre-

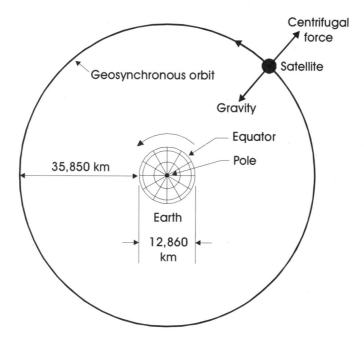

Figure 18.2 The geosynchronous orbit.

quency by a *transponder*. The signal is retransmitted to earth by the downlink, covering an area known as the satellite's *footprint*.

The boundary of the footprint is the contour of the *effective isotropic radiated power* (EIRP), which must be high enough to provide satisfactory reception on the earth. The EIRP is the ratio in decibels between the actual signal strength and the signal from an *isotropic antenna* (a physically unrealizable antenna that would radiate equally in all directions) radiating one watt.

18.3 The Geosynchronous Orbit

Satellites may be placed in orbit at any altitude above the earth's atmosphere and at any orientation to the equator. However, different altitudes have different uses. One special case is the geosynchronous orbit.

18.3.1 Orbit location

The *geosynchronous orbit* is a circle in the *equatorial plane*,[1] 35,850 km (22,300 miles) above the earth's surface (Figure 18.2). It has the property that the centrifugal force acting upward on objects located on the orbit and rotating with the earth equals the gravitational force acting downward. Under these conditions,

---

1. Sometimes the orbit becomes inclined slightly with respect to the equatiroal plane as a result of fuel saving for the N–S station keeping of the satellite (see Section 18.5.1).

satellites in the orbit appear to be stationary when viewed from the earth.

18.3.2 Orbital slots

Satellites are assigned fixed angular positions on the geosynchronous orbit known as *orbital slots* that are identified by their longitude. By international agreement, each country is allocated an arc of the geosynchronous orbit for its satellites, and the slots within this arc are assigned to individual licensees by the country's regulatory authority, the FCC for the United States. Since there is only one geosynchronous orbit, and there are requirements for satellite spacing, the total number of geosynchronous satellites in use is severely limited.

The United States is allocated the arcs 62°–103° and 120°–146° W. Long. (west longitude) for C-band satellites and 62°–105° and 120°–136° W. Long. for Ku-band satellites. (See Section 18.4.2 for definitions of the C and Ku bands.) The slots allocated to Canada are located in gaps in the U.S. arcs.

The spacing between orbital slots is critical, and it should be as small as possible to maximize the number of satellites that can be accommodated in the geosynchronous orbit. The minimum spacing is determined by the directivity of uplink and downlink antennas, since all satellites in a given band operate on the same frequencies and must depend on antenna directivity for isolation. Initially, C-band satellites were spaced 4°, and Ku band 3° on the orbital arc. (The Ku-band spacing was smaller because a narrow antenna beam can be produced more easily at its shorter wavelength.) Later, the spacing in both bands was reduced to 2°.

The spacing between DBS satellites in the Ku band is wider (9°) because smaller receiving antennas with wider beam angles are used in this service.

18.3.3 Elevation and azimuth angles

The *elevation* or *look angle* of a satellite is the angle above the horizontal made by the line of sight from a point on the earth to the satellite. The azimuth angle is the longitudinal position of the line of sight measured in degrees of west of Greenwich from 0° to 360°.

To calculate the elevation and azimuth angles, from a point on the earth's surface, first calculate the path length L and the angle β:

$$\beta = \cos^{-1}[\cos(\delta long)\cos(lat)] \tag{18.1}$$

$$L = (18.2 - 5.4 \cos\beta) \times 10^{-4} \text{ km} \tag{18.2}$$

where $\delta long$ is the difference between the earth station and satellite longitudes and lat is the latitude of the earth station.

The elevation and azimuth angles are then:

$$elevation = \cos^{-1}\left(\frac{4.22 \times 10^4}{L}\right)\sin\beta \tag{18.3}$$

$$azimuth = 180° + \tan^{-1}\left(\frac{\tan(\delta long)}{\sin(lat)}\right) \tag{18.4}$$

Assuming there is a choice, it is desirable that an orbital slot be chosen that results in as high an elevation angle as possible when the satellite is viewed from receiving locations. With low elevation angles there may be difficulty in clearing obstacles, the path through the troposphere is longer, which may result in more rain attenuation (especially in the Ku band), multipath distortions, and electrical noise generated by the earth's heat.

It is considered good design practice to utilize a minimum elevation when possible of 5° for C band and 10° to 20° for the Ku band.

18.3.4 Prime orbital arcs

The *prime orbital arc* for a location is the range of orbital slots that have an elevation angle in excess of 5°. The prime C-band arcs for the *continental United States* (CONUS), Alaska, and Hawaii are:

	Degrees west longitude
CONUS	55–138
CONUS plus Hawaii	80–138
CONUS plus Anchorage	88–138

The prime arcs for Ku band are shorter because of the higher minimum angle. The arcs are also shorter at more northerly latitudes (geosynchronous satellites are below the horizon at the north pole).

18.3.5 Solar eclipses

Solar eclipses occur when the earth comes between the satellite and the sun. They are not of direct interest to satellite users, but they have a major effect on satellite design. Satellites depend on solar cells for electrical power, and they must be equipped with on-board batteries to supply power during the eclipses.

Eclipses occur near midnight at the satellite's longitude for a period of 21 days centered on the spring and autumn equinoxes. Their duration is a maximum of 70 minutes.

18.3.6 Sun outages

Sun outages occur when the sun and satellite are in line as seen from the receiving location. The high temperature of the sun causes it to transmit a high-energy noise signal to earth station receiving antennas whenever it passes behind the satellite. The increase in noise is so great that a signal outage usually results.

The length and duration of sun outages depends on the latitude of the earth station and the size of the receiving antenna. At an average latitude of 40° in the United States and with a 10-m antenna, outages occur for a period of 6 days with a maximum duration of 8 minutes. Outages are longer with smaller receiving antennas because their beamwidth is wider and the sun remains within the beam for a longer time.

Outage times for specific locations can be calculated or can be obtained from the satellite operator. Sun outages cannot be avoided, and if continuous service is to be maintained the traffic must be switched to another satellite.

18.4 Satellite Classifications

Satellites are classified by their usage and by their technical characteristics.

18.4.1 Usage Classifications

The FCC defines two usage classifications for communication satellites used for television: *Fixed Satellite Service* (FSS), a broad classification that includes most commercial communication satellites, and *Broadcast Satellite Service* (BSS), a specialized FSS classification for satellites designed for broadcasting directly to homes.

The BSS may use either analog or digital transmission. If analog transmission is used, it is usually described as DBS, direct broadcast service. If digital transmission is used, it is described here as digital DBS, although all DBS is rapidly becoming digital, so the qualification will soon be moot.

18.4.2 Technical classifications

The most important technical classifications of satellites are their spectrum bands and service areas.

18.4.2.1 Spectrum bands

Three spectrum bands are used for commercial television satellite communications: C-band FSS, Ku-band FSS, and broadcast (B/C) satellite service. The uplink and downlink frequencies allocated to these bands are:

Service	Uplink (GHz)	Downlink (GHz)
C-band FSS	5.925–6.425	3.700–4.200
Ku-band FSS	14.0–14.5	11.7–12.2
B/C satellite service (DBS)	18.3–18.8	12.2–12.7

Direct-to-home broadcasting is also permitted on the FSS bands—in fact the C-band FSS spectrum was the first to be used for this purpose—but the wider spacing (see above) and higher power authorized for satellites in the broadcast service band permit the use of much smaller receiving antennas.

The characteristics of satellite service in these three bands are quite different with respect to propagation effects, equipment performance, and FCC regulation.

C and Ku bands have both advantages and disadvantages. C-band satellites share spectrum space with microwave systems, and the location and directional patterns of C-band earth stations must be coordinated or "cleared" with nearby microwave stations. It is sometimes impossible to clear a site in crowded metropolitan areas, and it is often not practical to use C-band uplinks for SNG where the uplink location cannot be planned in advance. In addition, the EIRP of C-band downlinks is limited to prevent interference to microwave systems. Finally, C-band uplink earth stations require larger antennas because of their longer wavelength.

Offsetting these problems, the C band has significant advantages, the most important being virtual immunity from signal outages resulting from rain attenuation. C-band components are generally lower in cost, in part because of larger mechanical tolerances, although this difference is diminishing as the production volume of Ku-band components is increasing.

18.4.2.2 Service areas

Satellite service areas vary widely in size, ranging from a metropolitan area to most of a hemisphere. The size of the service area depends on the size of the downlink antenna on the satellite—the larger the antenna the narrower the beam and the smaller the service area. One classification defines four service area sizes: global, regional, national, and spot.

Global beams cover the entire area that is visible from the satellite. They are used for international communications.

Regional beams cover a group of countries, for example, western Europe.

National beams cover all or a significant portion of a single country. Most domestic U.S. satellites are in this category.

Spot beams cover a limited area and are used principally for point-to-point voice and data communications. They are seldom used in television service.

18.5 Satellite Construction

The construction of a communication satellite is shown in Figure 18.3. It includes two primary subsystems: the bus and the payload. The *bus* includes the satellite housing and the subsystems that provide power and maintain the satellite in its orbital slot.

The *payload* includes the communications subsystems—the antennas, the receivers, and the transponders that shift the frequency and amplify the uplink signals for retransmission.

18.5.1 Satellite bus

The satellite housing can be rectangular as shown in Figure 18.3 or cylindrical, depending on the choice of the stabilizing mechanism that maintains the attitude of the satellite. The satellite in Figure 18.3 uses internal gyroscopes and *three-axis stabilization*. An alternative is spin stabilization in which the housing is a slowly rotating cylinder and the orientation of the antenna is held fixed through a rotating joint.

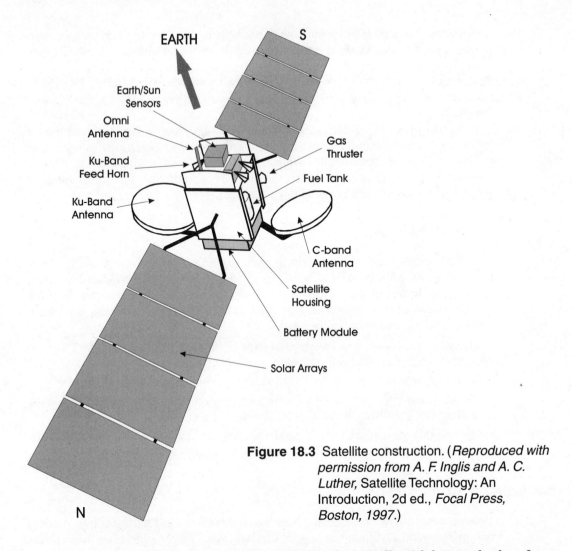

Figure 18.3 Satellite construction. (*Reproduced with permission from A. F. Inglis and A. C. Luther,* Satellite Technology: An Introduction, 2d ed., *Focal Press, Boston, 1997.*)

The power system consists of arrays of solar cells with battery backup for use during solar eclipses. The solar cells can be mounted on flat panels as in Figure 18.3 or on the cylindrical surface of a spin-stabilized satellite.

Although the forces on a geosynchronous satellite are nominally in balance, there are minor perturbations, e.g., the effect of irregularities on the earth's surface, that would cause it to leave its slot if uncompensated. Maintaining the satellite in its slot is known as station keeping, and it is achieved by ejecting small amounts of gas through *reaction thrusters*. The amount of gas that can be stored in the satellite is one of the elements that limits its life.

Substantial savings in fuel and a corresponding increase in the useful life of the satellite can be achieved by omitting N–S station keeping, thus allowing the satellite to depart from the equatorial plane. The result is a buildup in the inclination of the satellite's orbit with respect to the equatorial plane of about 0.85° per year. This causes cyclic annual variations in the azimuth and elevation of the satellite

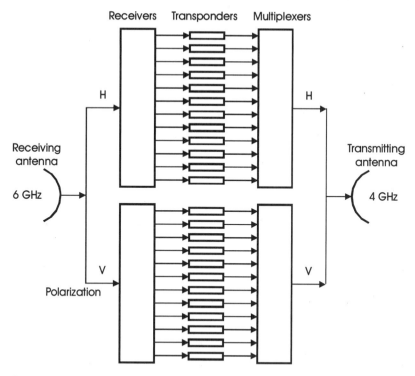

Figure 18.4 Satellite payload.

as seen from the earth that may be great enough to require periodic repointing of earth station antennas.

18.5.2 Satellite payload

A functional diagram of the satellite's payload, its reason for being, is shown in Figure 18.4. It includes the transmitting and receiving antennas, the receivers, and the transponders. Separate channels are provided for vertically and horizontally polarized signals or, alternatively, circularly polarized antennas with opposite rotations.

18.5.3 Frequency reuse

Frequency reuse nearly doubles the satellite's communications capacity by using both horizontal and vertical polarization (or right- and left-handed circular polarization for digital DBS systems). The de facto standard for the utilization of the downlink spectrum of a C-band satellite is shown in Figure 18.5. It includes 24 40-MHz channels—12 in each of the two polarities. Adequate isolation between channels sharing the same frequency is achieved by a combination of cross-polarization and offsetting the channel centers.

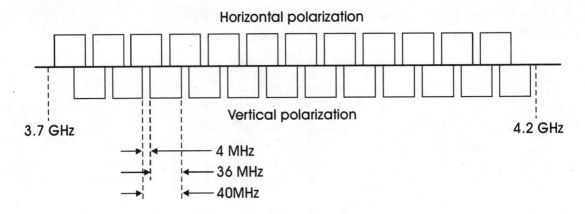

Figure 18.5 C-band spectrum utilization.

Additional isolation between channels on adjacent satellites is also achieved through cross-polarization.

Ku-band satellites do not have a de facto or official channel bandwidth standard. For FSS service, it ranges from 36 to 72 MHz, with 54 MHz as a common choice. As with C band, cross-polarization is used to double the satellite capacity, but the lack of standardization of channel bandwidth makes it impractical to increase the isolation between adjacent satellites by cross-polarization.

Digital DBS satellites have 16 transponders (or 32 with cross-polarization), each with a bandwidth of 24 MHz as described in Section 18.9.

18.5.4 Satellite antennas

The size and shape of the satellite's footprint is determined by the directivity of its transmitting antenna.

18.5.5 Satellite receivers

The satellite normally receives a strong signal from the earth station, and the demands on its receiver's sensitivity are not extreme. There are usually two pairs of receivers in each satellite, two (one spare) for each group of channels with like polarization.

18.5.6 Satellite transponders

The transponder receives an IF signal for a single channel (sometimes two channels in Ku-band systems) from the satellite receiver, shifts its frequency to the downlink channel, and amplifies it for retransmission to the earth. Transponders originally used traveling-wave tubes, but, as with much of the electronics industry, a transition has been made to solid-state power amplifiers, SSPAs.

Most transponders exhibit the property of saturation, a power level above which an increase in input causes no further increase in output. The output of the receiver should drive the transponder to saturation.

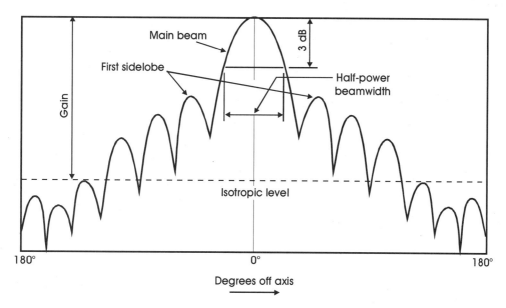

Figure 18.6 Earth-station antenna directivity.

Transponder power is determined by FCC regulations, economics, and technical feasibility.

In the C band, the FCC (and similar international organizations) specification on downlink power density limits the transponder power to about 8 watts. The power density limit is stated in terms of power per kHz of bandwidth per unit area. For FM signals with energy dispersal (see Section 18.7), it translates to EIRPs ranging from 38 dBw at elevations below 25° to 48 dBw at higher elevation angles.

The energy density limit is much higher in the Ku band, and economics and technical feasibility usually establish the actual transponder power. Forty watts is a typical choice for FSS service. Even higher power is required for DBS and transponder powers up to 250 watts have been proposed (see Section 18.9).

18.6 Earth Stations

Earth stations come in a wide range of sizes, costs, and performance. Digital TVROs (television receive-only), small earth stations that sell for less than $300 and are designed for home reception, are at one end of the range (although they use technically sophisticated technology). At the other end are multimillion-dollar installations owned by satellite operators and program suppliers.

18.6.1 Earth station antennas

The antenna is a basic component of both uplink and downlink earth stations. Most earth station antennas employ the familiar format of backyard TVROs, a parabolic or quasi-parabolic reflector illuminated by a feed horn. Their performance, however, differs widely depending on their size and design. The directional properties of earth station antennas are shown in Figure 18.6. They are rated by their gain with respect to an isotropic radiator (or alternatively with respect to a half-wave dipole), their half-power beamwidth (HPBW), their null beamwidth (NBW), and their side-lobe amplitude (see Figure 16.12 for definitions).

The shape of the pattern depends on the size of the antenna and the uniformity of the illumination of the antenna aperture by the feed horn. The width of the beam is inversely proportional to the antenna diameter and is a minimum when the illumination is uniform. Uniform illumination results in rather large side lobes, however, and these can be reduced at the expense of greater beamwidth by tapering the illumination from the center to the edge of the reflector.

The half-power beamwidth is given by Eq. (18.5):

$$\text{Half-power beamwidth} = C_{\text{HP}} \frac{\lambda}{D} \qquad (18.5)$$

where C_{HP} is a factor that varies from 50° for uniform reflector illumination to 70° for typical practical designs, λ is the wavelength, and D is the antenna diameter measured in the same units as λ.

The null beamwidth is given by Eq. (18.6):

$$\text{Null beamwidth} = C_{\text{NULL}} \frac{\lambda}{D} \qquad (18.6)$$

where C_{NULL} varies from 114° to 170° depending on the uniformity of the reflector illumination.

The gain of the antenna is inversely proportional to the square of the beamwidth (or the product of the vertical and horizontal beamwidths if different). It is also affected by the antenna efficiency, the fraction of the energy that is radiated after subtracting ohmic losses and energy intercepted by the feed horn. Table 18.1 lists specifications of representative uplink and downlink antennas assuming 70% efficiency and nonuniform illumination.

18.6.2 Uplink functional diagram

A functional diagram of a typical earth station uplink for FM carriers is shown in Figure 18.7. The functions of the components in an earth station with digital transmission are considerably different and the FM exciter is replaced by a series of components having functions that generate and process the bit stream for transmitting the audio and video signals (see Section 18.9).

The video and audio signals may be *scrambled* to prevent unauthorized reception.[2] They then pass to the exciter, the heart of the earth station. It generates the IF carrier, typically at 70 MHz for the visual carrier, adds preemphasis and a

---

2. For a summary of security techniques, see A.F. Inglis (ed.), *Electronic Communications Handbook*, Chap. 17, McGraw-Hill, New York, 1988.

Table 18.1 Antenna specifications

	Antenna diameter (meters)	HPBW (degrees)	Null BW (degrees)	Gain (dBi)
C-band downlink (4 GHz)	2	2.6	6.4	36.6
	5	1.05	2.5	44.5
	7	0.75	1.8	47.4
	10	0.52	1.3	50.6
C-band uplink (6 GHz)	5	0.70	1.7	48.0
	7	0.50	1.2	50.9
	10	0.35	0.85	54.0
Ku-band downlink (12 GHz)	1	1.75	4.2	40.0
	1	0.88	2.1	46.0
	5	0.35	0.95	54.0
	7	0.25	0.60	56.9
Ku-band uplink (14 GHz)	2	0.75	1.8	47.4
	5	0.30	0.72	55.3
	7	0.21	0.51	58.4

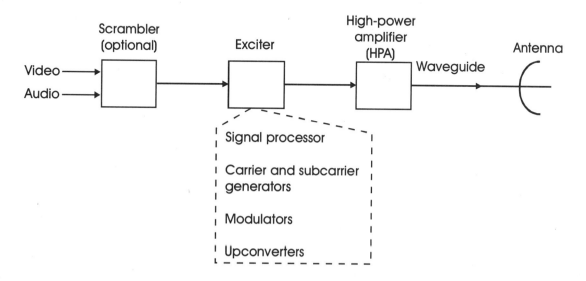

Figure 18.7 Earth station uplink.

Table 18.2 Uplink antenna maximum off-axis gain

Off-Beam Angle (deg)	Maximum Gain (dBi)
1	29.0
2	21.5
3	17.0
4	14.0
5	11.6
6	9.6
7	8.0

sideband energy dispersal waveform (see Section 18.7), modulates the IF carrier and *up-converts* it to the final carrier frequency and drives the high-power amplifier (HPA).

The power of the HPA in combination with the uplink antenna gain should be sufficient to drive the satellite transponder to saturation. This depends on the characteristics of the satellite, but a typical figure is 83 dBw. If the uplink antenna gain were 54 dB (see Table 18.1), the required power delivered to the antenna would be 29 dBw and a rated power of 1 kW would probably be chosen.

Traveling wave tubes have been frequently used for HPAs, although there is now a trend to solid-state devices.

Uplink antennas must meet the FCC specification shown in Table 18.2, which limits the off-axis uplink antenna gain to avoid interference to adjacent satellites.

18.6.3 Downlink functional diagram

A functional diagram of a satellite downlink earth station for FM modulation is shown in Figure 18.8. Digital downlinks are more complex and are described in Section 18.9.

18.6.4 G/T, the downlink figure of merit

Other factors being equal, the SNR performance of a downlink earth station is proportional to the ratio, G/T, where G is the gain of the antenna (see Table 18.1) and T is the *equivalent noise temperature* of the system in Kelvin [see Eq. (18.7) below].

Most of the noise in a satellite system is random and has the same characteristics as thermal noise, the electrical signals generated by hot objects. The equivalent noise temperature, T, is a convenient measure of the noise contribution of a component or system, even though it is nonthermal in origin.

T is given by the equation:

$$T = 293° (F_N - 1) \tag{18.7}$$

where F_N is the *noise figure* expressed as an arithmetic (not logarithmic) ratio

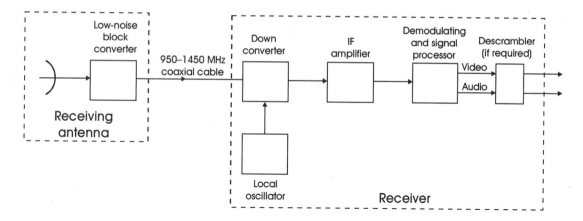

Figure 18.8 Earth station downlink.

and is equal to the quotient of the SNR at the output of a system to the SNR at the input normalized to an input temperature of 293K.

Because of its universality, noise temperature can be used to specify the noise contribution of all the components of a system. The system noise temperature is then equal to the sum of the noise temperatures of its components.

The major sources of noise in practical systems are the antenna and the input stage of the receiver. The antenna noise is electrical energy radiated from the earth because of its warmth and picked up by the antenna. The input stage of the receiver is a *low-noise amplifier* (LNA), chosen for its low equivalent noise temperature.

The amount of electrical noise power received from the earth by the antenna and its equivalent noise temperature depends on its directivity and its elevation angle. Noise also originates in heavy rainfall, which is a source of noise in the Ku band.

Typical antenna noise temperatures are shown in Table 18.3. Important progress has been made in recent years in the development of reasonably priced LNAs. Among these developments are the *field effect transistor* (FET) and the *parametric amplifier*. For the most demanding applications, the parametric amplifier can be artificially cooled. Equivalent noise temperatures for typical LNA types are shown in Table 18.4. The LNA is mounted at the antenna and provides the initial amplification. Alternatively, as in Figure 18.8, a *low-noise block converter* (LNB)

Table 18.3 Antenna noise temperature

Elevation angle	C band	Ku band
5°	58K	80K
10°	50K	58K
30°	30K	38K

Table 18.4 LNA noise temperature

LNA Type	C band	Ku band
Field effect transistor	85K	150K
Parametric amplifier	60K	120K
Cooled parametric amplifier	29K	43K

amplifies and translates all the incoming signals to a lower frequency range, often 950 to 1450 MHz, for transmission to the receiver by coaxial cable.

The remainder of the receiver functions are conventional.

18.7 Transmission Formats

The transmission format used by satellite systems may be either analog or digital. Video signals are inherently analog, and the format initially used for satellite transmission of these signals was frequency modulation. This format still predominates in the FSS service, but with recent developments in digital technology the use of digital transmission is enjoying increasing usage, particularly in the digital DBS services (see Section 18.9).

The advantages of frequency modulation as compared with amplitude modulation are relative noise immunity, insensitivity to system nonlinearities, and the ability to spread the transmitted energy over the entire channel by sideband energy dispersal (see below). Its major disadvantage as compared with amplitude modulation is its greater bandwidth requirement. Its performance as compared with digital transmission is described in Section 18.9.

The baseband waveform and the sideband spectrum for a carrier that is frequency modulated with a video signal are shown in Figure 18.9. The sync tip is clamped to a reference frequency so that the dc component of the signal is retained.

Dispersal of the sideband energy from the FM signal across the channel bandwidth is required in the C band to meet the FCC regulatory limits on EIRP per kHz. It is achieved by adding a low-frequency, 30-Hz (25-Hz for PAL), sawtooth waveform to the video signal to cyclically move the range of the frequency deviation of the carrier. This spreads the sideband energy over a larger portion of the spectrum and prevents the heavy concentration around the carrier that would severely limit the amount of downlink power because of the FCC limitation on energy density. The added sawtooth is removed at the receiver by the dc restoration circuit.

18.8 FM Performance Characteristics

The defining specifications of FM transmission systems are bandwidth and SNR.

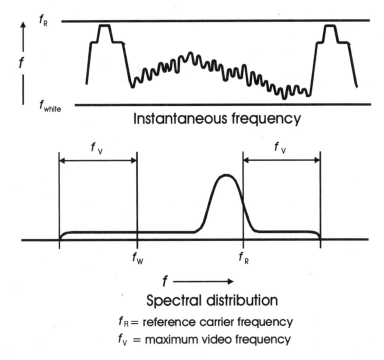

Figure 18.9 Video FM sideband spectrum

18.8.1 Bandwidth

The bandwidth of an FM signal is greater than the peak-to-peak frequency deviation and is determined by the spectrum of its sidebands. In theory they extend to infinity in both directions, but almost all of the sideband energy is confined to a frequency band extending from the extremes of the instantaneous frequency deviation waveform by an amount approximately equal to the video baseband bandwidth. The approximate total IF (or RF) bandwidth B is:

$$B_{IF} = \delta f_V + 2B_V \tag{18.8}$$

where δf_V is the peak-to-peak frequency deviation and B_V is the video baseband bandwidth. The same frequency units should be used for all terms.

A complete 40-MHz C-band transponder is usually used for each analog video signal, although two could be compressed into this spectrum space at the expense of SNR. It is common practice to transmit two analog video signals in a single 54-MHz Ku-band transponder.

18.8.2 Signal-to-noise ratio

The procedure for determining the SNR of a satellite circuit that employs frequency modulation begins by performing a *link analysis* to determine the carrier-to-noise ratio (CNR) of the unmodulated carrier as described in Section 18.8.2.1.

SNR is then calculated by adding the *FM improvement*, *preemphasis*, and *noise weighting* factors as shown in Eq. (18.9):

$$\text{SNR} = \text{CNR} + F_{\text{FM}} + F_{\text{DEEMP}} + F_{\text{W}} \tag{18.9}$$

18.8.2.1 Carrier-to-noise ratio

The CNR is calculated by means of Eq. (18.10):

$$\text{CNR} = (\text{EIRP} - 20\log f_{\text{GHz}} - 10\log B_{\text{Hz}} + 45) + G/T \tag{18.10}$$

where EIRP is the effective radiated power expressed in decibels relative to 1 W from an isotropic radiator, i.e., one radiating equally in all directions, f_{GHz} is the carrier frequency in gigahertz, and B_{Hz} is the bandwidth expressed in hertz.

The CNR and hence the fade margin (see Section 18.8.3) decrease with increasing frequency and bandwidth. As a result, CNR rather than SNR is sometimes the limiting factor in the performance of Ku-band systems. CNR can be increased by reducing the receiver bandwidth.

18.8.2.2 FM improvement factor

The presence of the FM improvement factor, F_{FM}, is a major advantage of the FM transmission mode. It is the ratio (in dB) of the SNR of the demodulated receiver output to the CNR of the unmodulated carrier—usually measured at the receiver IF. It is calculated by Eq. (18.11):

$$F_{\text{FM}} = 20 \log \left(\frac{\delta f_{\text{V}}}{2B_{\text{V}}}\right) + 10 \log \left(\frac{B_{\text{IF}}}{B_{\text{V}}}\right) + 7 \tag{18.11}$$

where F_{FM} is the FM improvement factor in dB, B_{IF} is the bandwidth of the radio-frequency channel, f_{V} is the video frequency, δf_{V} is the video peak-to-peak frequency deviation, and B_{V} is the video bandwidth. The ratio $\delta f_{\text{V}}/f_{\text{V}}$ is the *modulation index* F_{FM} can be quite large, sometimes more than 20 dB.

18.8.2.3 FM noise weighting factor

The noise weighting factor, F_{W}, for video signals, which converts unweighted to weighted noise, is described in Section 4.15.3. The noise spectrum for frequency-modulated video signals is triangular, i.e., the amplitude of the noise components increases with frequency. Since the acuity of the eye is less for fine-grained high-frequency components, the noise weighting factor is greater for FM noise than for flat noise. The weighted SNR for an NTSC signal transmitted by a frequency-modulated carrier, calculated by the EIA standard noise weighting curve, is about 12 dB higher than the unweighted value.

18.8.2.4 Frequency preemphasis/deemphasis

The SNR of an FM transmission system can be improved further by the use of

Table 18.5 CCIR rec. 405 preemphasis standard

Baseband Frequency	Relative Response (dB)
10 kHz	-10
20	-10
50	-9.5
100	-8.8
200	-6.8
500	-2.0
1 MHz	+1.4
2	+2.8

frequency preemphasis/deemphasis. The amplitude of the higher frequency components of the video baseband is increased or emphasized for transmission and deemphasized at the receiver. The deemphasis (F_{DEEMPH}) offsets the increase in amplitude of the high-frequency signal components and reduces the amplitude of the high-frequency noise components.

Standardization is required to obtain the full benefits of preemphasis/deemphasis and the CCIR has issued Recommendation 405 (see Table 18.5) for this purpose.

18.8.3 Fade margin

The *fade margin* is the difference in dB between the receiver input power under nonfading conditions and the receiver's *threshold*. The threshold is a property of FM receivers and is the point at which the SNR of the output signal begins to drop much more rapidly than the CNR. The fade margin is given by Eq. (18.12):

$$\text{Fade margin} = P_R - T_S \tag{18.12}$$

where P_R is the power in dBm at the terminals of the receiver under nonfading conditions.—it is directly proportional to the CNR and a high CNR is required for a good fade margin, and T_S is the threshold sensitivity in dBm of the receiver—the point at which the SNR at the output of the receiver begins to drop much more rapidly than the CNR.

Since the signal quality decays rapidly at power levels below the threshold, it is important that the fade margin be sufficient to maintain service, even during fading from refractive effects or heavy rainfall.

The length of the path of satellite signals through the troposphere is relatively short, and tropospheric fading is not a severe problem except for Ku-band signals in regions of torrential rainfall. For C-band and many Ku-band satellite circuits, a 3 dB fade margin is adequate.

Many receivers have *threshold extenders* that reduce the threshold power, typically from 12 to 8 dBm. These work by reducing the receiver bandwidth as the signal level approaches the normal threshold value.

Table 18.6 LNA noise temperature

	Trunk	CATV	ENG	Homesat	Homesat	DBS
Link parameters						
Band	C	C	Ku	C	Ku	DBS
Downlink EIRP	38	38	43	38	43	54
Antenna						
Diameter (m)	11	4.5	5	2	1	0.5
Gain (dB)	55	42	54	36	40	34
Noise Temp. (°K)						
Antenna	40	40	50	40	50	50
LNA	50	85	150	125	150	150
G/T (dB/°K)	35	21	31	14	17	11
B_{IF} (MHz)	32	32	24	24	24	27
B_V (MHz)	4.2	4.2	4.2	4.2	4.2	4.2
δf_V (MHz)	21	21	15	21	21	20
R_{CV} Threshold (dBm)	12	8	12	8	8	8
Link Performance (all in dB)						
F_{FM}	24	24	20	25	23	23
$F_{DEEMP} + F_W$	13	13	13	13	13	13
CNR	31	17	24	11	10	14
Weighted SNR	68	54	56	46	45	48
Fade Margin	19	9	12	3	2	6

18.8.4 FM satellite system performance

This section illustrates the principles that establish the SNR and fade margin of FM satellite links by applying them to representative examples:

1. A C-band trunk circuit of high quality and reliability such as would be used for interconnecting two major switching centers.
2. A C-band downlink to a CATV head end.
3. A Ku-band ENG circuit. One-half of a 54 MHz transponder is assumed.
4. A C-band homesat (earth station used for home reception) receiving transmissions intended for CATV systems.
5. A Ku-band homesat operating with a higher EIRP than is permitted on the C band-sometimes called medium-power DBS. As with the ENG circuit, one-half of a 54 MHz transponder was assumed.
6. A high-power analog Ku-band DBS system (see Section 18.9) operating in the broadcast satellite service band.

The circuit parameters that have been assumed for these examples are shown in Table 18.6 together with the calculated weighted SNRs and fade margins. Thirteen dB was assumed for $F_{DEEMP} + F_W$. A receiver threshold of 8 dB was assumed except for the CATV and homesat downlinks where 4 dB of threshold extension was assumed.

18.8.4.1 Trunk circuits

Trunk circuits are designed to add negligible noise to the signal since they are in series with other system elements, e.g., CATV downlinks, which for economic reasons leave very little margin for noise in the trunks. This is recognized in EIA/TIA-250-C performance standards, which specify a minimum unweighted SNR of 56 dB for point-to-point satellite transmission circuits. The example in Table 18.6 with a weighted SNR of 68 dB meets this standard.

Since the EIRP is limited by FCC rules, a high SNR must be achieved with a receiving system having a high G/T—in this example with an 11-m antenna and a cooled parametric amplifier for the LNA. This not only increases the SNR but permits a full bandwidth IF amplifier to be employed without adversely affecting the CNR and the fade margin.

The fade margin is 19 dB even without a threshold extender.

18.8.4.2 CATV downlinks

Downlinks are a significant part of the cost of CATV systems, but their G/T is usually chosen so that their SNR exceeds current industry standards for system performance. A weighted SNR of 54 dB, the performance of the CATV downlink in the example in Table 18.6, exceeds these standards.

The fade margin of 9 dB is satisfactory, even without the use of a threshold extender.

18.8.4.3 ENG circuits

An examination of the CATV and ENG system specifications in Table 18.6 illustrates some of the differences between C-band and Ku-band performance parameters.

The gain of the Ku-band antenna is 5 dB higher because of the shorter wavelength, but this is offset by a greater space loss and a higher antenna noise temperature. The higher CNR of the Ku-band system results from the greater EIRP of its footprint, and its smaller IF bandwidth, the latter chosen to increase the fade margin to compensate for rain attenuation.

The fade margin of 12 dB is adequate except in areas of extremely heavy rainfall.

18.8.4.4 C-band homesats

C-band satellite systems are not primarily designed (or even intended) to provide service directly to homes, and Table 18.6 clearly shows the compromises that must be made to operate a C-band homesat with a small antenna.

The IF bandwidth is reduced to increase the fade margin but at the expense of frequency response. Even with this increase, the use of a threshold extender would be desirable.

18.8.4.5 Ku-band analog FM homesats

The FSS Ku band has been proposed for analog direct-to-home transmission, but it has not been widely used for this purpose. The example shown in Table 18.6 assumes a 1-m antenna and two video channels in a 54 MHz transponder. The 45 dB weighted SNR is marginal as is the 2 dB fade margin.

18.9 Digital DBS Services

The technical and economic practicality of direct-to-home broadcasting from satellites was increased enormously by the introduction in 1994 of a digital transmission format for DBS services. Its growth has been extremely rapid (digital DBS receivers have been constantly in short supply), and the rate shows no sign of diminishing.

The following reasons for using digital transmission for DBS service have been advanced in filings with the FCC:

1. More efficient use of the spectrum by transmitting more video channels per transponder as the result of bandwidth compression (see Section 18.9.9).
2. A major improvement in the achievable SNR (see Section 18.9.11).
3. Smaller receiving antennas.
4. More flexibility in the assignment of channels.
5. More easily integrated with audio and other services.

The high technical quality of digital DBS broadcasts is encouraging public interest in high definition television, and it will probably be a major factor in the future acceptance of DTV systems. The following sections describe the basic technologies of digital satellite broadcasting.

18.9.1 Spectrum allocation

Digital DBS shares the spectrum (see Section 18.4.2.1), 12.2 to 12.7 GHz, that has been allocated for broadcast satellite services.

18.9.2 Orbital slots

Eight orbital slots have been allocated to the United States for DBS:

Deg. W.Long.
61.5
101
110
119
148
157
166
175

Only three of these slots—101°, 110°, and 119°—provide coverage for CONUS. 61.5° covers only the east while 148°, 157°, 166°, and 175° cover the west. Transponders on the 61.5° satellite can be paired with transponders on the four western satellites, however, to provide complete CONUS coverage from these satellites.

18.9.3 Transmission mode

It was initially expected that DBS satellites would use analog FM transmission, the standard format for the FSS services. A number of different combinations of transponder EIRP and downlink receiver G/T were proposed. Table 18.6 shows an example in which the EIRP is 52 dB (as compared with 43 dB for a typical Ku-band FSS satellite). With the assumptions used in the example, the fade margin would be 4 dB and the weighted SNR would be 48 dB. This is acceptable although not excellent performance, and it would be satisfactory in the absence of competition from signal sources with better performance. The plan to use FM for this service, however, was overtaken by the demonstrated advantages of digital transmission, which have caused it to be the preferred mode.

18.9.4 Satellite configuration

Thirty-two channels are allocated to each orbital slot as shown in Table 18.7. Alternate channels have right- and left-hand circular polarization. The transponder bandwidth is 24 MHz and the spacing between transponders of the same polarity is 29.16 MHz, thus providing a 5.16 MHz guard band. The spacing between transponders of opposite polarity is 14.87 MHz, which means that their spectra overlap as shown in Figure 18.10. The use of alternate left-hand and right-hand polarization and offsetting the transponder centers provide the necessary

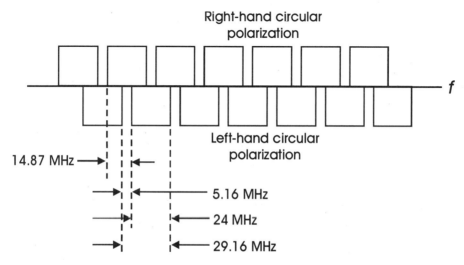

Figure 18.10 Digital DBS spectrum utilization.

Table 18.7 DBS channel allocations

Transponder Number	Polarization	Frequency (MHz)
1	RH	12,224.00
2	LH	12,238.80
3	RH	12,253.16
4	LH	12,267.74
5	RH	12,283.32
6	LH	12,296.90
7	RH	12,311.48
8	LH	12,326.06
9	RH	12,340.64
10	LH	12,355.22
11	RH	12,369.80
12	LH	12,384.38
13	RH	12,398.96
14	LH	12,413.54
15	RH	12,428.12
16	LH	12,442.70
17	RH	12,457.28
18	LH	12,471.86
19	RH	12,486.44
20	LH	12,501.02
21	RH	12,515.60
22	LH	12,530.18
23	RH	12,544.76
24	LH	12,559.34
25	RH	12,573.92
26	LH	12,588.50
27	RH	12,603.08
28	LH	12,617.66
29	RH	12,632.24
30	LH	12,646.82
31	RH	12,661.40
32	LH	12,675.98

isolation between adjacent transponders in the overlap area (equivalent to the use of vertical and horizontal polarization in FSS satellites).

18.9.5 Satellite locations

The two initial digital DBS satellites launched in 1994 are nominally collocated at 101° W. Long., but their precise locations are at 100.8° and 101.2°—approximately

700 mi apart in space-to avoid a collision. Each satellite has sixteen 120-W transponders, one-half of those assigned to the orbital slot, to reduce the satellite's power requirements to a reasonable level. From the standpoint of frequency allocation, the two are treated as though they were a single satellite.

18.9.6 System capacity

With 8 orbital slots and 32 transponders per slot, the allocation described in Section 18.9.4 provides spectrum space for 256 transponders. Only three of these slots have full CONUS coverage, but the equivalent of an additional full coverage slot can be obtained by pairing the eastern slot with one of the western slots. Full coverage can then be achieved with the equivalent of 128 transponders, 96 single transponders and 32 transponder pairs. With video compression, these could transmit more than 1,000 NTSC signals simultaneously, the total depending on the amount of compression used (see Section 18.9.8). The paired slots, of course, require the use of dual TVROs.

18.9.7 Transponder power

Each of the initial digital DBS satellites has a total power capacity of approximately 2 kW. This can be configured to power sixteen 120-W or eight 240-W transponders, which compares with 40 watts for the transponders in typical Ku FSS satellites. With the 120-W option, the maximum downlink EIRP in the United States with complete CONUS coverage is 54 dBW.

18.9.8 System architecture

The architecture of the uplink of a digital satellite broadcast system is shown in Figure 18.11. The video signal is first converted to a digital format by sampling, quantizing, and compression (see Chapter 3). Compression is followed by encryption—scrambling or otherwise making the signal unusable by all but authorized viewers—there are many ways to do this once the signal is digital. The signal is then divided into packets to provide the mechanism for error correction and multiplexing. The resulting video packets are interleaved with similar-sized digital audio packets.

The result is a complete audio-video bit stream for a single program source. It is then multiplexed with the bit streams for other program sources by further packet interleaving and the complete bit stream modulates the transmitter.

The architecture of the receiver is the converse of the transmitter.

The receiver is tuned to the desired program, a process that includes selecting the antenna polarization, the transponder, and the channel. The audio and video packets are separated, errors are corrected in the video bit stream, the encryption is defeated for the desired signal, the signal is decompressed and finally reconverted to the analog format.

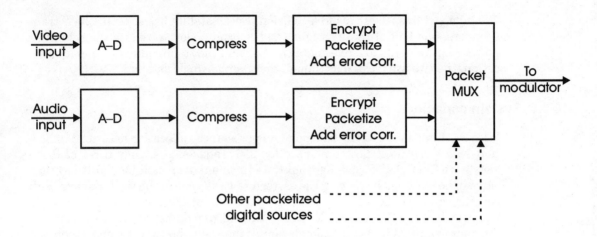

Figure 18.11 Digital DBS uplink architecture.

18.9.9 System parameters

Typical parameters of a satellite digital transmission format for NTSC signals are shown in Table 18.8.

The 4 f_{SC} sampling rate with 8 bits per sample, giving a net data rate of 114.56 Mbs (for signal information only), are common choices for these parameters.

The gross data rate, which includes bits for error correction and "overhead" for the packet headers, (see below), can be computed from the packet structure. A packet includes 147 bytes—127 for the audio and video payload, 17 for error correction, and 3 for the header (see Figure 18.12). The header includes 12 bits to identify the program, 4 bits to identify the encryption, 4 bits to identify the use of the packet, and a 4-bit continuity counter.

The MPEG signal compression algorithm is the same as recommended for broadcast use. The system designer has a wide variety of choices for the compression ratio. There is no sharply defined limit, and as it is increased there is a gradual

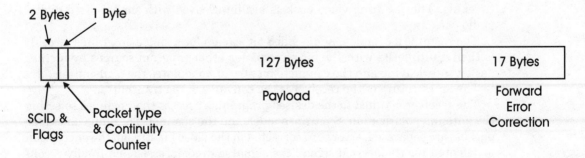

Figure 18.12 Digital DBS data packet.

Table 18.8 Digital signal transmission parameters—525 lines

Parameter	Value		
A–D sampling rate (4 f_{SC})	14.32 MHz		
Bits/sample (2^8 brightness levels)	8		
Single-channel video net data rate (uncompressed, no audio or error correction)	114.56 Mb/s		
Single-channel gross data rate (includes audio and overhead)	132.60 Mb/s		
Packet length (see Section 3.7.1 and Figure 18.12)	147 bytes		
Compression algorithm (see Section 3.12)	MPEG-1		
Transponder data capacity (QPSK modulation, 24 MHz bandwidth, see Table 3.4)	45.6 Mb/s		
Compression ratio (amount is optional. Maximum is determined by desired picture quality and image motion)	2:1 to 24:1		
Representative compression ratios:	9:1	12:1	24:1
Compressed bit rate (Mb/s)	14.73	11.05	5.53
Channels per transponder	3	4	8
Encryption for signal security (addressable)	"Smart Card"		
Error correction algorithm (see Section 3.8.3)	Reed-Solomon		
Multiplex format	Packetizing		
Modulation mode (see Section 3.7.2)	QPSK		

deterioration of image quality and the ability to handle rapid motion. Compression ratios up to 24:1 usually do not significantly affect image quality, and even greater compression is practical.

The encryption method must control access not only to complete program services but also to individual programs, as in pay-per-view. One technique is to use a "smart card" on which the authorized program services or program times are recorded remotely by the program supplier. To view an authorized program, the smart cart also stores the necessary information to control the receiver's digital processing to undo whatever digital scrambling may have been used at the uplink.

18.9.10 Digital DBS for HDTV

The high quality of digital DBS signals makes it an attractive potential medium for HDTV. In fact, there is one view that digital DBS will be a catalyst that will lead to the successful introduction and growth of HDTV service. No transmission standards have been adopted or even officially proposed for digital DBS transmission of HDTV signals, but it seems almost inevitable that they will be in time. Since the ATSC standards for broadcast DTV specify a digital format (see Chapters 7 and 16), the adaptation of this format to satellite transmission would be relatively straightforward. In fact, this has been done in the DVB standards, which already include a satellite version of their broadcast standards.

18.9.11 Digital DBS link performance

Noise in digital transmission circuits is the result of bit errors. Error correction can be so effective below a threshold BER that the noise contribution of the circuits is negligible. When the BER threshold is exceeded, error correction quickly becomes ineffective and the SNR of the signal deteriorates rapidly in a manner analogous to below-threshold FM signals. As with FM signals, the fade margin is the difference between the strength of the received signal under nonfading conditions and its level at the point where error correction fails. Most of the digital DBS receivers now in use provide a large fade margin even with small (18-in) receiving antennas.

Chapter 19

Fiber-Optic Transmission Systems

19.1 Overview

Fiber optics is one of the many new technologies that are revolutionizing the television industry. Because its transmission medium is infrared radiation having an extremely high frequency in electromagnetic terms, fiber-optic transmission systems have enormous bandwidths—20 GHz is typical. They are especially suited to digital signals, although analog transimission is also possible (see Section 19.5.2).

The cost of fiber optics now limits applications to point-to-point transmission and CATV trunks (see Chapter 17), but as its technology advances and cost reduces, it can be expected that it will be used for distributing voice, emssage, video, and audio service to individual homes.

Fiber-optic systems are simple in principle as shown in Figure 19.1. An optical energy source—a laser or LED—is modulated with the video signal. The optical source illuminates the end of an optical fiber, and the energy is transmitted by internal reflection within the fiber to its opposite end. Here it is detected and demodulated and the original baseband signal is recovered.

Unlike electromagnetic transmission systems, it is necessary to demodulate

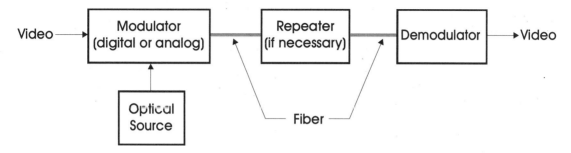

Figure 19.1 Fiber-optic transmission system.

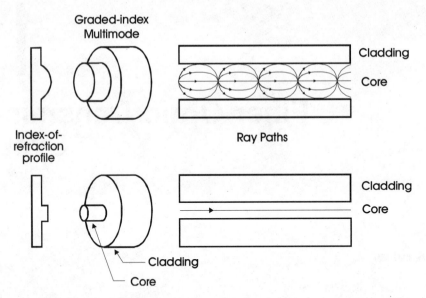

Figure 19.2 Optical fiber construction. (*From Ira Jacobs, "Fiber-Optic Transmission Systems," Chapter 8 in A. F. Inglis (ed.),* Electronic Communications Handbook, *McGraw-Hill, New York, Figure 8.1.*)

and remodulate the signal at repeater points in fiber-optic systems. This is not a severe problem for digital systems, but it causes analog systems generally to be used only for single hops.

19.2 Fiber-Optic Construction and Operation

The heart of a fiber-optic system is, of course, the fiber. This section covers some of the properties of optical fibers.

19.2.1 Construction

An optical fiber (see Figure 19.2) consists of an exceedingly fine glass filament that is surrounded by a concentric cladding, also made of glass but having a slightly lower index of refraction. The index-of-refraction profile for single-mode fiber is known as the step index. The index for the multimode fiber in Figure 19.1 is carefully shaped because it determines the bandwidth and frequency response of the fiber (see Section 19.4.2). It is known as *graded-index* multimode fiber.

The diameter of the fibers depends on the transmission mode (see Section 19.3). The cores of single-mode fibers typically have diameters of 7 to 10 μm while multimode cores range from 50 to 85 μm. One standard for multimode fibers is a 50 μm core and a 125 μm cladding. These diameters are comparable with those of human hair.

A fiber-optic cable consists of one or more fibers bundled together and sur-

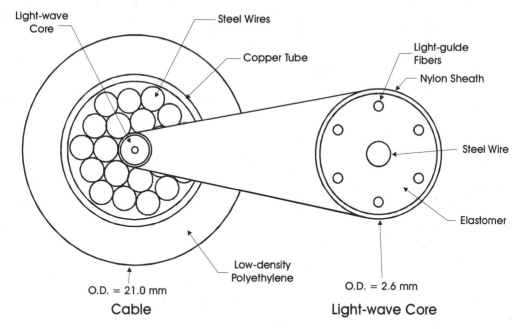

Figure 19.3 Fiber-optic cable. (*From Donald G. Fink (ed.)*, Electronic Engineers' Handbook, *McGraw-Hill, New York, 1989, Figure 22.25a.*)

rounded by a metallic or plastic sheath for mechanical protection. The bundle may include a steel wire for mechanical strength (see Figure 19.3). In one configuration, 12 groups, each containing 12 fibers, are bundled together. Needless to say, the bandwidth and hence the communication capacity of such a cable is enormous.

19.2.2 Operation

Infrared radiation is injected into the end of the fiber and is totally reflected when it strikes the cladding that has a lower index of refraction. It is contained within the core and would continue down the fiber until it is completely attenuated by losses in the glass, unless it is received at an earlier point along the fiber.

19.3 Single and Multimode Fibers

The operation of single-mode and multimode fibers is illustrated in Figure 19.2. The single-mode fiber has a smaller diameter, and only one mode, a single path down the fiber, is possible. The shortest wavelength that can be transmitted in a single mode is:

$$\lambda_{MIN} = 2.6r(n_1^2 - n_2^2) \tag{19.1}$$

where r is the radius of the core and n_1 and n_2 are the indices of refraction of the

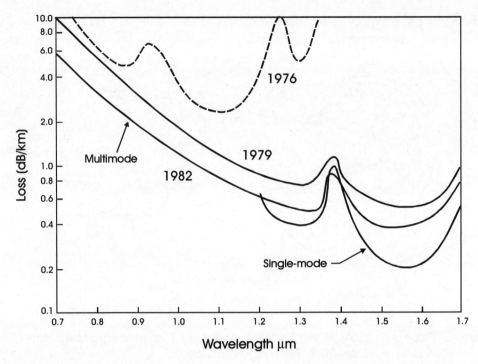

Figure 19.4 Fiber-optic attenuation. (*From Ira Jacobs, "Fiber-Optic Transmission Systems,"* Chapter 8 in A. F. Inglis (ed.), Electronic Communications Handbook, *McGraw-Hill, New York, Figure 8.2.*)

cladding and core. Larger-diameter fibers permit multiple reflections and multimode operation is possible. Each mode has advantages and disadvantages.

Single-mode fiber requires far greater mechanical precision in splicing. A lateral offset of only 1.5 μm in the ends of the fiber cores results in 0.5 dB attenuation. An offset of 7 μm would be required to produce the same loss in multimode fiber.

On the other hand, at wavelengths in excess of 1.1 μm, the attenuation of single-mode fiber is substantially less (see Figure 19.4). Also, single-mode fiber is not subject to *modal dispersion* (see Section 19.4), which would reduce its bandwidth. The practical result is that single-mode fibers are commonly used for wider bandwidth (higher speed) applications and for longer paths.

19.4 Attenuation and Dispersion

The principal signal loss effects in optical fibers are attenuation and dispersion.

19.4.1 Attenuation

Both single-mode and multimode fibers suffer from attenuation as shown in Figure 19.4 for a representative glass type. For this glass, the attenuation of single mode fibers reaches a low of about 0.2 dB/km at wavelengths in the 1.5 to 1.6μm infrared range. Long distance signal transmission by fiber-optic cables was made possible by the development of energy sources in this region of the spectrum together with the availability of glass with lower attenuation and the development of techniques for splicing the tiny single-mode fibers.

19.4.2 Modal Dispersion

Dispersion in fiber-optic circuits results from differences in the arrival time of different rays or different frequency components. It can be caused by differences in the path length of different modes, *modal dispersion*, or by differences in transmission speed for different wavelengths, *chromatic dispersion*.

The effect of dispersion is to cause pulses to spread, and this limits the maximum pulse rate. The effect can also be described as a bandwidth limitation. The relationship between bandwidth and pulse spreading is given by Eq. (19.2):

$$B = \frac{0.44}{i} \tag{19.2}$$

where B = bandwidth in MHz and i = the width in microseconds between the points on the edges of the pulse where it is one-half its maximum value.

The bandwidth of multimodal fiber, when determined by modal dispersion, is a function of the wavelength and the index-of-refraction profile. The profile can be adjusted so that the index of refraction is greatest and the transmission velocity is the least on the shortest paths. This partially equalizes the transmission time, reduces the pulse spread, and increases the bandwidth for a given circuit length. Figure 19.5 shows three examples of bandwidth due to the variation in the refractive index of the fiber with wavelength. The second is waveguide dispersion, which is caused by a graded index fiber in which the index of refraction varies across its diameter. Material and waveguide dispersion are in opposite directions, and the waveguide dispersion can be controlled so that the variation in delay is equal to zero at some wavelength. The wavelength at which this occurs can be adjusted within limits by appropriate choice of the refraction index profile.

Figure 19.6 compares the dispersion in picoseconds per km for each nanometer of source spectral width for two fibers, one with a uniform index of refraction across the core and the other with an index profile that has been adjusted to move the zero dispersion wavelength from 1.3 μm to 1.6 μm.

19.4.3 Chromatic dispersion

Both single-mode and multimode fibers are affected by chromatic dispersion, which is the principal bandwidth limiter for single-mode fibers. Chromatic dispersion occurs because fiber-optic light sources are not monochromatic, and the transmission velocity veries with wavelength.

Two effects are present that affect the variation of transmission velocity with wavelength. The first is *material dispersion* due to variations in the refractive

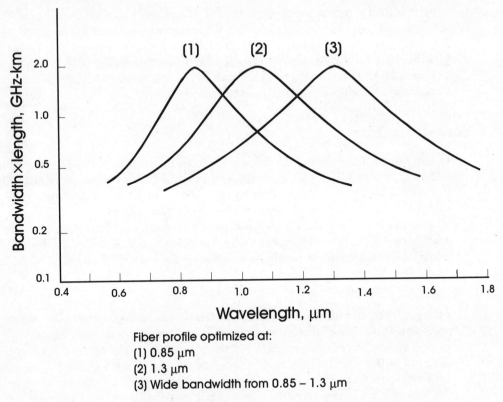

Figure 19.5 Fiber-optic bandwidth. (*From Ira Jacobs, "Fiber-Optic Transmission Systems," Chapter 8 in A. F. Inglis (ed.),* Electronic Communications Handbook, *McGraw-Hill, New York, Figure 8.3.*)

index of the fiber with wavelength. The second is *waveguide dispersion*, which is caused by a graded-index fiber in which the index of refraction varies acriss the diameter of the fiber. Material and waveguide dispersion are in opposite directions, and the waveguide dispersion can be controlled so that the variation in delay is equal to zero at some wavelength. The wavelength at which this occurs can be adjusted within limits by appropriate choice of the refraction index profile.

Figure 19.6 compares the dispersion in picoseconds per km for each nanometer of source spectral width for two fibers, one with uniform index of refraction across the core and the other with an index profile that has been adjusted to move the zero-dispersion wavelength from 1.3 to 1.6 μm.

19.5 Signal Formats and Modulation Modes

Fiber-optic systems may use digital or analog modulation. Both are covered in this section.

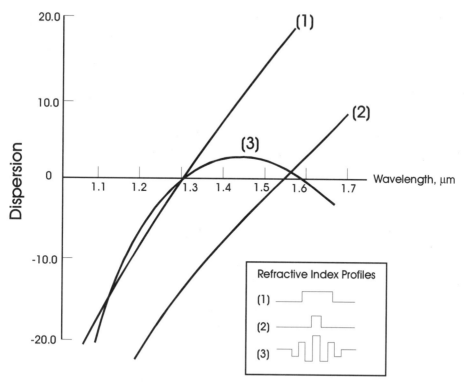

Figure 19.6 Spectral dispersion. (*From Ira Jacobs, "Fiber-Optic Transmission Systems," Chapter 8 in A. F. Inglis (ed.),* Electronic Communications Handbook, *McGraw-Hill, New York, Figure 8.4.*)

19.5.1 Digital format

The digital format is uniquely suited to the characteristics of fiber optics. Fiber-optic systems have a surplus of bandwidth but are limited with respect to power and receiver sensitivity. Pulse code modulation (PCM) trades bandwidth for SNR, and is almost universally used except for short single-hop paths. Present optical sources are not practical for frequency-shift or phase-shift keying.

Digital video signals with any of the standard bit rates described in Chapter 3 can be transmitted by fiber optics. They generally use a simple binary code (see Chapter 3) with *on-off keying* (OOK). Other modulation modes such as *quaternary pulse-position modulation* (QPPM), can be used to effect a different tradeoff between SNR and bandwidth. Modes involving phase-shift or frequency-shift modulation are not practical for use with an optical carrier.

19.5.2 Analog format

An analog format is more economical because ADC and DAC are not required, and it is practical for some single-hop systems. Amplitude modulation can be used

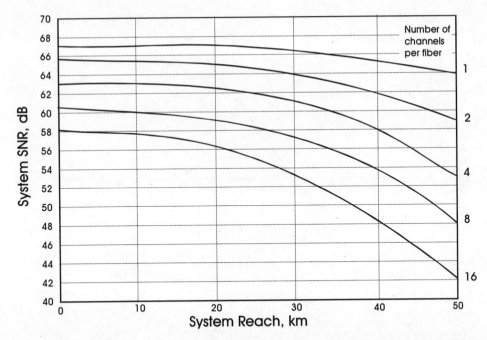

Figure 19.7 Range (reach) of analog systems. (*From E.D. Horowitz and V.D. Conanan, "Television Communication Systems Design," Chapter 17 in A. F. Inglis (ed.), Electronic Communications Handbook, McGraw-Hill, New York, Figure 17.19.*)

with intensity modulation of the optical source. Alternatively a subcarrier that is frequency modulated by the signal can be created and used to intensity modulate the light source. This eliminates problems of linearity and provides an increase in SNR by the FM improvement factor (see Section 18.8.2.2).

Frequency modulation has the additional advantage that several video signals can be multiplexed on the same fiber by the use of FDM-FM, frequency division multiplex.

The range of a single-hop system is sometimes known as its *reach*, which is established by the SNR. This calculation for multichannel FDM-FM systems is complex and beyond the scope of this book. The reach of these systems is indicated in Figure 19.7, which shows the SNR based on empirical results.

19.6 Path Performance, Digital Format

The measure of performance for a digital transmission system is the bit error rate rather than SNR as in analog systems. This is determined on a statistical basis and is expressed as a bit error probability.

Bit errors can be caused not only by system noise but also by signal-dependent noise including such noise sources as intersymbol interference and improper set-

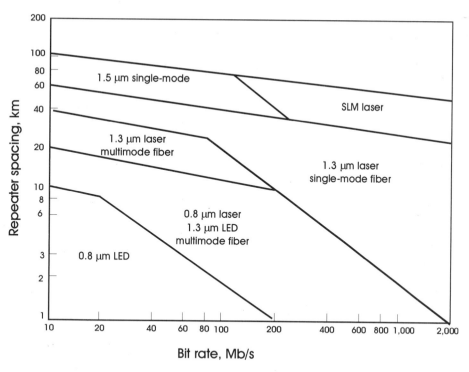

Figure 19.8 Repeater station spacing. (*From Ira Jacobs, "Fiber-Optic Transmission Systems," Chapter 8 in A. F. Inglis (ed.),* Electronic Communications Handbook, *McGraw-Hill, New York, Figure 8.17.*)

ting of decision thresholds. In total these are known as *eye degradation*, which refers to the *eye pattern*, an oscilloscopic method of viewing the operating margins of digital reception. The calculation of bit error probability in fiber-optic systems is beyond the scope of this book, and the reader should consult the references.

The ultimate criterion of the performance of fiber-optic digital transmission systems is the required repeater spacing for multi-hop systems. This is not an exact calculation because it must be based on statistical results and it will depend on the performance standards. Approximate maximum spacings for single and multi-mode fibers with laser and LED optical sources (see Section 19.7) are shown in Figure 19.8. This graph clearly shows the increase in range that has resulted from the development of laser sources in the 1.3 to 1.6 μm range and the introduction of single-mode fibers.

19.7 Transmitters for Fiber-Optic Systems

Two types of optical transmitters are used in fiber-optic systems, *lasers* and *light-emitting diodes* (LEDs). Both are semiconductors, but there is a fundamental difference in that the radiation from lasers is coherent, that is, it retains its wavelength and phase over a period of time.

Table 19.1 Lasers and LEDs

Parameter	Laser	LED
Light output	6 dBm	6 dBm
Coupling loss	3 dB	20 dB
Spectral width		
At 0.8 μm	2 nm	40 nm
At 1.3 μm	4 nm	100 nm
Maximum modulation rate (approx)	5 Gb/s	500 Mb/s
Feedback power control	Yes	No
Temperature sensitivity	Strong	Weak
Failure mechanisms	Many	Wears out

The use of lasers was limited at first because they were not available in the 1.5- to 1.67-μm wavelength range, the area of lowest attenuation (Figure 19.4). The development of lasers that operated at these wavelengths was a significant breakthrough in fiber-optic technology.

Some important comparisons of lasers and LEDs are summarized in Table 19.1. In both of these examples, the same output power is emitted but the coupling of the power into the fiber is far more efficient for lasers.

The spectral width of laser emission is much narrower, an important advantage because it minimizes the loss of bandwidth due to chromatic dispersion.

The maximum rate at which lasers can be modulated is greater by a ratio of about 10:1. However, the maximum rate for LEDs—approximately 500 MHz—is high enough for most purposes.

The output power of lasers can be stabilized by feedback, an operation that is not possible with LEDs. On the other hand, LEDs are more rugged, less sensitive to temperature variations, and probably have a longer life. As a result of these differences, LEDs are used in short-hop applications, while lasers are used for long paths and multihop systems.

Although the spectral bandwidth of laser emission is comparatively narrow, it is not monochromatic, and its fine-grained spectrum shows a number of discrete spectral lines. Figure 18.9 shows a representative example of the spectrum of a laser emitter at different power levels.

With a spectral distribution of this width, the bandwidth is reduced as the result of chromatic dispersion (see Section 19.4.3) and this can be a significant problem on long-distance, high-speed circuits. For these circuits, the output spectrum can be limited to a single spectral line, often 1.55 μm, by the use of an external cavity or a resonant structure within the laser. The result is a *single longitudinal mode* (SLM) laser.

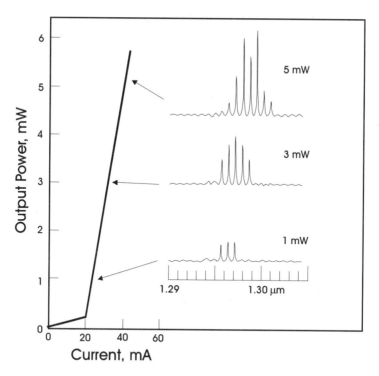

Figure 19.9 Laser emission spectrum. (*From Ira Jacobs, "Fiber-Optic Transmission Systems," Chapter 8 in A. F. Inglis (ed.),* Electronic Communications Handbook, *McGraw-Hill, New York, Figure 8.7.*)

19.8 Receivers for Fiber-Optic Systems

A fiber-optic receiver consists of a photodetector that converts an optical to an electrical signal, followed by stages of electronic amplification. The critical element in the receiver is the photodetector since it determines the receiver's sensitivity.

The sensitivity of a fiber-optic receiver is defined as the minimum received power, P_R, rwquired to achieve an error probability, P_E, at a bit rate B. P_R can be calculated from the minimum number of photons, n_P, that must be received per pulse:

$$P_H = \frac{n_P B h c}{\lambda} \tag{19.3}$$

where λ = wavelength, B = bandwidth, h = Planck's constant = 6.6252×10^{-34} joule-s, and c = velocity of light = 3×10^8 m/s.

n_P is calculated from the error probability:

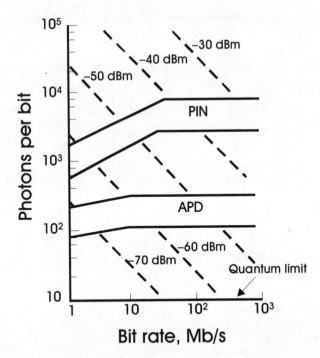

Figure 19.10 Receiver sensitivity. Photons per bit required at the receiver to achieve 10⁻9 bit error rate. Translation to received power level (in dBm) shown by dashed lines. (*From Ira Jacobs, "Fiber-Optic Transmission Systems," Chapter 8 in A. F. Inglis (ed.),* Electronic Communications Handbook, *McGraw-Hill, New York, Figure 8.8.*)

$$P_E = \frac{1}{2} e^{-n_p}$$

The most common photodetectors are the *avalanche photodetector* (APD), and the *PIN junction*. Their sensitivity in terms of the number of photons/bit and the received power level for a bit error probability of 10^{-9} for PIN and APD photodetectors is shown in Figure 19.10.

The demands on the performance of a fiber-optic receiver are severe. As an example, consider an APD receiver operating at a bit rate of 100 Mb/s. From Figure 19.10, it can be seen that it would require 10^2 photons per bit or 10^{10} photons per second to achieve a bit error probability of 10^{-9}. If each photon were converted to an electron by the photodetector (an optimistic assumption), the current would be 1.6×10^{-9} A, or 1.6 nA. This compares with an output signal of more than 100 nA for a typical photoconductive image tube in broadcast cameras. A current of this level requires a high degree of amplification in the electronics following the photomultiplier.

19.9 Passive Optical Components

A number of passive devices, including connectors, optical attenuators, couplers, and wavelength-division-multiplexing (WDM) filters, some necessary and some optional, are available to support the active devices and complete the system.

19.9.1 Connectors

The design of connectors that could place and maintain the ends of very tiny fibers in almost perfect alignment at joints has been a challenging engineering assignment. The success of single-mode fiber was made possible by the development of tools and techniques for splicing extremely thin fibers with precision. Permanent connections are bonded while demountable connectors utilize cylindrical bushings or conical plugs. Techniques have been perfected that have reduced the losses at splices to 0.05 dB or less.

19.9.2 Optical attenuators

Because of the wide amplitude range of received signals, a means of adjusting the optical input power is required for the operation of practical systems. This is accomplished by attenuators, both fixed and adjustable neutral density filters.

19.9.3 Couplers

Couplers divide the power in a single fiber between two or more paths. This can be accomplished with a Y junction or, if only a small amount of power is desired in one of the branches, by putting a sharp bend in the fiber so that reflection is less than total and the power that escapes is captured by a second fiber.

19.9.4 Wavelength-division-multiplexing filters

The purpose of WDM filters is to separate the radiation from the optical source into radiation bands, each of which forms a signal channel that is modulated separately. The channels are multiplexed and transmitted by a single fiber. This is a very effective technique for utilizing the large bandwidth of fibers to increase their information handling capacity.

The filters are thin-film dielectric materials that transmit one wavelength and reflect the otheras with the dichroic surfaces used in color television cameras.

Chapter 20
Video on the Internet

20.1 Introduction

With the growth of digital video systems, it was only natural to transmit video over digital networks designed for communications among computers. This is being widely done within broadcast and postproduction facilities, but most important in this regard is that worldwide network of networks—the *Internet*.

20.1.1 History

In the 1970s, the United States Department of Defense recognized the need for worldwide communication among its diverse types of computers and networks, and commissioned the *Defense Advanced Research Projects Agency* (DARPA) to develop the basis for such a network. Later, it was recognized that this super network would be valuable for use by organizations outside of the defense community as well, and it was opened to broader use; at this point the network became ARPANET.

Still later, nondefense use had grown so much that the United States government decided to turn the network over to private interests, and it became simply *the Internet*, which is the way it is today.

20.2 Internet Architecture

The Internet consists of software protocols that can be implemented on any type of computer and any type of digital transmission path or network. Some dedicated hardware supports network addressing and routing of messages to any location in the world.

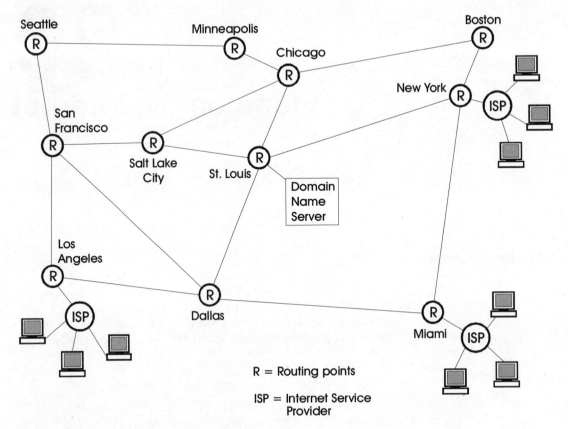

Figure 20.1 Diagram of a hypothetical Internet.

20.2.1 Hardware architecture

A simplified and somewhat hypothetical diagram of the Internet in the United States is shown in Figure 20.1. Computers at a particular location are connected to an *Internet service provider* (ISP), who maintains a connection (a *gateway*) to the Internet at the nearest routing point and offers a service to local customers that is usually connected by phone lines. The local computers can be any type, ranging from simple PCs to high-powered workstations or even mainframe computers. All must be equipped with software capable of performing Internet processes (see Section 20.2.2).

The network itself comprises various transmission paths between routing points, or *routers* (R in the diagram). The purpose of the routers, which are computers interfaced to the transmission paths and programmed specifically for this purpose, is to direct individual messages along an appropriate path to their destination. If a transmission path fails or is out of service, the routers may select an alternate path to the destination. The routing process is performed by *packet switching*, which depends on messages being packetized (see Section 3.7.1). Packets are

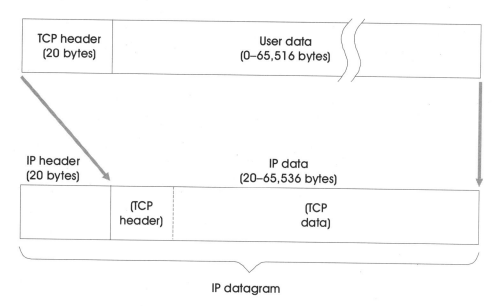

Figure 20.2 The TCP and IP packets. (*Reproduced with permission from [1].*)

self-contained so far as the network operation is concerned; each contains complete information for its routing and control.

Although the diagram only shows locations in the United States, this structure of transmission paths, routers, and ISPs is replicated throughout the world. Each part of the system is paid for by its users, who may be anyone. Thus, the Internet is completely public and open to all.

20.2.2 Software standards

The concepts of packetizing and packet switching are the keys to the effectiveness of the Internet. Packetizing standards were set early in the game and have been kept completely open so that all user computers can process their own packets. Another important part of the pie is an addressing scheme that is capable of worldwide identification and supports the routing scheme. These software standards are described below.

20.2.2.1 Internet protocol—IP

The Internet protocol is actually composed of two parts—the *transmission control protocol* (TCP) and the *Internet protocol* (IP). Each defines a packet and its contents, but they are always used together, the TCP packet being carried within the IP packet. This is referred to as TCP/IP, although it is increasingly being called simply IP.

The TCP/IP packet structure (called a *datagram*) is shown in Figure 20.2. Packets are variable-length, up to 65,516 bytes of user data. Data objects larger than that are spanned across multiple packets. The TCP software is capable of recognizing multiple-packet objects and assembling them in the proper sequence when received at their destination.

Each of the packet types has a 20-byte header. All the features of the header will not be explained here, but several important ones are:

Length field—This is a 16-bit field in each header that gives the length of the associated packet in bytes.

Source and destination fields—These are address fields for sender and receiver using the addressing method described in Section 20.2.2.2.

Sequence number—This is a 32-bit field in the TCP header that allows the packet data to be assembled in proper order at the receiver. TCP actually uses this field to convey both packet length and sequence.

Time-to-live field—This field specifies the maximum number of router hops the datagram can receive before being "killed." This recognizes that the routing system would endlessly route a packet around if no one ever accepted it. Time-to-live makes sure the network cannot become cluttered with endlessly routing packets.

Further information about TCP/IP can be found in [2].

20.2.2.2 Internet addressing

Every user around the world must have a unique Internet (IP) address. This is handled by two 32-bit fields in the IP header of every packet, one for source and one for destination. The address is structured into two levels. The "coarse" part of the address, the *netid*, identifies an individual LAN; the "fine" part, the *hostid*, identifies a particular computer on that LAN. Netids are assigned by a central authority to assure they are unique; this identifier is the only information used by the routers on the Internet. The rest of the IP address is the hostid, which is used at the destination LAN for routing to a specific computer on that nework.

The netid and the hostid are packed into a 32-bit number for use by TCP/IP. However, working with such numbers is impractical for end users, who need to have something that is easier to understand and work with. This is accomplished by *domain names*, which provides a system of text names for the IP addresses. In this system, a group of users, such as a business, is given a domain name, such as mcgraw-hill.com. Within McGraw-Hill, individual sites are named by prepending their name, delimited by a period. Thus, the sales department might be called sales.mcgraw-hill.com.

The .com ending on the domain name is the *root domain* name and provides further information about the domain. For example, .com means the domain is that of a business (commercial), .edu is a college or university, .gov is a government domain, .net is a network, and so on.

The Internet routers still must operate with the numerical addresses. Thus, any software accessing the Internet is responsible for resolving the user's request by a domain name into the actual numerical IP address. This is done by keeping tables of domain names and addresses at various places around the Internet. A site that specializes in this task is called a *domain name server* (DNS); it will respond to anyone's request for name resolution. Anyone needing an address that is not available in the local table can request a name server to help. This process, of course, requires a lot of communication, so it can slow down the initial access to a site that has not been requested before. Once the site is found, the local servers will keep the name and address on file for future accesses.

20.3 Accessing the Internet

Anyone desiring to access the Internet must have a gateway connected into the net. This is an expensive proposition and is generally not done except by large organizations who have a broad and nearly continuous need for access. However, Internet access for others has become a business, conducted by *Internet service providers* (ISPs), who maintain a gateway and provide means for users to connect to it, usually via telephone lines. Other types of services offer Internet access as part of other computer services they provide from their own computers. These are called *on-line services*.

20.3.1 Internet Service Providers

An ISP provides Internet connectivity for a fee, which can be charged on a monthly basis or on a connect-time basis. Individual users connect by telephone to modems at the ISPs site, or in some cases, they may have direct line connections. Most ISPs provide software to run on the user's computer to operate the system, and they also may provide Web hosting services (see Section 20.5.4).

An important element of cost to the user is telephone line cost. Most desirable is for the ISP to offer the user a local phone number that will be covered under his or her local phone service charge. Next best is to have an "800" number so the user has no telephone time charges. However, this can be more expensive to the ISP, resulting is somewhat higher monthly rates for this type of service.

20.3.2 On-line services

On-line services, such as America Online (AOL), Microsoft Network, or Compuserve, maintain their own computers containing an extensive service of databases, chat rooms, games, and other features. This is a style of business that developed before the Internet came to the fore. However, such services cannot compete with the volume and breadth of information now on the Internet and they have responded by including connectivity to the Internet in their portfolio of services.

As opposed to ISPs, who tend favor local customers, the on-line services are nationwide or even worldwide. Thus, they depend on long-distance telephone services for their user connections; an important part of their competitiveness is offering the best deals for the telephone connection. Because they have a large volume of traffic, they do have some leverage with the telephone companies that smaller ISPs may not have.

20.4 Connecting

Most individual users connect to the Internet over the standard dial-up telephone network (*plain old telephone service*—POTS), using a modem in their computer to communicate with a similar modem at the ISP's location.

20.4.1 Dial-up telephone lines

The POTS voice telephone network can be applied to data communications by using a modem at each end of the circuit. Modems are necessary because POTS connections are configured for analog voice communications, even though much of the telephone network is actually digital.

The dial-up network is *circuit-switched*, which means that an end-to-end connection (circuit) is established at dial-up and maintained for the duration of each call. This is advantageous for digital audio or video communication because the circuit is dedicated to the current communication and theoretically will not be interrupted or delayed.

Another major feature of the telephone system is that it is *full-duplex*, meaning that there is equal, simultaneous communication in either direction.

20.4.1.1 ISDN

As the market for point-to-point digital communications has grown, telephone companies have been developing better ways to use their networks for digital communications. This has spawned several other varieties of telephone service that provide higher speeds and other features. None of these has so far become a large factor in the market, but the most successful one seems to be the *integrated services digital network* (ISDN), which has been offered by most of the major companies.

ISDN can be used for simultaneous voice and data communications over the same lines, if desired. It is basically a digital service and must, therefore, use a modem for voice communication. Service can be provided in multiples of 64 kb/s blocks; the most common configuration uses just two such blocks, giving a total data rate of 128 kb/s. ISDN terminal hardware is considerably more expensive than POTS terminals, which is probably one reason it is not being widely used.

20.4.2 Modems

A modem is used to match digital communications to an analog voice telephone line. Phone lines provide a bandwidth of about 4 kHz, but various multilevel digital coding methods have expanded this small bandwidth to data rates as much as 56 kb/s[1]. This capability has been slowly growing over the years, but it now seems to be near its limit and no further breakthrough in phone line data rates is expected. Further increases of communication speed will have to come from the use of different lines or different channels (see the rest of this section).

20.4.3 Dedicated lines

Since most telephone systems are already digital except for the end connections, faster digital service can be provided by making special connections into the network. This has been available for some time for large users who can afford the expense of the special connections and usage fees.

The most common of the dedicated-line services is the *T-1 line*, which provides an uninterrupted data rate of 1.5 Mb/s. Such a rate can support MPEG-1 video, although an expensive T-1 line is seldom used only for that. Even higher data rates are available to those who can pay for them.

20.4.4 Cable TV connections

Cable-TV service is widely available in many areas of the world and provides a shared bandwidth of 450 MHz or more to all users. As explained in Chapter 17, this bandwidth is normally divided into 6-MHz TV channels for downstream distribution of TV programs.

However, using the digital modulation standards of the ATSC for cable-TV, a data rate of up to 38 Mb/s can be achieved on a single TV channel. Thus, digital distribution is quite feasible. The difficulty is that this is *half-duplex* (one-way) transmission—most cable-TV systems have very little, if any, upstream capability. Therefore, to implement an interactive digital service over cable-TV, other channels must be used for upstream communication. Usually, telephone lines are used, and their limited data rate capability is often sufficient for many purposes.

20.4.5 Satellite connections

Another service that is widely deployed is satellite communications, both in the form of direct-to-home (DTH or DBS) broadcasting, and point-to-point satellite communications. Both of these use geosynchronous satellites (see Section 18.3), which provide reliable communication, but they have a significant delay because of the transmission distance up to the satellite and back. This amounts to about 0.24 s, which is significant for voice or interactive digital communication. Also, most satellite links are half-duplex—down from the satellite; a satellite uplink earth station is quite expensive and is not in the range of most users.

Driven by the impetus of voice communication, new satellite systems are being deployed using arrays of satellites located at lower altitudes than the geosynchro-

---

1. Note that the 56 kb/s data rate capability is downstream only; the maximum upstream data rate of phone modems is more like 33.6 kb/s.

nous orbit to reduce the amount of transmission delay. These can provide worldwide communication between users carrying portable phones. These systems are internally digital, and digital communication service is also a possibility.

20.4.6 Special networks

Because of the opportunity for expansion of digital communications for the Internet and other uses, many different networks, both hardware and software, have been developed and some are being deployed. One of the most important of these is the *asynchronous transfer mode* (ATM), which is a packet-switching method that offers high data rates and various levels of quality of service (see Section 20.6.1).

20.5 The World Wide Web

One of the keys to the current popularity of the Internet is the *World Wide Web* (WWW). This is not another network, but rather an application-level protocol running on the Internet. Instead of the text-based communication common on most of the Internet, the WWW uses a text-based language that converts the underlying text communication to a graphical user interface (GUI). That language is the *hypertext modeling language* (HTML), which is capable of a full graphical and multimedia interface, using hypertext for navigation (see Section 15.6.1.3).

20.5.1 WWW addressing

A Web document is called a *page*, which can be displayed by *browser* software in the receiving computer. Pages are delivered to users from a *Web site* server (see Section 20.5.5). As the user moves about in the Web document, he or she may be directed from page to page on the same site, or he or she may be sent to another distant site. This is done through the hypertext links, using the *universal resource locator* (URL) addressing system. URLs are a simple extension to the Internet addressing system described in Section 20.2.2.2), where the Internet address defines the site and extensions after the address define the page within the site. The URL is always prepended by `http://www.`, which indicates that the address is that of an HTML page on the WWW. The prefix tells both the server and the receiver what software to use.

Thus, one can move directly from a Web page on a local computer to a different page on a different computer halfway around the world. This happens transparently, with a simple click on a hypertext link with the mouse button. This extreme ease of navigation is an important reason for the success of the WWW.

20.5.2 HTML

An example of an HTML page is shown in Figure 20.3 The HTML text is at the left and the page displayed by a browser is at the right. The HTML code contains the text of the page with a series of *tags*, contained within angle brackets (< >).

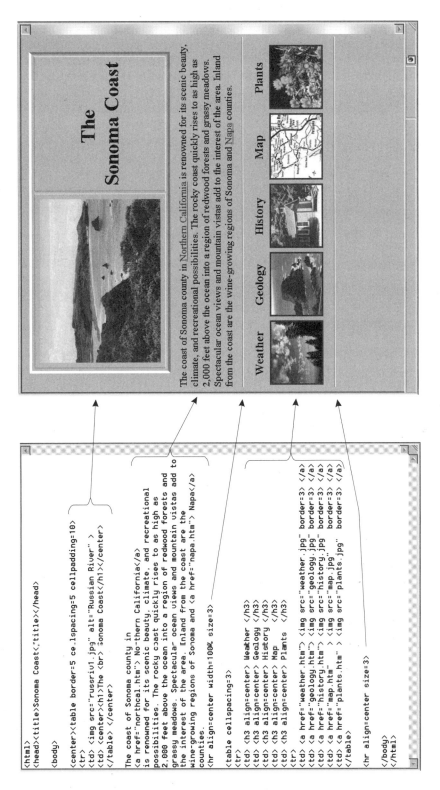

Figure 20.3 An example of HTML source code (left) and the page displayed by a browser (right). (*Reproduced with permission from [1].*)

The tags tell the browser software how to display the page. The arrows in the figure show what code controls what parts of the screen. The tags that contain the notation `href=` define the URLs for items that are hypertext links. Each of the five pictures at the bottom of the screen is actually a hypertext link to other pages on this site.

HTML is a developing language; it has gone through numerous versions as the feature set has been expanded and enhanced. It now contains audio and video features as well as the text and graphics features shown in the figure.

HTML is relatively simple and easy as far as programming languages go, but it is made even easier by various tools that are available on the market. For example, many document-creation tools, such as word processors, now contain a feature to export the document as an HTML page. The user simply creates the document normally in the word processor and, at the click of the mouse button, the page is transformed into HTML for use on the Web. Other tools are dedicated to HTML creation from scratch, but they still operate graphically and hide the actual HTML code from the user. It is reaching the point that anyone familiar with computer use can create Web pages.

20.5.3 Browser software

The receiving computer of a Web page must run a browser program to display the page correctly. Several browsers are on the market (Microsoft *Internet Explorer*, Netscape *Navigator*, etc.). They all do the same thing in displaying a particular version of HTML, because that is standardized, but they may have additional nonstandardized enhancements. This causes confusion in the market, so that many Web sites find it necessary to specify which browser correctly displays their pages.

Browsers also may have additional features beyond the displaying of pages, such as e-mail, support for animation (see Section 20.5.4), and so on.

20.5.4 JAVA

Web pages created in HTML are static; they are simply displayed and show no movement or animation. Something else must be used to create dynamics on a Web page screen. Various approaches have been used to accomplish this, but the most popular of these is the *Java* language developed by Sun Microsystems.

Java is a text-based language that can be embedded in HTML pages. It is run by being interpreted by a special module in the browser. Various dynamic effects are easily accomplished as well as actions within the Web page that are achieved without requiring the loading of a new page from the Web server. For example, if the user is required to enter information into the Web page, that can be verified by a Java program before it is sent back to the Web server for further action.

As with HTML, development tool programs are available for Java. Because it involves dynamics, Java programming is more complex than HTML, but it is still something that is easily used by a programmer of general experience.

20.5.5 Web hosting

Putting a Web site on the Internet requires an Internet connection and a dedicated computer that will respond to requests for display of the site's pages. Large users have their own equipment and software for this purpose, but smaller users can get their pages on line through ISPs who offer *Web hosting* services. They provide the Internet connection and the hosting computer; the user simply delivers a set of HTML files that define the site's content. The ISP offers a specified amount of storage for the site's files and charges a monthly fee for the service. By this route, almost anyone can have a Web site. It is another of the features that have made the Web grow so rapidly.

20.6 Internet Video Architectures

The Internet is capable of distributing any type of digital data and, with the Web, it provides comvenient means for interactivity. Thus, it could be used for presentation of video or audio on demand from users. However, there are several important caveats.

20.6.1 Streaming video

Real-time video playback requires a continuous flow of the video bit stream. This is called *streaming*. If the bit stream is interrupted or changes speed, the video playback can be affected. A certain amount of buffering at the receiver can help accommodate such disturbances, but it costs in RAM and it introduced a delay equal to how many frames are buffered.

In the extreme case, the entire video bit stream can be buffered before it is played back. This is called *store-and-forward* and it requires hard disk storage but, of course, it is not affected by any transmission glitches as long as the entire bit stream eventually is received. However, there is a delay at least equal to the transmission time of the entire video clip.

20.6.2 Quality of service

In digital communications, the system performance is often described as *quality of service* (QoS). This may include the following factors relative to video service:

Data rate—With compression, the higher the data rate, the better the pictures will be. Internet data rates are limited by the type of connection one has to the ISP and also the loading on the Internet itself. What matters for real-time video transfer is the sustained rate that can be achieved. At the 56 kb/s speed of a phone line and modem connection, video is severely limited to very low resolution and very low frame rates. As faster connections come into use, the problems of the Internet itself will become more important.

Error rate—Compressed video is more sensitive to errors than uncompressed video because every bit in a compressed stream is doing much more and all redundancy has been removed. BERs of better than 10^{-6} are necessary.

Delay—There is a delay in transmission caused by the transmission time through all the cables and connections involved. This is a fraction of a second and would not be a problem so long as it is constant. However, the packet switching of the Internet can cause a different delay for each packet because they may not all get the same routing. Transmission may be interrupted by other messages. Packets can even arrive out of order at the receiver. Of course, the TCP software will assemble them back in the correct order, but a delay will occur while TCP waits for a missing packet to appear.

Present-day Internet connections generally do not deliver good enough QoS for real-time video or audio playback. However, it is the objective of many Internet initiatives to improve the QoS to the levels required by video or audio. This is a challenge to achieve at affordable costs for most users and in the face of continually increasing volume of service.

20.7 References

1. A. C. Luther, *Principles of Digital Audio and Video*, Artech House, Norwood, MA, 1997.
2. M. J. Riley and I. E. G. Richardson, *Digital Video Communications*, Artech House, Norwood, MA, 1997, Ch. 4.

Chapter 21

The Future of Video Systems

21.1 Introduction

The goal of video engineers is to create systems that speak to the human senses of hearing and vision to entertain, educate, or convey information. The accomplishments of the first 50 years, described in the preceding chapters, are ample proof of their success. But what lies beyond that?

DTV, including HDTV has just been introduced—its technical capabilities have been proven—but success depends on many other factors that are not yet concluded. Will consumers buy HDTV? Will programs be produced for it? Will it replace present TV? How soon will it happen? These questions and many others probe the technology and marketing factors that determine a new product's success or failure. This chapter explores those questions and other issues that may affect the path of video engineering into the next century.

A warning to the reader: This chapter contains forecasts of the future, which are inherently shaped by the writer's opinions since there is nothing factual about the future until it becomes present or past. Therefore, the reader is free to disagree as much as he or she pleases. However, this discussion should prove stimulating and may trigger some surprising new thoughts on the part of the reader.

21.2 Technologies in Video Engineering

One of the interesting aspects of video engineering is the breadth of technology that it involves (see Table 21.1). No single engineer can be master of all these skills, but a senior video engineer must be familiar with them all at a level sufficient to understand their interactions and when more in-depth expertise may be required. This list of technologies is also useful when considering trends for the purpose of forecasting what may happen in the near future. The following sections discuss such considerations.

Table 21.1 Video engineering technologies

Technology	Areas of applicability
Audio	Creation, recording, receiving
Colorimetry	Cameras, displays
Communications	Transmission, reception, recording
Digital	Everywhere
Ergonomics	Cameras, recorders, control panels, user interfaces, receivers
Materials	Cameras, recording, transmission
Mechanical	Cameras, tape transports, enclosures, antennas
Optics	Cameras, recorders, displays, transmission
Solid-state electronics	Custom LSI, everywhere
Systems	Everywhere

21.2.1 Trends in solid-state devices

The progression of solid-state technology described in Figure 8.1 is expected to continue well into the 21st century. More circuits per chip, larger chips, lower cost per device, all result from this. General-purpose devices such as microprocessors and RAM chips will continue to become more powerful and lower in cost. There also will be progress in improving and simplifying the techniques for making custom chips so that more discrete functions can be integrated on custom chips. This will continue the trend of reducing size and cost of the electronic portions of all products. The result will be that products can include more sophisticated electronic functions while, at the same time, they become less expensive. The current introduction of DTV depends on this trend for its market success.

The trend of solid-state technology underlies much of the progress in video systems and their use. As systems simultaneously become more powerful and less expensive, the range of practical application of video is broadened. Applications which could not afford video a few years ago may now embrace it simply because it has become a cost-effective approach. One class of application that may increase its use of video is education. However, the development of courseware is an important element in educational applications and the cost of doing that must also be brought in line with the lower hardware costs. That is a technical and marketing challenge.

An important component of video systems is the imager. The trends of solid-state technology also apply to CCD imagers and higher resolutions at lower costs will continue to occur. This is a driver that will result in improved image quality of video systems at all cost levels.

21.2.2 Trends in digital technology

Most features of digital video systems are based on principles that were developed in the computer industry years ago. They have simply become practical for digital video because of continued cost reduction of digital functionality. However, because digital video products are new, there is a lot of room for improvement and the process of refinement will go on for some time. An overwhelming breakthrough is unlikely in the near future, but there will be new applications introduced and essentially all analog functionality in video systems will be replaced by digital devices over the next five years.

In the digital era, hardware engineers are even more likely to say "we can do *anything*," but the practicality and success of new products will depend more on their market suitability and presentation as much as or more than their technology (see Section 21.3.2).

21.2.3 Trends in optical technology

Communication, cameras, recorders, and displays depend on optical technology. Great improvement in materials and devices has been made, but optical technology is still expensive for many uses. The following paragraphs discuss each of the application areas listed above.

21.2.3.1 Optical communication

As digital systems demand higher bandwidths, the alternatives to optical transmission technology begin fading away. This is resulting in many applications for optical communication despite the cost and it is increasing the momentum of optical development. Costs for fiber-optic cable and transducer devices will continue to reduce and their volume of use will increase. This is one of the drivers of the information superhighway (see Section 21.4).

21.2.3.2 Camera optics

Camera lenses and prisms are mature components that are simply undergoing slow refinement and cost reduction as the volume of use increases. CCD imagers are adopting their own optical features, such as the on-chip lenses that increase sensitivity of interline-transfer devices (see Section 9.3.3).

21.2.3.3 Optical recording

Although the laser optical videodisc has been on the market for more than 20 years, it never became a mainstream product. The world waited until the digital audio Compact Disc (CD) to embrace an optically-recorded product. Since then, various spinoffs have been developed (see Section 10.18.1), mostly for computer applications. The latest CD spinoff, for digital video program distribution, is the DVD disc. This will compete with prerecorded videotapes for the home video play-only market.

In any case, the DVD brings with it a substantial advance in optical recording density, which is applicable to other uses such as a high-capacity CD-ROM, which does not require the massive consumer-market infrastructure to get started. DVD drives are aailable for all new PCs, and the early market acceptance is very promising. The recordable version of the DVD-ROM (DVD-RAM) is not far behind.

It is easy to predict that optical recording density will increase by as much as another 10-fold in the next five years, but it's harder to say where it will make the greatest market penetration.

21.2.4 Trends in display technology

In both computers and television the trend is to larger, higher-resolution displays. Although CRT technology is currently dominant, flat-panel displays have highly desirable features for the market and much research and development is going into this area. The portable computer market has been an effective proving ground for this work. LCD displays for desktop computers are also beginning to appear, but their prices are still high. The products are attractive and prices should fall as volume builds.

As HDTV comes on line in the next few years, the pressure on new display technologies will increase even further because the HDTV market will need displays that are at or beyond the practical limits of CRT size. This may be the opportunity for flat panels to move into the television market in a big way. At this point, many laboratories are addressing the problems of large, high-resolution flat panels; at this point the challenge is not the technology so much as it is low-cost manufacturing. If the developments are successful, it could provide the breakthrough that the display market needs.

When display components are readily available in the HDTV 16:9 format, they also will be used for personal computers. Thus, the market for 16:9 flat panel displays will be even larger than the TV market.

21.2.5 Trends in audio technology

The audio market was the first consumer application of digital technology. The audio CD has taken over the prerecorded audio market from records and cassettes. This success is a result of the superb quality and reliability of digital technology and the convenience and durability of the optical medium.

In audio production and postproduction, digital audio is also taking over as both digital tape and hard disk digital recorders become available. In the long run, hard disk recorders will probably prevail because of their fantastic editing features. Tape will become a backup and archiving medium, much as has happened in the computer business.

21.2.6 Trends in recording technology

Magnetic recording is the primary technology for video recording, but optical recording has become important for audio and may soon be an important video medium as well. As shown in Figure 10.9, magnetic recording density has steadily improved over the years, resulting in recorders that are smaller and less expensive to operate. Magnetic video recording has also made the transition from analog to digital and new video recorders in all fields should be digital from now on.

Further advances in both magnetic and optical recording are certainly expected—the density curve should continue to go up and optical recording costs should continue to come down.

21.2.7 Technology conclusions

The bottom line of technology trends is that all the technologies of video engineering have potential for improvement and none are up against brick walls. In several places (recording and displays), there are even possibilities for a breakthrough. Thus, the deployment of video systems in the foreseeable future should not be limited by technology considerations. But there are other factors...

21.3 Video Systems

A video system is a set of equipment, software, and connections that provide a specified end-to-end service to its users. Examples of video systems are:

television broadcasting
cable television
prerecorded video programs and movies
multimedia program distribution
satellite broadcasting
picture-phone

Each of these systems includes all the ingredients shown in Figure 1.1 for creation, transmission, and reception. These elements are called the *infrastructure* of the system, which provides program content material, transmission paths, consumables (such as tapes), and the means whereby users acquire the necessary equipment. Systems also require standardization so that signals and equipment will be compatible between different sources and manufacturers. Video system considerations are discussed below.

21.3.1 Video systems engineering

There are two levels of video systems engineering. The first is the process of creating an entirely new system using available technology to provide a new service.

This is discussed in the following sections. The second class of systems engineering is the specification of a video system using existing standards and specifications, for example, the design of a new broadcasting station. This type of systems engineering will not be discussed further.

21.3.2 Developing a new video service

The first step of creating a new video service is the specification of what that service will do and who the users will be. This is actually the most crucial part of the entire development project because it determines where the market is and the potential market value of the service. These two things have to be defined and understood at the outset and all subsequent steps of the development must be consistent with these objectives both in content and cost. Many projects are doomed to failure because this market focus doesn't exist or it gets lost; it's too easy to let the technological issues overwhelm the project to the detriment of the market objectives.

Once the service objective makes sense in market terms, technological choices can begin so that the entire development can be planned and cost estimated. All of the considerations of infrastructure must enter at this stage and be included in the projected cost. The cost of infrastructure can often be greater than the equipment development cost, so much so that no single entity can handle it. This leads to the "chicken and egg" syndrome, where equipment cannot be sold until the necessary infrastructure is in place, but no one wants to invest in infrastructure until equipment is selling. Many new systems face this dilemma.

A classic example was NTSC color television, which was standardized in 1953 but it took more than 10 years after that before receivers began selling at a level that generated any profit. During that 10-year gestation period, investors in color TV were acting on faith because there were no profits. If they hadn't stuck it out, color TV as a worldwide service might never have happened.

Most new systems require a good strategy to overcome "chicken and egg." For example, the home digital cassette system (DV—see Section 11.9) was introduced first in camcorders, which are usable by consumers even though there are no prerecorded tapes available. When the value of the system has been established by camcorder sales, DV players could be introduced to play tapes made in camcorders and then prerecorded tape suppliers could begin offering the DV format. Eventually, DV might replace all analog home recorders. However, this scenario already has competition from the DVD disc for prerecorded program distribution. I believe this is already a lost cause and a market for prerecorded DV tapes will never develop.

21.3.3 Standards

The purpose of standards is to remove technology decisions from the marketplace. When a product is properly standardized, products, signals, and media from all manufacturers or sources will work interchangeably. The user does not have to worry about compatibility issues and he/she can base buying decisions on other considerations such as features, performance, and price. The value of the user's investment in equipment and media is also increased because there is a wider range of use for the system.

With standards, manufacturers also know exactly the requirements their product must meet to guarantee compatibility with other products in the field. Well-conceived standards define only what is necessary for compatibility and still allow manufacturers to develop their own competitive edges within the standard.

However, standards for new systems are not easy to achieve because the developers of new systems do not want to compromise their competitive positions by sharing information with potential competitors in order to achieve standardization. But history has shown time and again that offering incompatible products to the market for the same service will, in the long run, delay market acceptance, alienate customers, and produce one or more losers in the manufacturing community. The battle in the initial VCR marketplace between Beta and VHS is a good example. When preparing to introduce a new service, it makes no sense to proceed without achieving a standard among potential competitors.

Fortunately, there are several industry organizations who develop standards and the industry will also form an ad-hoc organization to create standards for an important new service. A recent example of the latter approach is the Grand Alliance DTV system. The SMPTE has been very active in video recording standards since the original quadruplex days, and essentially all video recording formats are standardized by the SMPTE.

A new aspect of standardization comes with the emergence of software-driven video products, which is something the computer industry faced long ago. With software, it is primarily necessary to standardize the interfaces—between components and between software modules. If a product offers a standard API for its software, writers can create any software they like to deliver new features, and their software will run on all hardware that provides the standard API. This offers a new degree of flexibility to video products.

21.4 Communications

Video systems all require means to move video or other information from one location to another—communication. This may be by means of cable, broadcast, fiber optics, or by physically carrying a recorded medium package from one location to another. In the digital era, communication consists simply of moving bits and performance is equated to a sufficient bit rate and bit error rate for the intended service.

A digital communication path can be used for many types of information simply by changing the meaning of the bits being communicated. This opens the possibility of general-purpose digital communication services that will convey bits for any purpose and for anyone, just like the present telephone network offers voice communication to anyone who is connected. The idea of a high-speed two-way digital communication network connecting homes and offices around the world has been given the popular name *information superhighway* (ISH).

The ISH would be the universal medium for worldwide communication of voice, video, business data, and anything else that can be represented by bits. It is obvious that whoever controls such a network has tremendous political and economic power and for this reason many companies and countries are pursuing the development of this concept. The investment required is enormous and it is unlikely that a single entity can pull it off—it takes an international consortium.

One system already in place that fits the model of the ISH is the Internet, which was originally developed as a computer network for military and academic purposes but is now a public network. Some have proposed that the Internet might evolve into the ISH, and that is a distinct possibility. However, a massive upgrading of Internet speed and traffic capacity is required. Many initiatives by the telecommunications companies are being pursued to achieve this.

Another contributor to the ISH is CATV service. Cable systems have analog bandwidths of up to 300 MHz, which would be capable of handling high-data-rate digital signals. The ATSC DTV standard contains means for transmitting 38.6 Mb/s in each 6-MHz TV channel of a cable system, which would give a 50-channel cable system a total digital transmission capacity of 1.93 Gb/s and it is already connected to homes! But again, that is only part of the solution because cable systems are local entities that are not interconnected worldwide, protocols are not developed for general digital communication via cable service, and most cable systems are only one-way.

Still other potential ISH providers are satellite communication companies, who have existing capability for high-speed digital communications worldwide. Again, however, the protocols for using satellite distribution for general digital communication are not developed, and the cost of connecting everyone to a satellite system is high, especially when two-way communication is considered.

The above discussion shows that no one has a complete handle on the ISH today—it is an exciting idea that one can already observe in use with the Internet. There is much activity in industry and government to expand this concept and a solution will probably evolve in the next five to ten years.

21.5 New Video Systems

Although the march of technology will help the existing video systems to improve and become less expensive, it is more interesting to explore what new systems might become possible. Many possibilities have been discussed in the press and the literature, usually based on the technology aspects only. As mentioned above, the business and market considerations are much more difficult to develop and these will probably prevent some of the new ideas from ever becoming commercial successes. However, here are some of the ideas.

21.5.1 HDTV

DTV, including HDTV modes, has been standardized in the United States and in Europe, although the standards are somewhat different. As described in Chapter 7, DTV offers exciting features and it can operate over existing broadcast and cable channels. Although it is offered as an eventual replacement for present analog television systems, it is not compatible in that the public must buy new receivers to enjoy the service. Likewise, program producers must buy new origination equipment to digitize existing programs and to produce HDTV programs. This is worse than the chicken-and-egg situation that existed at the start of color TV in 1953, where at least there was enough compatibility that existing receivers could view color programs in monochrome. With digital HDTV, existing receivers will receive nothing on the DTV channels, so DTV broadcasters are addressing only viewers who have new receivers or digital adapters boxes for old receivers.

Existing analog program material can be converted to DTV format for broadcasting but the picture and sound quality will not be much better than analog broadcasting. This does not create much incentive for consumers to buy new, more expensive, DTV receivers. What will sell DTV receivers is HDTV. DTV manufacturers will surely be very creative in building features into digital receivers that will be usable even without much actual HDTV program material. But the receiver base may grow very slowly.

Program producers have been improving their picture quality recently by using digital component video systems in production and postproduction. This is advantageous to them because it allows programming to be converted to and released in various formats, such as NTSC, PAL, videotape, even film. That increases the market value of their programming. Similarly, programs produced in existing digital component formats can be converted to HDTV with improved picture quality. The problem with this is the aspect ratio, which cannot be fully converted. Material produced for 4:3 viewing must either be shown on HDTV with black areas at the edges or it must be enlarged to full width with loss of information at the top and bottom. Either way, some of the quality advantage of the component origination is lost.

A source of program material that is valuable for HDTV is motion picture film that has been produced for wide-screen theater viewing. With a digital HDTV telecine, this material converts beautifully to HDTV. To the extent that motion pictures can be shown on TV (due to rights considerations), films are a source of programming that can fully exploit the performance of HDTV. Some TV programming is actually produced in wide-screen format for possible showing in theaters; in this case, the 4:3 aspect ratio TV viewers have to take the disadvantage of aspect ratio conversion, but such material will show perfectly on HDTV.

HDTV is exciting, but it suffers from initially high costs and poor availability of full-quality program material. It is now a marketing challenge to bring this system up to broad usage.

21.5.2 DVD

Another new service that is now on the market is the *Digital Versatile Disc* (DVD), which is a spinoff of the CD-ROM for video program distribution. (In the previous edition, I called this *Video CD*.) DVD is in the early stages of market acceptance, but a large volume of prerecorded discs is now available, which alleviates much of the chicken-and-egg problem. It appears to be on its way to replace VHS videotapes for video distribution to the home. The principal deterrent is the large number of videotape libraries in existence—this is a major investment that cannot be replaced overnight. However, I think DVD will be the principal video distribution medium within five years.

As expected, DVD has spawned DVD-ROM players in the computer market, which provides additional outlets for players and discs. Since the computer market usually adopts new technologies faster than the consumer market, this further speeds the acceptance of DVD in the consumer market.

Recordable DVD drives are also entering the computer market, which is happening much faster than it did with the CD-ROM, but in this case, the technology is already available and it is just a matter of standards. At this writing, there is still competition for recordable-DVD standards, and this must be settled before recordable drives will begin selling in large quantities. That will happen in the next few years.

21.5.3 The Information Appliance

With television becoming digital in the form of DTV, a combined television-computer product is technically feasible. The same product could be used either for computing or TV viewing or, through on-screen windowing, both at the same time. This would be valuable to some customers, but for the bulk of the TV market, it is probably not desirable. That is because of the basic difference between the viewing environment for computers and television. Computers are operated at viewing ratios as low as 1:1 because the single user sits close to the screen. The typical television viewing situation has several people sitting across a room from the display, so viewing ratios are much higher—4:1 or more. This means that a TV display must be much larger so the viewers can see all the picture detail from across the room. (This statement assumes that HDTV and computers have similar resolution performance, which is already true at the high end of the PC market.) Because serious computing usually has only one user at a time and other materials such as books and papers are often required, computers are typically set up on a desktop rather than in a living room environment. A single display configuration does not readily serve both uses.

There is another, broader, version of the information appliance that may have more potential, although it will be more difficult to develop and sell. That is the *Home Information System* (HIS), which is a digital network in the home that integrates video, audio, computer data, and home control into a single system.

Home networks have been marketed before with little success because they were proprietary systems to one manufacturer, they did only one of the above functions, and they were too difficult for the consumer to understand, install, and operate. The HIS faces the same problems, but its value to the user is greater because it integrates all home communications and, with standardization, it can be nonproprietary so multiple manufacturers can compete for the parts of the system. The understanding, installation, and use problems can be overcome with proper design of hardware and software.

The HIS concept is not new; in fact, exactly the same idea was described in the previous edition of this book four years ago. Much progress has been made in the technology needed, but a system is still not ready for deployment. Figure 21.1 shows an example of an HIS, configured for a family of four.

21.5.3.1 The HIS server

The heart of the system is a powerful PC that provides interface to the outside world, storage of local data, and network control—this is the HIS server. The server has its own large-screen flat-panel display—probably located in the kitchen. All displays are in HDTV 16:9 aspect format, although they may have different resolutions. The kitchen display might be fairly high resolution and it could serve as the family bulletin board (like the notes stuck on the front of the refrigerator), a local TV display, and the entry point for home control, system control, door and phone answering, etc. The server has a multitasking operating system, so it keeps the rest of the system going while family members interact with its display; it could even support more than one person using it at a time by having multiple mice with different cursors operating on the same screen. If the server display is full HDTV resolution (1920 × 1080) and roughly 30 inches wide, separate windows could be simultaneously displayed for each of its functions.

21.5.3.2 ISH input/output

The server accepts the home's connection to the ISH and processes data between it and other locations within the home. If telephone connections are separate from the ISH, they also are handled by the server. For telephone, the server performs answering service, voice mail, intercom, and routing of phone connections. If future phone connections are via the ISH, the server provides the same services from that source.

Video coming in over the ISH is either stored in the server or routed to another location for decompression and immediate display. Because the home network handles only compressed video, displays that show video must have decompression capability built in. The displays generally also have capability to display data windows along with video, so the communication and control capabilities of the system can be accessed from any location.

At least one video digitizing input is available on the server. It has a video switcher so that the one digitizer can support cameras for door answering, room surveillance (baby's room, for example), and local camcorder use. The server stores the digital video or routes it over the home network to any location that requests display.

21.5.3.3 HIS displays

Small video displays are used everywhere except the server and the living room. The small displays are lower resolution, such as 854 × 480, but the server and living room displays are 1920 × 1080 for showing full-resolution HDTV or multiple computer windows. The living-room display also has full surround-sound

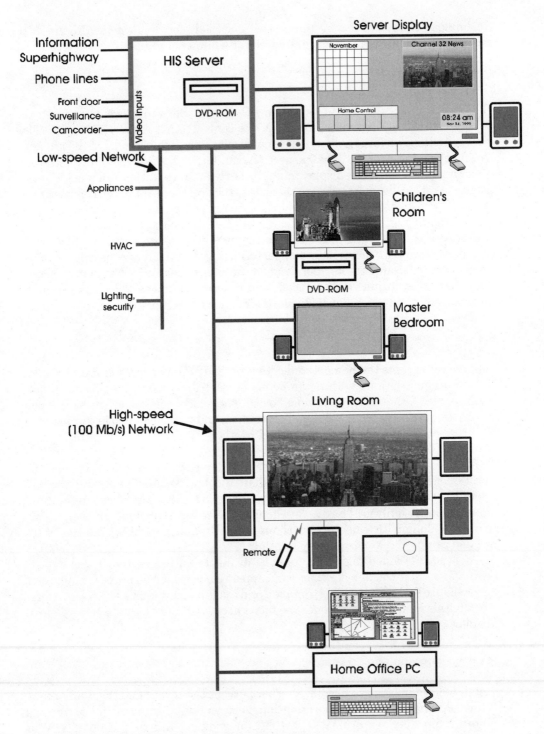

Figure 21.1 Home information system of the future.

speakers.

All displays have a button that can be touched or clicked to gain access to the server menus for program selection, communication, or control. Displays also have touch-screen, wired mouse, infrared remote mouse, and/or keyboard as required. To support these operations, all displays must have a local CPU, I/O, and memory, but they rely on the server for mass storage.

If the home includes a home office computer, that is probably a standalone PC, but it should be connected to the home network to utilize the I/O and storage capabilities of the HIS server.

21.5.3.4 Developing HIS

The HIS concept is something that requires concerted industry effort to accomplish because of the number of different products that are involved. That is unlikely to happen all at once, but a number of industry groups are working on the interfaces and components needed by such a system. With that done, manufacturers can make their own decisions about providing those interfaces in their products, either built-in or as options. Software developers can work on approaches to create the server device using a standard computer as the platform and system marketers can begin promoting the system concept. This way, the system is achieved in an evolutionary way.

A computer network like the HIS is an extremely difficult technology from the point of view of the people installing and operating it. Bringing such a complex concept into the home and making it usable by unskilled consumers is a tremendous challenge. Successfully solving this problem is essential for the HIS to be accepted by consumers.

In today's terms, the HIS looks like an expensive concept and it is. However, with the steady cost reduction of system components that can be counted on in coming years, it won't always be so expensive. Also, homeowners can begin with the server and add the peripherals over time to spread the cost over a number of years. In the end, it still depends on whether homeowners will buy the concept. Another marketing challenge.

21.6 Conclusion

As shown in these examples, the digital video era offers many possibilities, including super-quality television (HDTV), high-quality home video production (DV), worldwide video communication (ISH), high-quality prerecorded video (DVD), and computer and TV integration (HIS). The next five or ten years should be a very exciting and challenging time for video engineers as they work to bring these and other new developments to fruition.

Glossary

3:2 pull down A technique for presenting 24 frames/second motion picture film on a 30 frames/second interlaced TV system. One film frame is shown for three TV fields and the next film frame is shown for two TV fields.

4:1:1 In a digital component system, refers to subsampling of color-difference components by 4:1 in the horizontal direction only.

4:2:0 In a digital component system, refers to subsampling of color-difference components by 2:1 in both horizontal and vertical directions.

4:2:2 In a digital component system, refers to subsampling of color-difference components by 2:1 in the horizontal direction only.

A-B roll editing An editing method that uses two source recorders (A and B) playing synchronously and combining their outputs through a video switcher to a third recorder in the Record mode.

active-matrix LCD A liquid-crystal display that has a transistor array behind the panel to substantially improve the speed and contrast of the display.

acuity, visual The ability of the human eye to distinguish fine detail. It is usually expressed in seconds of arc.

adaptive differential PCM (ADPCM) A form of differential PCM that adjusts its parameters dynamically according to the characteristics of the signal being encoded.

ADC See analog to digital conversion.

additive color A color reproduction system that mixes colored lights (usually red, green, and blue) to display an image. (See also subtractive color.)

address In computers, the use of a digital word to specify a location in a memory array. For example, 16-bit addressing can specify 65,536 locations.

ADPCM See adaptive differential PCM.

Advanced Television Systems Committee A committee in the United States charged with completing and maintaining the DTV standards.

algorithm The specification of the steps that perform a process, such as drawing a circle or compressing an image.

aliasing An artifact produced by sampling a signal that contains frequency components higher than one-half the sampling frequency. In horizontally sampled images, aliasing produces jagged lines.

alpha wrap In a helical-scan tape recorder, a configuration where the tape wraps completely around the scanner in the manner of the Greek letter alpha (α).

amorphous silicon A noncrystalline form of silicon.

amplitude modulation (AM) Modulation that is accomplished by varying the amplitude of a carrier signal in accordance with the information to be transmitted.

analog The expression of a variable as a continuous function of time, amplitude, etc.

analog to digital conversion (ADC) The process of converting an analog signal to a digital bit stream. (See also sampling and quantization.)

aperture (1) In image scanning, the size of the sensitive spot that moves over the image. (2) In an antenna, an imaginary cylinder in space surrounding the radiating elements.

aperture correction In video cameras, an electronic process that corrects for some of the high-frequency losses due to aperture response.

aperture response In video cameras, the frequency response characteristic due to the effect of a finite-sized scanning aperture. It is usually specified in terms of the modulation transfer function.

application programming interface (API) In computers, a software interface for use by programmers writing applications for that system.

application software In computers, the software that performs a specific end-user task, such as word processing or drawing.

artifact Any unnatural component. With respect to images, distortions of all types are called artifacts.

aspect ratio The ratio of the width to the height of an image. For example, analog TV uses a 4:3 (1.33) aspect ratio, while high-definition TV has an aspect ratio of 16:9 (1.78).

assemble editing In video postproduction, single-recorder editing that is done by adding new material onto the end of a previous recording.

asynchronous transport mode (ATM) A data communication transport system based on packetization.

ATM See asynchronous transport mode.

ATSC See Advanced Television Systems Committee.

azimuth recording In video recording, the technique of offsetting the azimuth of adjacent tracks to reduce cross talk between the tracks at short wavelengths. (The azimuth is the angle between the head gap and a perpendicular to the track edge.)

basic input/output system (BIOS) In computers, built-in software (usually in ROM) that performs the interface between the system's permanent hardware and the operating system software.

beam tilt In broadcasting antennas, the technique of directing an antenna's vertical pattern slightly below the horizontal to improve fringe-area signal strength.

beam width In broadcasting antennas, the angle between the half-power points of a directivity pattern.

BER See bit error rate.

binary A number system based on powers of 2. A binary digit is a bit.

BIOS See basic input/output system.

bit A digit of a binary system. It has only two possible values: 0 or 1.

bit error rate (BER) In digital transmission or recording systems, the average rate at which bit errors may occur. For example, a BER of 10^{-4} means that, on the average, there will be a bit error for every 10,000 bits received.

bitmapped graphics Graphics or images that are defined by specifying every pixel value in the image. For example, a 640 × 480 bit map will contain 307,200 pixel values. (See also vector graphics.)

black matrix In cathode-ray tubes, a technique to make the screen appear black when no pixels are illuminated. This improves the contrast by reducing the reflection of incident illumination.

blanking In an analog video signal, the periods between active lines and frames, intended to provide time for retrace of the display scanning beam. Blanking intervals are normally at black level or blacker-than-black and they also are used for transmitting synchronizing information. Blanking may be eliminated in digital video signals.

blooming An artifact of a camera imager where an extreme highlight in the scene causes a spreading white area in the reproduced picture.

Boolean algebra A branch of mathematics named for the English mathematician, George Boole, that deals with the relationships between and manipulation of binary numbers.

browser Application software designed for viewing information from the World Wide Web. Its key feature is an interpreter for the HTML language.

burned-in time code In videotape recording, the technique of placing time code numbers in the picture for off-line editing.

bus In computers, a parallel interconnection of all the internal units of the system. It provides for transfer of data, address, and control information.

C-band The frequency band from 3.0 to 10.0 GHz.

cable television (CATV) A television system where programs are distributed to users over coaxial cable or fiber-optic cable. It provides for multichannel service in areas where broadcast service is limited or to augment broadcast service in other areas.

cache memory In computers, a block of fast memory used to temporarily hold data for use by the CPU. It provides a significant speedup when information from main memory or mass storage can be moved into cache memory before the CPU needs it because the CPU does not have to wait for access to slower memory.

camcorder An equipment that combines a video camera and recorder in one unit.

cassette A box that holds two reels of tape for use in an automatically threading tape recorder.

cathode-ray tube (CRT) A video display device that operates by electron-beam bombardment of an electroluminescent phosphor in a vacuum. There are monochrome and color CRTs.

CATV See cable television.

CCD See charge-coupled device.

CD-ROM An acronym for Compact Disc-Read-Only Memory. (See also Compact Disc.)

central processing unit (CPU) In a computer, the main processor that performs data movement and manipulation. In a personal computer, the CPU is usually a microprocessor chip.

characteristic impedance In a transmission line, the ratio of the voltage to the current in the line. If the load impedance at the end of the line equals the characteristic impedance, all the energy coming down the line is absorbed. If the load impedance does not match the characteristic impedance, some energy is reflected at the load and travels back the line. (See also voltage standing-wave ratio.)

charge-coupled device A solid-state imaging device that stores an electrical charge representation of the optical image by means of photoconductivity. A readout mechanism converts the charge image to a video signal.

chroma keying A means for combining two video signals where a color in the foreground image specifies whether to show a second (background) image. For example, an announcer standing in front of a blue screen can have a second image replace all blue areas, giving the appearance that he is actually standing in the second image.

chromaticity diagram A standard diagram that shows the chromaticity coordinates of the visible spectrum and on which any color can be plotted. It was originally developed by the CIE (Commission Internationale de l'Eclairage) in 1931. (See also trichromatic.)

chrominance In color video systems, the signal component(s) that describe color-difference information.

CISC See complex information-set computer.

clamping In analog video systems, the process of setting some portion of the video signal to a fixed level to restore the dc component. Usually part of the blanking interval is used for clamping.

clock frequency (1) In a computer, the operating frequency of the CPU. (2) In digital transmission systems, the basic rate of data transmission.

CLUT See color lookup table.

coded orthogonal frequency-division multiplexing (COFDM) A modulation method where a multiplicity of orthogonal (noninterfering) carriers are modulated by the information signal. It can be designed to be very immune to multipath

coercivity A measure of threshold magnetizing force for a magnetic material.

COFDM See coded orthogonal frequency-division multiplexing.

color burst In an analog color video signal, a burst of the color subcarrier frequency placed in the blanking interval for subcarrier synchronization purposes.

color difference In a color video system, signals that represent the difference between specified colors and the luminance component. They have the property that they go to zero for monochrome images.

color lookup table (CLUT) In digital bit-mapped images, the bit map data value for each pixel is an index into a color lookup table that reads out the actual pixel value. For example, an 8 bits/pixel bit map can have only 256 different values in its bit map, but by using a CLUT containing 18-bit values, the 256 colors in the image can be selected from a palette of 262,144 colors. This improves color rendition if the CLUT colors are carefully selected for each image.

color-under In analog video recording, a system of color encoding that encodes the chrominance signal in a frequency band below the frequency range of the luminance FM signal. It is used in home recording systems.

colorimetry (1) The science of color. (2) In a video system, the specific colorimetric parameters of the system.

comb filter A transversal filter designed to produce a frequency response that has a series of nulls at equally-spaced frequencies. For example, a comb filter can be used to separate the interleaved chrominance and luminance sidebands in an NTSC video signal.

comet tails In video cameras, an overload artifact that shows tails behind moving highlights.

Compact Disc (CD) A 12-cm diameter digital optical disc used for audio or computer data storage and distribution. The computer format is called CD-ROM.

compiler In computer software, a program that reads a high-level language such as C, Pascal, or BASIC and generates microprocessor instructions in the form of an executable file for later use.

complex instruction-set computer (CISC) A central processing unit that has a large instruction set of both low- and high-level operations. Typically, different CISC instructions take different numbers of CPU clock cycles to execute, proportional to their complexity. (See also reduced instruction-set computer.)

component signal Signals in a color video system where separate signals represent aspects of the picture—typically three. Examples are R,G,B; Y,I,Q; and so on.

composite signal A color video signal that multiplexes all the picture information (components) into a single signal, such as NTSC, PAL, or SECAM.

compression In digital systems, the process of data manipulation to reduce the amount of data needed to represent information, such as a text file, a still image, motion video, or audio.

constellation diagram In a modulated digital transmission system, a diagram that shows the allowable modulated-signal states, usually shown in terms of modulated-signal amplitude and phase.

contrast In an image, the ratio of the brightest to the darkest areas. Sometimes called "contrast ratio."

contrast compression In a video camera, special processing to improve reproduction of a scene that has a contrast ratio higher than can be handled by the video system.

CONUS an acronym for "Continental United States."

CPU See central processing unit.

cross color In a composite color system, distortion caused by luminance components appearing in the chrominance channels.

cross luminance In a composite color system, distortion caused by chrominance signals appearing in the luminance channel.

crossword code In digital error protection, the application of error protection algorithms to a two-dimensional data block. This is also called product coding.

CRT See cathode-ray tube.

cyclic redundancy check (CRC) In digital error protection, a form of Hamming code for error detection of data blocks.

DAC See digital-to-analog conversion.

dark current In an imager, the signal current that flows in the output when no light is present on the sensitive surface (lens capped).

DCT See discrete cosine transform.

decode The process of undoing an encoding algorithm to restore the original signal.

definition In an image, the property of sharpness. This is different than the resolution of an image.

dichroic surface An optical surface that has a special coating of a thickness that reflects light in one region of the optical spectrum and transmits the rest.

differential gain In a composite video system, a nonlinear effect that causes the luminance level to affect the subcarrier amplitude. It shows up as a variation of color saturation with image brightness.

differential phase In a composite video system, a nonlinear effect that causes the luminance level to affect the subcarrier phase. It shows up as a variation in color hue with image brightness.

differential pulse code modulation (DPCM) A form of pulse code modulation where the transmitted values represent the differences between adjacent sample values of the input signal.

digital A signal system where signals have values chosen from a specified set. For example, a binary digital system has only two valid signal values; similarly, an octal digital system has eight valid signal values.

digital-to-analog conversion (DAC) The process of converting digital values into analog values. It is a necessary step to display digital video signals on an analog display such as a CRT.

digital signal processor (DSP) A special microprocessor that has an instruction set optimized for signal processing tasks.

digitization Another word for analog-to-digital conversion.

dipole A basic radiating element of a broadcasting antenna.

discrete cosine transform (DCT) In image compression, a transform that converts blocks of pixel data into a form that is more suitable to run-length and statistical compression. It is a variation of the Fourier transform.

display refresh In digital displays, the process of continuously reading from display memory to deliver signals to the monitor.

dithering In a digital system, the introduction of small random variations that mask the effect of quantization errors. In representing digital images with a limited number of colors (such as with a color lookup table), dithering varies the colors of adjacent pixels to give the appearance of more colors.

dot pitch In a color display device, such as a CRT or LCD panel, the distance between adjacent dots of the same color. This parameter determines the limiting resolution of the display.

downlink In satellite communications, the transmission path from the satellite to the ground. See also uplink.

DPCM See differential pulse code modulation.

driver In computer software, a software module that interfaces custom hardware to the system.

drop-out In magnetic video recording, a signal interruption caused by a surface defect or foreign particle that momentarily affects the head-to-tape contact.

drops In CATV systems, the cables that connect individual homes to the feeder cable.

DSP See digital signal processor.

edit decision list (EDL) In audio or video editing, a list of time code locations and commands that can control the editing of a program.

editing In audio or video production, the process of assembling individual shots or clips into a complete program.

EDL See edit decision list.

effective radiated power (ERP) In a broadcast transmission system, the specification of radiated power that takes into account the transmitter power output, the transmission line losses, and the antenna gain.

EIA See Electronic Industries Association.

Electronic Industries Association (EIA) An organization of electronics manufacturers in the United States.

electronic news gathering (ENG) The use of portable video and audio equipment for collecting news stories for television.

encode The process of converting information to a different form for transmission or storage. (See also decode.)

ENG See electronic news gathering.

envelope delay distortion In an analog system, distortion caused by the system delay varying with frequency.

ERP See effective radiated power.

error protection In digital systems, the provisions made to detect and correct for errors.

exciter In a broadcast transmission system, the equipment that generates the modulated carrier.

fade margin In FM communication systems, the difference in dB between the receiver input power under nonfading conditions and the receiver's threshold (the point where the signal-to-noise ratio of the receiver begins to deteriorate rapidly with reduction of input power).

FDM See frequency-division multiplex.

feeders In a CATV system, the cables that run in front of homes, from which individual homes are connected by drops.

field In interlaced scanning, one vertical scan.

field sequential A composite color TV system that transmits different colors on successive scanning fields. In the 1950s, such a system was briefly adopted in the United States, but because it was incompatible with existing monochrome receivers, it was quickly replaced by the NTSC color system.

field strength In broadcasting, the measure of electric field strength at a receiving location, defined in millivolts/meter or μvolts/meter.

flash memory In computers, a form of nonvolatile solid-state memory that can be updated by applying a special procedure.

flying-spot scanner A form of image sensing that uses a scanning spot of light and a photocell pickup. In a light-transmission mode with a CRT as the light source, flying-spot scanners are sometimes used for motion picture film pickup. They can also be used in a reflective mode to scan physical objects, but this is not practical for television.

FM improvement factor In frequency-modulated (FM) transmission systems, the ratio (in dB) of the system signal-to-noise ratio to the signal-to-noise ratio of the unmodulated carrier.

Fourier transform A mathematical process that converts an electrical waveform (time domain) into its frequency components (frequency domain).

frame In video systems, the scanning of a complete picture. To convey motion, frames are repeated at a rate high enough that the eye sees a continuous image, typically 30 frames/second or higher.

frequency-division multiplex (FDM) A technique for combining multiple signals in the same transmission path by sending each signal at a different frequency. For example, a CATV system uses FDM.

front porch In an analog video signal, the part of the horizontal blanking interval ahead of the sync pulse. It exists to prevent video crosstalk into the sync.

gamma correction In a video camera, the process of adjusting the amplitude transfer characteristic to obtain optimum reproduction of the brightness levels in a scene.

genlock In video systems, the process of synchronizing a video source to the signal from a remote source.

geosynchronous orbit A satellite orbit having the property that the orbital rotation is the same as the earth's rotation, causing the satellite to appear stationary relative to the earth. It is at an altitude of 35,850 km (22,300 miles).

ghost In television broadcasting, interference due to multipath transmission that produces multiple images.

graceful degradation In analog systems, the name given to the accumulation of distortions that occurs as system elements or processes are cascaded. The signal becomes steadily worse as the system is extended, but it may remain useful even though seriously impaired.

Grand Alliance An organization formed in the United States in 1993 to develop a single proposal for an HDTV system. This work was completed in 1995.

graphical user interface (GUI) In personal computers, a system for user-computer interaction that includes a video display and a pointing device such as a mouse. The user controls the system by pointing at objects (icons) on the screen and clicking a button.

graphics In computers, the name given to computer-generated drawings and pictures. (See also vector graphics and bit-mapped graphics.)

graphics accelerator In a PC video display adaptor, a special-purpose processor that speeds up common graphical operations such as displaying windows or drawing objects.

gray scale Refers to the amplitude characteristic of the luminance channel of a video system.

GUI See graphical user interface.

Hamming code A digital error protection method that is able to detect single errors and point to the specific bit that is in error. It is easy to implement, but lacks the power of better algorithms, such as the Reed-Solomon code.

hard disk In a computer, a mass storage device utilizing a magnetic coating on a rotating disk and moving heads to achieve random access.

HDTV See high-definition TV.

head end In a CATV system, the equipment that creates the signals and modulates them to drive the cable.

header In a digital data stream or data file, a specified block that defines the characteristics of the data that follows.

helical scan In video recording, a rotating-head scanner that writes diagonal tracks across the tape to achieve the high head-to-tape speed needed for video recording.

hexadecimal A digital number system based on 4-bit words, giving 16 values per word.

high-definition TV (HDTV) Television systems of higher resolution, wider screens, and better sound compared to present NTSC or PAL systems. In the United States, the ATSC all-digital system meets these goals.

horizontal detail In an image, the scene features that are affected by resolution in the horizontal direction, i.e., vertical lines.

HTML See hypertext modeling language.

hue The "color" of a pixel, such as red, blue, yellow, violet, etc.

Huffman coding In digital systems, a form of statistical coding for data compression that assigns word lengths in proportion to the frequency of occurrence of values. For example, frequently occurring values get short words, seldom-used values get long words.

hypertext In computers, a system for presenting text that allows words or phrases to be marked so the user can select them and cause the display of additional or related information.

hypertext modeling language The programming language used on the World Wide Web.

I/O See input/output.

icon In computers, a graphical symbol that represents objects or actions on the screen. In graphical user interface, the user selects an icon to go to the represented object or perform the represented action.

illuminance A measure of the illumination incident at a point on a surface. The units are lux or foot-candles.

image enhancement The act of modifying or processing an image to make it look better. Examples are aperture correction in video cameras or adjusting the brightness and contrast of an image in a computer.

image scanner A hardware device for digitizing a hard-copy document or picture into a computer.

imager A device that converts optical images to electrical signals, such as a charge-coupled device.

impedance In electrical circuits or devices, the ratio of voltage to current. The units of impedance are ohms.

incidental phase modulation In a TV transmitter, an undesirable variation of carrier phase with carrier amplitude. Depending on the type of modulation, this may cause distortion of the transmitted video signal.

index of refraction In a material used to transmit electromagnetic waves (including light), a parameter that defines the refractive properties of the material.

input/output (I/O) In computers, the part of the system that deals with interfacing to external devices for input or output, such as keyboards or printers.

insert editing In audio or video editing, a technique for replacing a segment of a previously recorded program.

instructions In microprocessors, the digital words that tell the microprocessor what actions to perform.

interactivity In any system, provision for the user to control the action of the system and for he/she to view the results of his/her action on the system.

interframe In television systems, any action that makes use of the content of more than one frame. For example, a compression system that was based on how much change there was between adjacent frames would be interframe compression. (See also intraframe.)

interlaced scanning In video systems, the technique of scanning part of the lines of the image in a first vertical scan and adding the in-between lines in a subsequent vertical scan or scans. In standard analog television, 2:1 interlace is used: half of the lines are scanned in each field to create a frame out of two vertical scans.

interleaving In digital systems, a technique that rearranges the data in a specified way, usually to cause bursts of transmission errors to be spread out in the data.

International Standardizing Organization (ISO) A worldwide cooperative organization responsible for standards in many fields.

Internet A public worldwide computer network of networks. See World Wide Web.

interpreter In computer software, a program that reads commands from a file written in a programming language such as BASIC, and converts the language statements to instructions for the system's microprocessor, which are immediately executed.

intraframe In television systems, any action that uses only information from a single frame. (See also interframe.)

ISO See International Standardizing Organization.

isotropic radiator In antennas, a theoretical radiator that radiates uniformly in all directions. Such a radiator cannot be constructed.

Java language A programming language that is capable of running on any type of computer by using an interpreter program that converts Java code into the computer's native language. It is used extensively on the World Wide Web.

Joint Photographic Expert Group (JPEG) A working group of a committee of the ISO. They developed the JPEG worldwide standard for compression of still images that is applicable to nearly all fields of electronic imaging.

JPEG See Joint Photographic Expert Group.

Kell factor In raster-scanned video systems, a factor that expresses the ratio of the observed vertical resolution to the total number of scanning lines. For conventional analog television systems, the Kell factor is about 0.7.

kinescope The name given to a cathode-ray tube designed for TV viewing.

lag In video cameras, an artifact where moving objects in the image tend to smear as they move.

LCD See liquid-crystal display.

limiting resolution In an image, the highest resolution that can be seen in the reproduced image. Limiting resolution is often measured by observing a pattern of converging lines (called a "wedge") and looking for the place on the wedge where the lines just fade out.

linearity In any system, the degree to which a property approximates a straight line. In video systems, scanning linearity and amplitude linearity are important.

lip sync In a system involving the simultaneous presentation of sound and motion video, the degree to which the audio and video are in sync. The name "lip sync" comes from the observation of a talking-head scene, which is one of the most critical cases for audio/video synchronization.

liquid-crystal display (LCD) An electronic display panel that is based on the properties of a liquid-crystal material sandwiched between two transparent plates and illuminated from behind. The transmittance of the liquid-crystal material changes with the applied electric field.

local bus In computers, a special bus that connects parts of the system that require the fastest data transfer, such as the CPU, RAM, and video display adaptor.

luminance In color video systems, a component signal that represents the brightness of the image. (See also luminosity function.)

luminosity function A function that expresses the apparent brightness of the spectral colors. It is used in video systems to calculate the luminance signal.

MAC See multiplexed analog components.

masking In color video systems, the process of mathematically mixing the color channels to produce controlled changes in the system colorimetry.

mass storage In computers, the nonvolatile, large-capacity storage elements of the system, such as hard disks, CD-ROM, or floppy disks.

mastering In optical recording, the process of creating a master recording to be used for replication, usually by pressing.

meander gate In a PAL video system, a circuit that controls the insertion of the color burst during the vertical blanking interval.

memory-mapping In computers, the technique of accessing hardware by assigning the hardware to memory addresses in the system memory space.

millions of instructions per second (MIPS) In microprocessors, the average speed at which the microprocessor executes instructions.

MIPS See millions of instructions per second.

mixing In audio or video systems, the process of combining controlled proportions of two or more source signals to create a new signal.

modem A contraction of the words "modulate-demodulate." It is most commonly used as the name of a device that connects a computer to a telephone line for data communication.

modulation In communication systems, the process of modifying the channel signal (often called the "carrier") to represent information to be transmitted. Typical analog modulation methods involve the frequency, amplitude, or phase of the carrier.

modulation index In frequency-modulation (FM) systems, the ratio of the frequency deviation to the highest modulating frequency. This parameter determines the extent of the spectrum generated by the FM process.

modulation transfer function (MTF) In video systems, the system's response to line patterns of varying spacing. The cross-sectional darkness of the lines varies sinusoidally. This is one measure of aperture response.

motion compensation In video compression systems, a method of compression that determines the degree of motion between adjacent frames and encodes only the changes.

Motion Picture Expert Group (MPEG) A working group of a committee of the ISO. They developed the MPEG worldwide standard for motion video compression and transport.

MPEG See Motion Picture Expert Group.

MTF See modulation transfer function.

multimedia The combining of different media (text, audio, graphics, video, animation, etc.) to create a presentation. This is done with computers or video systems.

multipath In radio transmission, the effect of signal reflection or refraction. This causes receivers to see several components of the original signal coming at different times due to the different path lengths of direct, reflected, or refracted signals. (See also ghosts.)

multiplexed analog components (MAC) In video systems, a system of component transmission where the components are transmitted sequentially in time, usually at TV line rate.

multitasking In computer operating systems, the ability to run two or more programs concurrently by rapid time-switching of the microprocessor between programs.

multithreading In computer operating systems, a form of multitasking that allows multiple tasks (called "threads") within the same program.

National Television Systems Committee (NTSC) The committee that decided on the compatible color television system for the United States, adopted by the FCC in 1953. This system is called the NTSC system.

N_e In a television system, a single-number representation of aperture response determined by defining a rectangle that has the same area as the area under the aperture response squared curve. The N_e is the width of the defined rectangle.

nonvolatile memory In computer memory devices, a memory that retains its stored data when power is removed.

nonlinear editing In video editing, a technique for viewing edits in real time by rapidly selecting video sources on the fly. This is usually done with computers and hard disk storage-it cannot be done with magnetic tape.

NTSC See National Television Systems Committee.

null fill In broadcast antennas, a technique that partially fills in the signal nulls that would otherwise appear in the vertical pattern. This improves the close-in coverage.

Nyquist frequency In a sampled system, the sampling frequency needed to avoid aliasing. It is twice the highest signal frequency.

Nyquist limit In a sampled system, the highest signal frequency that can be sampled without aliasing with a given sampling frequency. It is one-half the sampling frequency.

octal A digital system that has eight valid values. In binary digital systems, octal representation uses 3-bit words.

off-line editing In audio or video editing, an editing system that operates with copies of the master recordings (usually in a different format) in order to facilitate or lower the cost of edit decision making. As edits are chosen, an edit decision list is created with commands and time codes that can be subsequently used to perform the final edits using the master recordings.

offset binary In digital systems, a binary representation where signal zero is offset to the center of the word value range. This allows positive and negative signal values to be represented. For example, in an 8-bit offset-binary representation, signal zero would be at binary 128, maximum-positive signal would be at binary 255, maximum-negative signal would be at binary 0. (See also two's complement.)

omega wrap In a helical-scan tape recorder, a scanner design where the tape wraps partially around the scanning drum, entering and leaving the drum and its guides along a radius. This is like the Greek letter omega (Ω).

operating system In computers, the program that provides the basic functions needed to run programs.

orbital slot In satellite communication using geosynchronous satellites an arc of the orbit allocated to a single satellite. Orbital slots are identified by their longitude.

packetization In digital transmission systems, the technique of breaking up a data stream into blocks to facilitate transmission, multiplexing, and error protection.

PAL See phase-alternating line.

passive-matrix LCD A liquid-crystal display panel that uses a pattern of row and column lines to provide access for refreshing the display. This is the lowest-cost type of LCD panel.

pay-per-view In video distribution systems, the technique of controlling access so that the customer can be charged on the basis of what programs he/she watches.

PCM See pulse code modulation.

permeability The property of a magnetic material that defines how easily it can be magnetized.

phase-alternating line (PAL) The analog composite color video system developed in Europe and used by certain countries around the world. It is similar to NTSC but it uses a subcarrier phase alternation technique that makes certain kinds of transmission errors less visible.

phase-lock loop A circuit that synchronizes a local oscillator to an incoming signal. Phase-lock loops are used, for example, to synchronize an analog television receiver to the horizontal scanning and color subcarrier of the incoming signal.

photoconduction The process whereby a material exhibits light sensitivity by changing its electrical conductivity with light level. For example, photoconductive materials are used in vidicon and Plumbicon pickup tubes and in CCD imagers.

photoemission The process whereby a material exhibits light sensitivity by emitting electrons in the presence of light. This was used in the early pickup tubes such as the iconscope and the image orthicon.

pixel In an electronic image, a single point in the picture. Each pixel has unique values of luminance and color. In a digital image, pixels are individual points of the image, each represented by a certain number of bits.

polarization In electromagnetic waves, describes the orientation of the electric field vector. It may be horizontal, vertical, or elliptical. With elliptical polarization, the electric field vector rotates as the wave passes a fixed point.

polysilicon Polycrystalline silicon. Its electrical properties are between single-crystal silicon and amorphous silicon. Unlike single-crystal silicon, which must be grown, polysilicon can be deposited on a substrate.

postproduction In the creation of audio and video programs, the process of assembling previously recorded source material into the finished program. Postproduction includes tasks such as image enhancement, audio sweetening, special effects, titling, and editing.

primary color In a trichromatic color system, one of the three colors that are mixed to create all other colors in the system.

print-through In magnetic tape recording, the tendency for demagnetization of recordings to occur between layers of the tape due to magnetic crosstalk when tape is rolled up on a reel.

product code See crossword code.

progressive scanning A scanning process where all the lines of the image are scanned by every vertical scan. (See also interlaced scanning.)

pulse code modulation (PCM) In digital audio and video systems, the representation of an analog signal by its direct digitized values.

quantization In analog-to-digital conversion, the process of fitting the sampled values into the digital bits/sample range. For example, if there are 8 bits/sample, the sample values must be quantized to no more than 256 levels.

quantum efficiency In an imaging device, the percent of incident light quanta that create useful electrical output.

radiating pattern In antennas, the variation of radiation intensity in different directions, both horizontally and vertically.

RAM See random-access memory.

random-access memory (RAM) In computers, the main system memory, usually consisting of (volatile memory) solid-state chips.

randomizing In digital systems, a process that ensures a pseudo-random distribution of ones and zeros. This improves the spectral distribution and clock content of the data stream.

raster In video scanning, the pattern made by the scanning spot as it covers the entire picture. The usual raster is a pattern of equally-spaced horizontal lines.

read-only memory (ROM) In computers, nonvolatile memory that stores permanent programs. ROM usually consists of solid-state chips.

Rec. BT.601-5 The document of the ITU-R that spcifies sampling rates and structures for SDTV component digital video.

recording density A figure of merit for recording systems based on the amount of information recorded in a given area of medium. For example, bits/cm^2.

reduced instruction-set computer (RISC) A microprocessor architecture based on a simplified instruction set that allows all instructions to be executed in a single clock cycle or less. Currently, the fastest microprocessors use this approach.

Reed-Solomon code A digital error protection code based on blocks of data with added redundancy. It is capable of correcting burst errors up to a limit set at design time.

reflection coefficient In electrical transmission lines, a parameter that describes the degree of mismatch between the characteristic impedance of the line and its load impedance.

rendering In computer graphics, the process of creating a bit-mapped representation of a vector graphics image.

resolution (1) In an electronic image, a measure of the reproduction of fine detail. See also limiting resolution. (2) In a digital image, the total number of pixels in the horizontal and vertical directions. For example, 640 × 480.

RISC See reduced instruction-set computer.

ROM See read-only memory.

SNR See signal-to-noise ratio.

sampling In an electronic system, the process of taking equally-spaced readings of the instantaneous value of an analog signal. It is the first step of converting the analog signal to a digital representation. (See also analog-to-digital conversion.)

saturation (1) In color video systems, the measure of the intensity or purity of a color. (2) In a magnetic material, the maximum flux density that can exist in the material.

SAW filter An electrical filter using the surface acoustic wave (SAW) technology. SAW filters are capable of transfer functions that are not possible in lumped-element circuits.

scanning In electronic image reproduction, the process of reading or displaying all the points of the image by passing a moving spot across the image. (See also raster, interlaced scanning, and progressive scanning.)

SECAM See Sequential Coleur et Memoire.

Sequential Coleur et Memoire (SECAM) The analog composite color television system developed in France and used in several other countries. SECAM uses a luminance signal and two FM subcarriers that transmit the color-difference components.

serial transmission A digital transmission method using a single wire or channel with the bits organized into a sequential stream.

server On a digital network, a computer that is devoted to providing network control, data storage, and/or input/output interfaces.

service area In broadcasting, the geographical area covered by a broadcast transmitter at a specified level of received signal quality.

set-top converter An electronic unit that connects a CATV system drop cable to a subscriber's TV receiver. It provides frequency conversion so the receiver can access all the cable channels and it also may include features for pay-per-view or other premium services.

shift register An electronic digital circuit that stores data that can be clocked out in sequence. If the shift register has a parallel input, it can function as a parallel-to-serial converter.

signal-to-noise ratio (SNR) In an electrical channel, the ratio of the signal amplitude to the random noise in the channel. For video signals, the ratio is between the peak-to-peak video signal and the root-mean-square (RMS) noise. For audio signals, the ratio is between the RMS maximum signal and the RMS noise.

SMPTE Society of Motion Picture and Television Engineers.

soft-decision decoder In a digital transmission or recording system, a technique for converting the channel signal into proper digital values by taking account of the values of adjacent symbols. This is known as "Viterbi" decoding.

software In computers, the modules of information that either already are or result in instructions to the system CPU. The software tells the computer what to do. Software modules are called "programs." (See also interpreter, and compiler.)

special effects In video postproduction, the creation of dynamic transitions between scenes or other dynamic modification to video signals.

streaking In an electronic image, an artifact where objects in the image create black-level shifts that go entirely across the image.

streaming data The data flow required by a system that must display the data as it is received. This is typical of video displays, because the video data rate is too high for the data to be stored for later display.

sub-Nyquist sampling In a sampled system, sampling at a rate lower than the Nyquist frequency.

subcarrier In an electrical transmission system, a carrier embedded in the baseband signal to transmit additional information.

subsampling In a sampled system, the process of deleting certain samples for the purpose of data reduction. For example, every other sample may be deleted. Proper use of subsampling requires filtering to prevent aliasing from the subsampling.

subtractive color A color reproduction system that uses pigments or dyes added to a white background. The pigments or dyes subtract certain colors from the light reflected from the white background. (See also additive color.)

super VGA (SVGA) In personal computers, a generic name for scanning and color standards that are an extension of the VGA standard.

SVGA See super VGA.

sweetening In audio postproduction, any process applied to the signal to make it sound better.

symbol In digital transmission systems employing modulation, the time period for transmitting a single unique state of the carrier. Depending on the modulation method, each symbol can transmit one or more bits of data.

tape transport In magnetic recording, the mechanism that moves the tape and presents it to the heads.

telecine A video system specifically designed for transferring motion picture film to television signals.

tempo In music, the beat rate, which determines the speed of playing notes.

timbre In music, the dynamic harmonic and amplitude properties of a note. This is the reason different instruments playing the same note have different sounds.

time code In audio and video recording, a system that records a number with each video frame or section of audio. This is used to facilitate precise control during postproduction. (See also edit decision list.)

total harmonic distortion In audio systems, a measure of distortion obtained by testing the system with a pure sine wave input and measuring the harmonic components generated at the output of the system.

transfer characteristic In electrical circuits, the property that specifies the ratio of output to input signal amplitude. (See also linearity.)

transfer function In electrical circuits, the complete relationship between input and output signals, usually expressed mathematically as a function of frequency, amplitude, or other parameters.

transitions In audio or video editing, the aural or visual actions that occur while changing the program from one scene to another. Examples are wipes, dissolves, fades, etc. (See also special effects.)

transmission line A hard-wired electrical circuit that transmits a signal from one point to another. It may be a coaxial cable, a waveguide, twisted-pair wires, etc., depending on the system requirements for frequency response, distance, and power level.

transponder In satellite communications, the equipment on the satellite that receives signals from the uplink and retransmits them for the downlink.

transport layer In the Grand Alliance digital HDTV system, the part of the system that packetizes the data streams to prepare them for transmission.

transversal filter An electrical filter produced by sending a signal down a transmission line and creating the output by combining signals tapped from various points along the line.

trellis coding In digital transmission using multilevel modulation, a technique for error protection that improves error performance without increasing the required channel bandwidth. Among other things, it makes use of soft-decision decoding.

trichromatic A color system that uses three color components to represent all colors. Also called a "tristimulus" system.

troposphere The lower portion of the earth's atmosphere, from about 4 to 12 miles altitude at the equator. In this region, temperature generally decreases with altitude.

TRS-ID In serial digital transmission systems, a code for synchronization and identification. The acronym means "timing reference and identification signal."

trunks In CATV systems, the cables that carry the signals between the head end and the feeder break-out points.

truth table In digital systems, a table that specifies the outcome of a process for all possible combinations of values of the inputs to the process.

TV lines (TVL) In video systems, a unit of measurement of resolution based on a pattern of equally spaced lines. The TVL number is obtained by counting all lines and spaces if the pattern is extended to reach a distance equal to the picture height.

TVL See TV lines.

two's complement In binary digital systems, a method of representing signed values that keeps signal zero at binary zero. (See also offset binary.)

universal resource locator (URL) The method of addressing different sites on the World Wide Web.

uplink In satellite communications, the transmission path from the ground to the satellite. (See also downlink.)

URL See universal resource locator.

vector graphics In computer graphics, the representation of objects by mathematical means. It makes possible the rendering of the object to be displayed at any size or format. (See also bit-mapped graphics.)

vertical detail In an image, the scene features that are affected by resolution in the vertical direction, i.e., horizontal lines.

vertical interval time code In video systems, a time code that is transmitted during the vertical blanking interval of the signal.

vestigial-sideband modulation A modulation system where one of the sidebands resulting from modulation is partially removed. The sideband is only partially removed in order to ease the filtering requirements in signal generation.

VGA See video graphics array.

video (1) The technology for the electronic capture, storage, transmission, and reproduction of images and motion pictures. (2) The signals generated by video technology.

video adaptor In personal computers, a hardware module including memory and control circuits to drive a video display monitor. It performs the tasks of video display refresh and graphics acceleration.

video graphics array (VGA) In personal computers, a video adaptor standard originally developed by IBM for the PC/AT. It has been greatly extended by others. (See also super VGA.)

video overlay In personal computers, the technique of combining analog video from a VCR or laser disc with computer information on the same screen. This often is accomplished by a special video adaptor.

video RAM (VRAM) In personal computers, a special form of RAM chip that has a separate serial-output port for display refresh operations. This architecture speeds up video adaptor performance.

viewing ratio When viewing a reproduced image, the ratio between the distance between viewer and screen to the height of the screen.

VITC See vertical interval time code.

Viterbi decoder See soft-decision decoder.

volatile memory In computers, memory that loses its data when power is removed.

voltage standing-wave ratio (VSWR) In a transmission line, the ratio between the voltage maximum to the voltage minimum in the standing wave which may exist on the line due to reflections. It is a fundamental measurement of transmission line performance.

VRAM See video RAM.

VSB See vestigial-sideband modulation.

VSWR See voltage standing-wave ratio.

waveform monitor An oscilloscope specifically designed for viewing television video waveforms.

waveguide In broadcast transmitters, a transmission line that consists of a hollow rectangular or circular tube within which an electromagnetic wave transmits power. It is sometimes used to connect a UHF transmitter to the antenna.

wavelength For periodic signals transmitted through a medium, wavelength equals the speed of propagtion through the medium divided by the frequency of the signal.

weighted noise In recognition of the fact that high-frequency noise is less visible in a picture than low-frequency noise, a high-frequency rolloff is sometimes used during S/N measurement. This rolloff is called a "weighting filter" and it makes the noise measurement more indicative of how the picture might look to an observer. Weighted-noise measurements give somewhat larger numbers than unweighted measurements.

wireless cable A multichannel broadcast service where the multiple channels are transmitted with frequency-division multiplex in a manner similar to cable television.

World Wide Web (WWW or the Web) The protocol on the Internet for multimedia. The Web is based on HTML and, using URLs, allows a user to jump to information sources located anywhere on the worldwide Internet.

wow and flutter In audio systems, the manifestation of time base instability. It appears as spurious frequency modulation of audio signals.

WWW See World Wide Web.

zigzag ordering In video compression systems based on the discrtet cosing transform, a method of arranging the output coefficients from the transform so they are more efficiently compressed by run-length and statistical coding.

Bibliography

General References

ANSI/EIA/TIA *Standard Electrical Performance for Television Transmission Systems*, Electronic Industries Association, Washington, D.C., 1990.

Benson, K. Blair, and J. Whittaker, *Television Engineering Handbook*, McGraw-Hill, New York, 1992.

Broadcast Engineering Magazine, published monthly by PRIMEDIA Intertec, 9800 Metcalf, Overland Park, KS, 66212-2215.

Dorf, Richard C., *Electrical Engineering Handbook*, CRC Press, Boca Raton, FL, 1993.

Fink, Donald G., and Donald Christiansen, *Electronic Engineers' Handbook*, McGraw-Hill, New York, 1988.

Fisher, D. E., and M. J. Fisher, *TUBE: The Invention of Television*, Counterpoint, Washington, D.C., 1996.

Gibson, Jerry D., *The Communications Handbook*, CRC Press, Boca Raton, FL, 1997.

Inglis, Andrew F., *Behind the Tube: A History of Broadcasting Technology and Business*, Focal Press, Boston, 1990.

Inglis, Andrew F., *Electronic Communications Handbook*, McGraw-Hill, New York, 1990.

Meyer-Ahrendt, Jurgen R., *Introduction to Classical and Modern Optics*, 4th ed., Prentice Hall, Englewood Cliffs, N.J., 1995.

Rorabaugh, C. Britton, *Communications Formulas and Algorithms*, McGraw-Hill, New York, 1990.

Various authors, *Television Technology in the 1990s*, SMPTE, White Plains, N.Y., 1981.

Various authors, *Tomorrow's Television*, SMPTE, White Plains, N.Y., 1982.

Various authors, *Television Technology: A Look Toward the 21st Century*, SMPTE, White Plains, N.Y., 1987.

Various authors, *Television Technology in Transition*, SMPTE, White Plains, N.Y., 1988.

Various authors, *Television-Merging Multiple Technologies*, SMPTE, White Plains, N.Y., 1990.

Chapter 1, Video System Fundamentals

Fink, D. G., *Television Engineering Handbook*, McGraw-Hill, New York, 1957.

Chapter 2, Color Video Fundamentals

FCC rules, *Part 73*.

CCIR, *Recommendations and Reports*, CCIR 15th Plenary Assembly, Vol. XI, Broadcasting Service (Television), ITU, Geneva, 1982.

EIA/TIA Standard, "Electrical Performance for Television Transmission Systems," EIA-TIA-250-C, Washington, D.C., 1990.

Hirsch, C. J., W. F. Bailey, and B. D. Loughlin, "Principles of NTSC Compatible Color Television," *Electronics*, Vol. 52, February 1952.

Loughren, A. V., "Recommendations of the National Television Systems Committee for a Color Television Signal," *J. SMPTE*, Vol. 60, April-May 1953.

Various authors, *Television Image Quality*, SMPTE, White Plains, N.Y., 1984.

Wentworth, John, *Color Television Engineering*, McGraw-Hill, New York, 1954.

Chapter 3, Introduction to Digital Technology

Luther, Arch C., *Digital Video in the PC Environment*, McGraw-Hill, New York, 1991.

Pohlmann, Ken C., *Principles of Digital Audio*, McGraw-Hill, New York, 1995.

Poynton, Charles A., *A Technical Introduction to Digital Video*, Wiley, New York, 1996.

Robin, M., and M. Poulin, *Digital Television Fundamentals*, McGraw-Hill, New York, 1998.

Rzeszewski, Theodore S., *Digital Video: Concepts and Applications Across Industries*, IEEE Press, New York, 1995.

Various authors, *Digital Video*, Vols. 1-2, SMPTE, White Plains, N.Y., Vol. 1 (1977), Vol. 2 (1979).

Watkinson, John, *The Art of Digital Video*, Focal Press, London, 1994.

Chapter 4, Elements of Image Quality

Ardito, Maurizio, "Studies on the Influence of Display Size and Picture Brightness on the Preferred Viewing Distance for HDTV Programs," *J. SMPTE.*, August 1994.

Herman, S., "The Design of Color Television Color Rendition," *J. SMPTE*, Vol. 84, 1975.

Various authors, *Better Video Images*, SMPTE, White Plains, N.Y., 1989.

Schade, Otto H., Jr., "Electro-Optical Characteristics of Television Systems," *RCA Review*, Vol. 9, March, June, September, December 1948.

Schade, Otto H., Jr., *Image Quality, A Comparison of Photographic and Television Systems*, RCA Laboratories, Princeton, N.J., 1975.

"Engineering Aspects of Television Allocations," TASO Report to the FCC, March 1, 1977.

Chapter 5, Audio Technology for Video

Benson, K. Blair (ed.), *Audio Engineering Handbook*, McGraw-Hill, New York, 1988.

Luther, A. C., *Principles of Digital Audio and Video*, Artech House, Norwood, MA, 1997.

Pohlmann, Ken C., *Principles of Digital Audio*, 3d ed., McGraw-Hill, New York, 1995.

Chapter 6, Analog Video Systems

RJG Ellis, "The PALplus Project: Conception to Introduction," *IBC Convention*, 16–20 September 1994.

See also references for Chapter 2.

Chapter 7, Digital Video Systems—DTV

Baron, Stanley N. (ed.), *Implementing HDTV: Television and Film Applications*, SMPTE, White Plains, N.Y., 1996.

Benson, K. Blair, and Donald G. Fink, *HDTV—Advanced Television for the 1990s*, McGraw-Hill, New York, 1991.

de Bruin, Ronald, and Jan Smits, *Digital Video Broadcasting: Technology, Standards, and Regulation*, Artech House, Norwood, MA, 1999.

Glenn, William E., and Karen G. Glenn, "High-Definition Transmission, Signal Processing, and Display," *J. SMPTE*, July 1990.

"Grand Alliance HDTV System Specification," Version 2.0, December 7, 1994.

Various authors, "HDTV and the New Digital Television," *IEEE Spectrum*, April 1995.

Whitaker, Jerry, *DTV: The Revolution in Electronic Imaging*, McGraw-Hill, New York, 1998.

Chapter 8, Digital Video Systems—Computers

DV: Digital Video Magazine, published monthly by Miller Freeman, 600 Harrison St., San Francisco, CA, 94107.

Hennessy, John L. and David A. Patterson, *Computer Architecture: A Quantitative Approach*, 2d ed., Morgan Kaufmann, San Francisco, 1996.

New Media Magazine, published monthly by Hypermedia Communications, Inc., P. O. Box 3039, Northbrook, IL, 60065-3039.

Ozer, Jan, *Video Compression for Multimedia*, AP Professional, Boston, MA, 1995.

Chapter 9, Video Cameras

Hamalainen, J., T. Leacock, and P. Westerlink, "Facts and Fiction: Some Aspects Regarding the Design of Digital Television Cameras Using CCD Image Sensors," Matsushita publication.

Luther, Arch C., *Video Camera Technology*, Artech House, Norwood, MA, 1998.

Thorpe, Laurence J., "HDVS CCD Camera—A Significant Advance in Real Time High Definition Imaging," Sony Publication.

Thorpe, Laurence J., "The Electronic Pursuit of Film Imaging," *IS&T 47th Annual Conference*, 1994.

Zettl, Herbert, *Television Production Handbook*, 6th ed., Wadsworth, Belmont, CA, 1997.

Chapter 10, Professional Video Recorders

Felix, M., and H. Walsh, "FM Systems of Exceptional Bandwidth," *Proc. IEEE*, Vol. 112, No. 9, September 1965.

Jorgensen, Finn, *The Complete Handbook of Magnetic Recording*, 4th ed., McGraw-Hill, New York, 1996.

Luther, Arch C., *Video Recording Technology*, Artech House, Norwood, MA, 1999.

Mee, C. Denis, and Eric D. Daniel, *Magnetic Storage Handbook*, 2d ed., McGraw-Hill, New York, 1996.

Watkinson, John, *The Digital Videotape Recorder*, Focal Press, London, 1994.

Chapter 11, Home and Semiprofessional Video Recorders

Videomaker magazine, published monthly by Videomaker, Inc., P.O. Box 4591, Chico, CA 95927.

Chapter 12, Video Postproduction Systems

Anderson, Gary H., *Video Editing and Postproduction: A Professional Guide*, 2d ed., Knowledge Industry Publications, 1993.

Ohanian, Thomas A., *Digital Nonlinear Editing*, Focal Press, London, 1993.

Chapter 15, Interactive Video Systems

Buford, John F. K., *Multimedia Systems*, Addison-Wesley, Reading, MA., 1994.

Laurel, Brenda, *The Art of Human-Computer Interface Design*, Addison-Wesley, Reading, MA, 1990.

Lu, Guojun, *Communication and Computing for Distributed Multimedia*, Artech House, Norwood, MA, 1996.

Luther, Arch C., *Authoring Interactive Multimedia*, AP Professional, Boston, MA, 1994.

Luther, Arch C., *Using Digital Video*, AP Professional, Boston, MA, 1995.

Chapter 16, Terrestrial Broadcasting Systems

de Bruin, Ronald, and Jan Smits, *Digital Video Broadcasting: Technology, Standards, and Regulation*, Artech House, Norwood, MA, 1999.

HDTV Tech Brief Series; published by Comark
  "Digital HDTV and RF Amplifier Dynamic Range," May 1992
  "Digital HDTV and Amplifier Class of Operation," June 1992
  "WRC-TV, Washington D.C., AD-ADTV test," October 1992
  "HDTV Transmitter Performance Criteria—Linearity Versus Range," December 1992
  "Linearity Correction for ADTV Transmitters—What's Best?" September 1993
  "8-VSB-The ATV Select Standard, What Is It?" March 1994
  "Is COFDM Dead?" May 1994
  "RFP/RFI—A Serious Consideration for IOT Transmitters," June 1994

McConnell, Chris, "MSTV Proposed Allocation Plan," *Broadcasting and Cable*, January 16, 1995.

McConnell, Chris, "ATV Comes Out on Top of NTSC Field Tests," *Broadcasting and Cable*, September 26, 1994.

Chapter 17, CATV Systems

Bartlett, Eugene R., *Cable Technology and Operations*, McGraw-Hill, New York, 1990.

Cicora, Walter S., "Inside the Set-Top Box," *IEEE Spectrum*, April 1995.

Inglis, Andrew F. (ed.), *Electronic Communications Handbook*, Chapter 17, McGraw-Hill, New York, 1988.

Large, David, "Creating a Network for Interactivity," *IEEE Spectrum*, April 1995.

Chapter 18, Satellite Video Communications

Baylin, Frank, *Miniature Satellite Dishes*, 2d ed., Baylin Publications, Boulder, CO, 1995.

Elbert, Bruce R., *The Satellite Communication Applications Handbook*, Artech House, Norwood, MA, 1997.

Ha Tri, T., *Digital Satellite Communications*, 2d ed., McGraw-Hill, New York, 1990.

Inglis, Andrew F., and A. C. Luther, *Satellite Technology: An Introduction*, 2d ed., Focal Press, Boston, 1997.

Miya, K., *Satellite Communications Technology*, KDD Engineering and Consulting, Inc., Tokyo, 1981.

McConnell, Chris, "The SNG Edge," [special report on satellites in 1994], *Broadcasting and Cable*, June 11, 1994.

Potts, James B., *Satellite Transmission Systems*, BASS II Publications, Lewes, DE, 1993.

Roddy, Dennis, *Satellite Communications*, 2d ed., McGraw-Hill, New York, 1996.

FCC Docket 86-496, [Reduced orbital spacings].

Chapter 19, Fiber-Optic Transmission Systems

Kao, C. K., *Optical Fiber Systems: Technology, Design, and Applications*, McGraw-Hill, New York, 1982.

Midwinter, J. E., *Optical Fibers for Transmission*, John Wiley & Sons, New York, 1979.

Personick, S. D., *Fiber Optics Technology and Applications*, Plenum Press, New York, 1985.

Chapter 20, Video on the Internet

Cohill, Andrew Michael, and Andrea Lea Kavanaugh, *Community Networks: Lessons from Blacksburg, Virginia*, Artech House, Norwood, MA, 1997.

Graham, Ian S., *HTML Sourcebook*, 2d ed., Wiley, New York, 1996.

Held, Gilbert, *The Complete Modem Reference*, 3d ed., Wiley, New York, 1997.

Kessler, Gary C., and Peter Southwick, *ISDN: Concepts, Facilities, and Services*, 3d ed., McGraw-Hill, New York, 1997.

Kientzle, Tim, *Internet File Formats*, The Coriolis Group, Scottsdale, AZ, 1995.

Chapter 21, The Future of Video Systems

Negroponte, Nicholas, *Being Digital*, Alfred A. Knopf, New York, N.Y., 1995.

Index

Symbols

1394 serial interface   68, 345
16-VSB   66, 184
3:2 pull down   248
4:2:2 components   48
8-VSB   66, 184

A

A-B roll editing   321
AC-3. *See* Dolby Digital audio
ACATS. *See* Advisory Committee on Advanced Television Service
acoustics   137
active-matrix LCDs   361
acuity   93
adaptive differential PCM   28, 144
ADC. *See* analog-to-digital conversion
add-on editing   320. *See* editing: videotape: assemble mode
additive color   32
ADPCM. *See* adaptive differential PCM
advanced graphics processing   208
Advanced Television Evaluation Laboratory   172
Advanced Television Systems Committee   172
Advanced Television Test Center   172
Advisory Committee on Advanced Television Service   172
AFC. *See* automatic frequency control
AGC. *See* automatic gain control
AGP. *See* advanced graphics processing
algorithm   75
aliasing   8, 23, 48, 90, 100
   CCD imagers   102, 231
   filtering for   107
   frame   102
   scanning line   101
alpha channel   207
alpha wrap   261
ambience   139
AML. *See* amplitude modulated link
AMLCD. *See* active-matrix LCDs
amplitude modulated link   444

AMSTV. *See* Association for Maximum Service Television
analog   2
analog video systems   151
analog-to-digital conversion   26, 47, 206, 239
animation   380
antennas   393, 411
   beam tilt   416
   beamwidth   411
   circularly polarized   414
   directivity   411
   half-power points   411
   height/power limitations   394
   helical   415
   isotropic radiator   393
   log periodic   438
   null fill   416
   panels   414
   radiation pattern   411
   slot radiators   414
   superturnstile   413
   YAGI   438
aperture correction   109
   vertical   110
aperture response   95
   CCDs   222, 227
   NTSC system   106
   sampling   100
API. *See* application programming interface
application programming interface   202
aspect ratio   13, 424, 513
assemble editing   320
Association for Maximum Service Television   425
AST. *See* automatic scan tracking
AT&T   172
ATEL. *See* Advanced Television Evaluation Laboratory
ATSC. *See* Advanced Television Systems Committee
ATTC. *See* Advanced Television Test Center
audio   27
   adaptive differential PCM   144
   analog signals   141
   bandwidth   27

551

552 Index

companding 146
differential pulse code modulation 144
digital 27, 143
Dolby Digital 144
equalizer 146
frequency response 141
home video recorders 303
intermodulation distortion 142
mixing 147, 326
multimedia 377
music 147
natural sound 137
phase distortion 142
signal properties 141
signal-to-noise ratio 142
stereo panning 147
stereophonic 139
sweetening 325
system 139
total harmonic distortion 142
trends 508
wow and flutter 143
authoring tools 380
automatic frequency control 273, 340
automatic gain control 340
automatic scan tracking 276, 305
azimuth recording 284, 302

B

B-frames 83
B-MAC 167
bandwith 14
basic input/output system 199
BER. *See* bit error rate
Betacam 279
Betacam SX 285, 292
bimorph mounting 276
binary 46
 addition 58
 multiplication 59
binaural 139
BIOS. *See* basic input/output system
bit 46
bit error rate 64, 125
bit streams 25
 headers 25
bit-flags 59
bit-mapped images 379
bits per pixel 16
black matrix 120
blanking intervals 17
block matching 83, 179
Boolean algebra 58
bpp. *See* bits per pixel
brightness 34, 92
Broadcast Satellite Service 456
broadcast video recorders 255

browser software 502
BSS. *See* Broadcast Satellite Service
burned-in time code 322
burst 156
burst errors 69
byte 46

C

Cable Television Relay Service 443
cable TV:
 cable-ready receivers 441
 channels 430
 digital transmission 446
 fiber-optic cables 447
 future systems 447
 head end 438
 history 429
 line amplifiers 434
 pay-TV systems 441
 performance standards 437
 scrambling 442
 set-top converters 439
 system elements 429
 trunk circuits 436
 wireless cable 442
cache memory 199
camcorders 312
 image stabilization 314
 professional 235
 professional features 315
 viewfinder 314
cameras. *See also* camcorders
 blanking 238
 color correction 239
 digital electronics 242
 film camera configurations 214
 Gamma correction 240
 highlight control 233
 lenses 236
 live camera configurations 213
 live pickup 213
 operating controls 245
 optical filters 237
 packaging 234
 primary colors 38
 prisms 236
 spatial offset 225
 specifications 246
 still-image 249
 taking curves 38
 telecine 213
 triax cable 242
 viewfinders 242, 314
carrier-to-noise ratio 193, 468
CARS. *See* Cable Television Relay Service
cassettes 262
cathode-ray tubes 343, 359. *See also* kinescope

Index 553

CATV systems. *See* cable TV
CCDs. *See* charge-coupled devices
CCIR. *See* Consultative Committee on International Radio
CCITT. *See* Consultative Committee on Telephone and Telegraph
CCU. *See* camera control unit
CD. *See* Compact Disc
CD-R. *See* CD-recordable
CD-recordable 200
CD-rewritable 200
CD-ROM 200, 291
CD-RW. *See* CD-rewritable
central processing unit 197
    backward compatibility 198
    clock frequency 199
    instructions 198
Chalnicon 216
characteristic impedance 402
charge-coupled devices:
    aliasing 231
    aperture response 222, 227
    architectures 222
    dark current 229
    frame-interline architecture 225
    frame-transfer architecture 223
    high-definition TV 112
    highlight control 233
    interline-transfer architecture 225
    lag 232
    limiting resolution 228
    on-chip lenses 220
    overflow gate 219
    potential well 219
    spectral response 230
    transfer function 231
    transfer smear 224
    tricolor imager 221
chicken and egg syndrome 510
chroma keying 325
chromaticity coordinates 34
chrominance components:
    NTSC equations 155
    PAL equations 156
CIE. *See* Commission Internationale de l'Eclairage
CIE Chromaticity Diagram 35
CinePak 209
CIRC. *See* cross-interleaved Reed-Solomon coding
circuit-switching 498
circular polarization 410, 414, 473
CISC. *See* complex instruction-set computer
clamping 391
clock frequency 199
closed captioning 345
CLUT. *See* color lookup tables
CMYK 33
coaxial cable:
    construction 432

    electrical characteristics 433
cochannel interference 91, 126
coded orthoginal frequency division multiplexing 192
coercivity 256
COFDM. *See* coded orthoginal frequency division multiplexing
color:
    burst 156
    correction in cameras 239
    correction in editing 324
    dithering 205
    illuminants 37
    matching functions 34
    optical filters 237
    temperature 37
color lookup tables 204
color systems:
    additive 32
    compatible 152, 172
    subtractive 32
color-difference signals 154
color-under system 298, 307
colorimetry 29
comb filters 23, 166
Commission Internationale de l'Eclairage 34, 92
Compact Disc 27, 291, 507
companding 146
compatibility 172, 386
complex instruction-set computer 198
component video 153
    analog formats 167
    digital recording 281
    sampling rate 50
composite video 24, 153
    analog 154
    sampling rate 50
compression 17
    algorithms 75
    discrete cosine transform 78
    DTV compression layer 178
    lossless 74
    lossy 74
    subsampling 76
    transformations 78
constant luminance principle 157
constellation diagram 65
Consultative Committee on International Radio 152
Consultative Committee on Telephone and Telegraph 152
continental United States 455
contours-out-of-green 239
contrast 92
contrast ratio 89, 118
contrast transfer function 96
CONUS. *See* continental United States
CPU. *See* central processing unit
CRC. *See* error protection: cyclic redundancy check
critical fusion frequency 128

cross color distortion 153
cross luminance distortion 153
cross-interleaved Reed-Solomon coding 145
CRT. *See* cathode-ray tubes
CTF. *See* contrast transfer function

D

D-1 recorders 283
D-2 recorders 283
D-3 and D-5 recorders 284
D-6 recorders 285
DAC. *See* digital-to-analog conversion
data:
 files 204
 interleaving 69, 186
 randomizing 186
 rates 17
  audio 28
David Sarnoff Research Center 172
DBS. *See* direct broadcast service
dc restoration 19
DCT. *See* discrete cosine transform
definition 89
 HDTV 113
dichroic surfaces 236
differential gain 153, 165, 401
differential phase 153, 165, 402
differential pulse code modulation 77, 144
diffraction 409
digit 45
digital 2, 45
 camera electronics 242
 modulation 64
 signal processing 60, 354
 still-image cameras 249
 synchronization 67
digital mirror device 363
digital recording:
 background 280
 Betacam SX 285
 D-2 standard 283
 D-3 and D-5 standards 284
 D-6 standard 285
 Digital Betacam 285
 DVCPro 288
 error protection 282
 HD-D5 format 285
 hybrid recorders 292
 optical recording. *See* optical recording
 standard formats 281
digital TV 4, 171
 ATSC system 174
 audio for 27
 compression layer 178
 display processing 365
 frame rates 177
 history 172
 layered architecture 175

modulation 66
objectives 173
picture formats layer 176
resolution 9
sampling rates 177
terrestrial broadcasting. *See* terrestrial broadcasting
transmission layer 183
transmission standards 423
transmitters 427
transport layer 180
Digital Versatile Disc 291, 508, 513
 on computers 210
digital video 45
Digital Video Broadcasting 173, 190, 478
 DVB-C 191
 DVB-S 191
digital-to-analog conversion 26, 73
digitization. *See* analog-to-digital conversion
direct broadcast service 452
discrete cosine transform 78, 180
displays. *See also* flat-panel displays
 limiting resolution 108
 primary colors 41
 refresh 352
 scanning 5
 shadow-mask kinescopes 343
 trends 508
dithering 16, 55, 205
DMD. *See* digital mirror device
Dolby Digital audio 144, 175
dominant wavelength 34
DPCM. *See* differential pulse code modulation
DRAM. *See* dynamic RAM
drop-out compensation 278
DSP. *See* digital: signal processing
DTV. *See* digital TV
DV:
 format 288
 system 310
DVB. *See* Digital Video Broadcasting
DVCam 288
DVCPro 288
DVD. *See* Digital Versatile Disc
dynamic RAM 353

E

earth stations 461
EBU. *See* European Broadcasting Union
editing 63, 318
 A-B roll 321
 decision list 322
 nonlinear 327
 off-line 321
 pre-roll 320
 SMPTE time code 327
 time code 326

video effects 325
   chroma keying 325
   split screen 325
video titles 325
video transitions 324
videotape
   assemble mode 320
   insert mode 321
EDL. *See* editing: decision list
EDTV. *See* extended-definition TV
effective isotropic radiated power 453
effective radiated power 394
EFM. *See* eight-to-fourteen modulation
EIA. *See* Electronic Industries Association
eight-to-fourteen modulation 145
EIRP. *See* effective isotropic radiated power
EL. *See* electroluminescent displays
electroluminescent displays 365
Electronic Industries Association 88, 103, 124
electronic news gathering 312
embedded microprocessor 374
encoding 48, 55
   NTSC 241
   offset binary 55
   pure binary 55
   two's complement 56
ENG. *See* electronic news gathering
envelope delay distortion 142, 165
equally tempered scale 147
equivalent noise temperature 464
ERP. *See* effective radiated power
error protection 64, 123
   audio CD 145
   concealment 69
   cross-interleaved Reed-Solomon coding 145
   crossword code 70
   cyclic redundancy check 71
   data interleaving 72
   effectiveness of 72
   Hamming code 71
   in recorders 282
   Reed-Solomon code 71
European Broadcasting Union 49
exposure 118
extended-definition TV 172
eye. *See* vision

F

fade margin 469
fast fourier transform 192
FET. *See* field effect transistor
FFT. *See* fast fourier transform
fiber-optic systems:
   digital performance 486
   fibers:
      construction 480
      performance 481
   in cable TV 447
   modulation modes 484
   optical receivers 489
   optical transmitters 487
   passive optical components 491
field effect transistor 465
field sequential color system 386
field-emission displays 362
fields 11
film 88, 317
film editing 63
Fixed Satellite Service 456
flat-panel displays 359, 508
   electroluminescent 365
   field-emission 362
   plasma 364
flicker 11, 90
   critical fusion frequency 128
floppy disk 200
flying spot scanner 211
FM. *See* frequency modulation
foot candles 92
foot lamberts 92
FPLL. *See* frequency and phase-lock loop
frame rate 11
frame-interline CCDs 225
frame-transfer CCDs 223
frequency and phase-lock loop 340, 365
frequency modulation 268, 466
   FM improvement factor 468, 486
   modulation index 468
frequency-shift keying 65
fresnel zone 445
front porch 20
FSK. *See* frequency-shift keying
FSS. *See* Fixed Satellite Service
full-duplex 498

G

GA. *See* Grand Alliance
gamma 89, 119
gamma correction 240
General Instrument Corp. 172
generations 318
genlocking 241
geometric distortion 91
geosynchronous orbit 453
ghosts 91
graceful degradation 125
Grand Alliance 172
graphical user interface 202
graphics 378
graphics acceleration 353
gray scale 89, 118
guard intervals 192

H

H & D. *See* Hurter and Driffield
Hamming code 70
hard disk technology 199, 288
HD-D5 format 285
HD-MAC 169
HDTV. *See* high-definition TV
hearing:
   amplitude range 138
   frequency range 138
   properties of 137
heat pipe 399
helical-scan recording:
   alpha wrap 261
   capstan drive 259
   omega wrap 261
   principles of 259, 295
hexadecimal 45, 47
Hi-8 recorders 302, 309
high-definition TV 4, 171
   digital DBS 477
   display trends 508
   future 512
   on computers 210
   recording of 285
   scanning standards 425
highlight brightness 118
highlight control 233
HIS. *See* Home Information System
Home Information System 514
home video recorders 255
   analog:
      audio recording 303
      color-under system 307
      frequency modulation 304
      record formats 299
      record parameters 301
      recording drum configurations 299
      VHS-C cassette 313
   automatic scan tracking 305
   azimuth recording 302
   background 295
   comparison with professional 295
   control interfaces 309
   digital:
      D-VHS system 312
      DV system 310
   time-base stability 298
horizontal blanking interval 17
horizontal detail 7
horizontal limiting resolution 105
HTML. *See* hypertext modeling language
hue 30
Huffman coding 80
hum 91, 126
Hurter and Driffield 118
hybrid recorders 292
hypertext modeling language 500

I

I and Q signals 155
I-frames 83, 179
I-Q axis sampling 51
iconoscope 212
icons 202, 373
IEC. *See* International Electrotechnical Commission
IEEE. *See* Institute of Electrical and Electronics Engineers
IF. *See* intermediate frequency
illuminance 92
illuminants 37
image:
   defects:
      aliasing 90
      cochannel interference 126
      flicker 90
      geometric distortion 91
      ghosts 91
      hum 91, 126
      lag 91
   definition 89, 94
   enhancement 109
   image orthicon 212
   quality 87
   scanner 205
   stabilization 314
imagers:
   CCDs. *See* charge-coupled devices
   Chalnicon 216
   flying spot scanner 211
   history 211
   iconoscope 212
   image orthicon 212
   instantaneous scanners 211
   Newvicon 216
   photoconductive 212
   photoemissive storage tubes 211
   Plumbicon 212, 216
   Saticon 212, 216
   vidicon 212, 216
images. *See* still images
incidental phase modulation 392, 402
index of refraction 408, 480
inductive output tube 399
information superhighway 511
infrared 30
input/output 200
insert editing 321
Institute of Electrical and Electronics Engineers 49
   1394 serial interface 68
integrated services digital network 498
Intel Indeo 209
interactive TV 381
interactivity 369
intercarrier sound 339
interlaced scanning 11
interleaving data 186

Index 557

interline-transfer CCDs 225
intermediate frequency 335
intermodulation distortion 142
International Electrotechnical Commission 80
International Standards Organization 152
International Telecommunications Union 49
Internet 493, 512
    addressing 496
    architecture 494
    cable TV connections 499
    history 493
    satellite connections 499
    transmission control protocol 495
    video architectures 503
Internet protocol 495
Internet Service Provider 497
intraframe coding 179
IOT. *See* inductive output tube
IP. *See* Internet protocol
ISDN. *See* integrated services digital network
ISH. *See* information superhighway
ISO. *See* International Standards Organization
isotropic radiator 393, 453
ISP. *See* Internet Service Provider
ITU. *See* International Telecommunications Union

J

Java language 502
Joint Photographic Expert Group 80
JPEG. *See* Joint Photographic Expert Group

K

Kell factor 103
kinescope:
    recording 318
    resolution 348
    shadow-mask 343
klystrode 399

L

lag 91, 232
lamberts 92
laser videodisc 291, 356, 507
lasers 487
LCD. *See* liquid-crystal display
LED. *See* light-emitting diodes
lenses 107, 236
    HDTV 111
light-emitting diodes 487
limiting resolution 89, 103, 228
linear time code 327
liquid-crystal display 360
live-pickup cameras 213
LNA. *See* low-noise amplifier
LNB. *See* low-noise block converter
lossless compression 74

lossy compression 74
loudness 138
loudspeakers 140
low-noise amplifier 465
low-noise block converter 465
LTC. *See* linear time code
lumens 92
luminance 17, 24, 29, 92, 155
    equation 152
luminosity function 92
luminous flux 92
luminous intensity 92

M

M-II recorders 279
MAC. *See* multiplexed analog components
magnetic recording:
    analog frequency modulation 268
    azimuth recording 284
    coercivity 256
    D-1 standard 283
    D-2 standard 283
    D-3 and D-5 standards 284
    D-6 standard 285
    digital. *See* digital recording
    frequency response 265
        aperture effect 265
        azimuth loss 266
        record depth effect 266
        separation loss 266
    HD-D5 format 285
    heads:
        design of 257
        materials 258
    helical scan 259
    magnetizing force 256
    principles of 256
    recording density 267
    SNR 269
    tape 263
    track guard bands 273
    trends 509
    Type C. *See* SMPTE Type C recorders
masking 240
mass storage 199
Massachusetts Institute of Technology 172
matrixing 38
MDS. *See* Multipoint Distribution Service
meander gate 163
memory 199
micromirror displays 362
microphone 140
Microsoft Windows 374
MIDI. *See* music: musical instrument digital interface
millions of instructions per second 199
MIPS. *See* millions of instructions per second
MIT. *See* Massachusetts Institute of Technology

mixing 326
MMDS. *See* Multipoint Multichannel Distribution Service
MMX. *See* multimedia: extensions
modems 498
modulation 64
   vestigial sideband 66
modulation transfer function 96, 227
moiré 167
monitors:
   computer display 358
   professional 349
   SMPTE Standard C primaries 42
motion compensation 81, 179
Motion Picture Experts Group 80
motion vector 83
motion video capture 208
motion-JPEG 81
Moviola machine 322
MPEG. *See* Motion Picture Experts Group
MTF. *See* modulation transfer function
multimedia 376
   audio 377
   development tools 380
   extensions 208
multimoding 402
multipath interference 192
multipath transmission 409
multiple sub-Nyquist encoding 172
multiplexed analog components 167
Multipoint Distribution Service 443
Multipoint Multichannel Distribution Service 443
multitasking 203, 380
MUSE. *See* multiple sub-Nyquist encoding
music:
   musical instrument digital interface 148
   pitch 147
   sampling synthesizer 149
   synthesizer 148
   tempo 148
   timbre 148

N

National Television Systems Committee 152, 386
natural sound 137
navigation 380
Ne 98, 227
nematic 360
neutral density filters 237
Newvicon 216
noise. *See also* signal-to-noise ratio
   figure 348, 464
   temperature 123
   weighting factor 124, 468
nonlinear editing 327
notch diplexer 393
NTSC. *See* National Television Systems Committee

NTSC decoder 341
NTSC system:
   chrominance components 155
   encoding 241
   signal spectrum 160
   subcarrier 156
number systems 45
Nyquist frequency 8, 48, 112
Nyquist limit 8, 48, 100

O

off-line editing 321
offset binary encoding 55
omega wrap 261
on-chip lenses 220
on-off keying 65, 485
OOK. *See* on-off keying
operating systems 201
   command-line interface 202
   file system 202
   graphical user interface 202
   multimedia 380
   multitasking 203
   multithreading 203
optical components 236
   color filters 237
   filters 107
   HDTV filters 112
   light splitters 236
optical recording 291
   Compact Disc 291
   Digital Versatile Disc 292
OS. *See* operating systems
overflow 58

P

P-frames 81
packet switching 494
packetization 63, 175, 180
PAL system:
   chrominance components 156
   subcarrier 160
   subcarrier equations 161
panel antenna 414
parallel transmission 57
parametric amplifier 465
parity 70
passive-matrix LCDs 361
pay-per-view 381
pay-TV systems 441
PC. *See* personal computers
PCM. *See* pulse code modulation
permeability 258
personal computers:
   architecture 196
   audio 27

Index 559

bus 197
cache memory 199
central processing unit 197
display monitors 358
graphics 378
graphics acceleration 353
HDTV 210
history 195
images. *See* still images
Input/output 200
main memory 199
mass storage 199
motion video formats 207
operating system 201, 380
pixel formats 206
pointing device 202
sampling rates 51
scanning standards 15
software 201
text 378
video adaptors 351
video standards 209
video subsystem 351
Philips Electronics North America 172
photoconductive imagers 212
photoemissive storage tubes 211
photographic film 88, 130
picture quality. *See* image quality
picture tube. *See* kinescope
picture-in-picture 347
piezoceramic materials 276
pitch 147, 215
pixels 9, 106, 206
   square 176, 352
plasma displays 364
Plumbicon 212, 216
Pointer colors 43
pointing device 202
polarization 410
postproduction 4, 63, 317, 513
pre-roll 320
preemphasis-deemphasis 141
primary colors 32
prism optics 236
production 4, 317
progressive scanning 13
projection receivers 344
pulse code modulation 28
pure binary encoding 55

Q

QAM. *See* quadrature amplitude modulation
QPPM. *See* quaternary pulse-position modulation
QPSK. *See* quaternary phase-shift keying
quadrature amplitude modulation 65, 191
quadrature phase distortion 392
quantizing 10, 46, 53
   errors 54

levels 53
quantum efficiency 230
quaternary phase-shift keying 65, 191
quaternary pulse-position modulation 485
QuickTime 209

R

randomizing data 186
raster 5, 106
read-only memory 199
Rec. BT.601-5 50, 281
receivers:
   analog 334
      IF amplifier 337
      IF frequency response 337
      sound IF channel 339
      tuner section 335
   cable-ready 441
   closed captioning 345
   detection 340
   digital 334
      DTV decoder 341
      IF frequency response 339
   GA prototype 365
   history 333
   IF Gain 339
   noise figure 348
   picture-in-picture 347
   projection 344
   remote control 346
   selectivity 348
   sync separation 340
   video inputs 345
recording density 267
reduced instruction-set computer 198
Reed-Solomon code 70, 184
refraction 408
rendering 378
resolution 6
retrace 5
return loss 433
reverberation 139, 146
RGB-16 16
RGB-24 16
RISC. *See* reduced instruction-set computer
RLL. *See* run-length limiting
ROM. *See* read-only memory
run-length encoding 79, 180
run-length limiting 282

S

S-Video 345
sampling 7, 99
   aperture response 100
   co-sited 51
   locations 51
   rates:

DTV 177
HDTV 177
SDTV 49
sub-Nyquist 8
satellite news gathering 451
satellites:
    background 451
    Broadcast Satellite Service 456
    communication system elements 452
    construction 457
    DBS services 472
    earth stations:
        downlink 464
        uplink 461
    elevation angle 454
    fade margin 469
    Fixed Satellite Service 456
    FM system performance 470
    frequency bands 456
    geosynchronous orbit 453
        orbital slots 454
        prime orbital arc 455
    link analysis 467
    sideband energy dispersal 464
    solar eclipses 455
    stabilization 457
    sun outages 455
    transponders 460
Saticon 212, 216
saturation 30
SAW filter. *See* surface acoustic wave filter
scanning 5
    aliasing 101
    bandwidth 14
    choosing standards 11
    fields 11
    HDTV 113
    interlaced 11
    pairing 12
    PCs 15
    progressive 13
    TV standards 14
scrambling 442
SDI. *See* serial digital interface
SDTV. *See* standard-definition TV
serial digital interface 68
serial transmission 57
set-top converters 439
SFN. *See* single-frequency network
shadow-mask kinescopes 343
signal-to-noise ratio 90, 123
    audio 142
    digital systems 125
    weighted 123
single-frequency network 193
single-sideband modulation 442
slope overload 77

Smith chart 404
SMPTE. *See* Society of Motion Picture and Television Engineers
SMPTE time code 327
SMPTE Type C recorders:
    automatic scan tracking 276
    playback electronics 274
    record electronics 273
    records 261, 273
    slow-motion and still-frame 278
    transport geometry 271
SNG. *See* satellite news gathering
SNR. *See* signal-to-noise ratio
Society of Motion Picture and Television Engineers 42
    Standard 125M 68
    Standard 240M 89, 122
    Standard 259M 68
    Standard 260M 89
soft-decision decoder 66
software 201
    application programming interface 202
    applications 202
    compilation 203
    drivers 202
    interpretation 203
    languages 203
sound. *See also* audio
    ambience 139
    binaural 139
    direction perception 139
    reverberation 139
    velocity 137
    wavelength 137
spatial offset in cameras 225
special effects 325
split screen 325
square pixels 176, 352
standard-definition TV 4
standardization 151
    new systems 510
standards conversion 62, 215
stereophonic 139
still images 379
    bit-mapped 379
    cameras 249
    capture 205
    compression 207
    formats 204
store-and-forward 503
streaming data 200, 503
studio cameras 234
subcarriers 156
subsampling 76
subtractive color 32
super VGA 16
Super VHS recorders 302, 309

superturnstile antenna 413
surface acoustic wave filter 338, 393, 398
SVHS. *See* Super VHS recorders
sweetening 325
symbols 64, 184
sync separation 20, 340
synchronization 67
synchronizing pulses 19
    equalizing pulses 20
    vertical serrations 20

T

tape 263
tape cassettes 262
TASO. *See* Television Allocations Study Organization
TBC. *See* time-base correction
TCP. *See* transmission control protocol
telecine cameras 131, 213
    history 246
Telecommunications Association 88
telephone service 498
Television Allocations Study Organization 125
tempo 148
terrestrial broadcasting 385
    aural modulation 393
    channel assignments 388
    current U.S. allocations 387
    designs of TV transmitters 394
    digital TV 421
        channel assignments 425
        transmitters 427
    DTV channel assignments 389
    DTV transmitter 186
    effective radiated power 394
    FCC field intensity curves 419
    history in the United States 385
    inductive output tubes 399
    klystrode 399
    negative modulation 391
    performance standards 399
    power amplifiers 398
    service area 418
    transmission lines. *See* transmission lines
    vestigial sideband filter 393
    visual carrier modulation 391
    visual modulator 397
text on computers 378
THD. *See* total harmonic distortion
Thomson Consumer Electronics, Inc. 172
TIA. *See* Telecommunications Association
timbre 148
time code 326
    burned-in 322
    linear 327
    SMPTE standard 327
    vertical interval 327
time-base correction 61, 277

title effects 325
total harmonic distortion 142
transfer characteristic 89
transfer function 118
transform coding 78
transient response 98
transmission:
    modulation modes 64
    parallel digital 57
    serial digital 57
transmission control protocol 495
transmission lines 402
    characteristic impedance 402
    multimoding 402
    waveguides 406
transparent 2
transponders 460
trellis coding 66, 187, 340
trellis decoder 367
triax camera cable 242
trichromatic theory 32
tricolor imager 221
trinitron 112
tristimulus values 34
tropospheric transmission 409
TRS-ID 67
Truemotion S 209
truth table 59
tuners 335
TV:
    audio 27
    scanning standards 14
    waveforms 17
TV lines of resolution 6
TV receivers. *See* receivers
TV transmitters. *See* terrestrial broadcasting
TVL. *See* TV lines of resolution
two's complement encoding 56

U

U-Matic recorders 301
ultraviolet 30
unexcited field brightness 118
universal resource locator 500
URL. *See* universal resource locator
user interfaces 370

V

varactors 336
VBI. *See* vertical blanking interval
VCR. *See* video cassette recorder
vector graphics 378
velocity of sound 137
vertical blanking interval 17
vertical detail 7
vertical interval time code 327
vestigial sideband filter 393

vestigial sidebands 66, 183, 391
VGA. *See* video graphics array
VHS. *See* Video Home System
video:
   analog 2
   cameras. *See* cameras
   digital 2
   NTSC waveforms 23
   overlay 356
   RAM 353
   recorders:
      broadcast 255
      history 253
      home 255
      quadruplex 253
   scanning 5
   server 290
   signal standards 3
   signals:
      composite 24
      dc component 17
      digital 25
      luminance 24
      spectral content 20
   storage 3
   systems 2
   transition 324
   transmission 4
video cassette recorder 262
video graphics array 16
Video Home System 297
video on demand 291
videotape recorder 246
vidicon 212, 216
viewfinders 242, 314
viewing ratio 6, 94

vision:
   acuity 93
   aperture response 96
   characteristics 92
   color 32, 34
   cones 32
VITC. *See* vertical interval time code
Viterbi decoder 66
VOD. *See* video on demand
voltage standing wave ratio 403
VRAM. *See* video: RAM
VSB. *See* vestigial sidebands
VSWR. *See* voltage standing wave ratio

W

waveform monitor 24
waveforms:
   NTSC TV 17
waveguides 406
wavelength 33, 258, 408
Weber's law 93
weighted SNR 123, 468
WFM. *See* waveform monitor
white field brightness 118
wireless cable 442
World Wide Web 500
   browser software 502
   universal resource locator 500
wow and flutter 143
WWW. *See* World Wide Web

Z

Zenith Electronics Corp. 172
zigzag ordering 81, 180
zoom lenses 236

ABOUT THE AUTHORS

Arch C. Luther is a noted expert in the field of video technology and is the author of *Principles of Digital Audio and Video*, *Video Camera Technology*, *Video Recording Technology*, and *Satellite Technology*.

Andrew F. Inglis, deceased, the originator of this book, was a consultant to the communications industry and a former president and CEO of RCA American Communications.